End of Life and People with Intellectual and Developmental Disability

Roger J. Stancliffe · Michele Y. Wiese ·
Philip McCallion · Mary McCarron
Editors

End of Life and People with Intellectual and Developmental Disability

Contemporary Issues, Challenges, Experiences and Practice

palgrave
macmillan

Editors
Roger J. Stancliffe
Centre for Disability Research and Policy
University of Sydney
Sydney, NSW, Australia

Michele Y. Wiese
School of Psychology
Western Sydney University
Penrith, NSW, Australia

Philip McCallion
School of Social Work
Temple University
Philadelphia, PA, USA

Mary McCarron
Trinity Centre for Ageing and Intellectual Disability
Trinity College Dublin
Dublin, Ireland

ISBN 978-3-030-98696-4 ISBN 978-3-030-98697-1 (eBook)
https://doi.org/10.1007/978-3-030-98697-1

This Palgrave Macmillan imprint is published by the registered company Springer Nature Switzerland AG
The registered company address is: Gewerbestrasse 11, 6330 Cham, Switzerland

Foreword

Everybody dies, but is death truly a great leveller? To what extent is there equity for people with intellectual and developmental disabilities at the end of life? Since the 1990s, this question has been first whispered and then asked out loud. With people's deaths increasingly occurring within communities, evidence was beginning to emerge around the challenges for services and the inequities around end-of-life care for this group of people. They include lack of knowledge and experience among healthcare staff in caring for people with intellectual and developmental disabilities, problems with symptom recognition and management, severe communication challenges, patients' difficulties in understanding complex concepts including death and dying, multiple co-morbidities and complex social situations.

People with intellectual and developmental disabilities need specific attention about end-of-life issues. While many of their needs may be no different from those of the general population, it can be more difficult to meet those needs, because of their often disadvantageous circumstances and unique challenges. People with intellectual and developmental disabilities have poorer health outcomes than the general population, and evidence has emerged over the past decades that this may be for reasons unrelated to the causes of their impairment, but rather, due to failings within healthcare systems. This is compounded by persistent negative attitudes and assumptions about quality of life of people with life-long cognitive impairments.

The specific challenges for palliative care services have been well described. In 2015, the European Association of Palliative Care published a White Paper titled *Consensus Norms for Palliative Care of People with Intellectual Disabilities in Europe*, which was produced with the involvement of 92 experts from 15 European countries (Tuffrey-Wijne et al., 2016). Its recommendations included that there should be ongoing exchange of experiences, expertise and best practice at local, national and international levels, and that there was a need for further research, as empirical evidence in this field remained limited.

Now that we are well into the twenty-first century, it is time to take stock. There is a small but steadily growing number of researchers who are helping to fill the many knowledge gaps. There are also growing numbers of organisations, practitioners and caregivers who are engaged with the end-of-life topic, trying out different ways of doing things, adapting support to fit a person's needs and speaking up about doing it better. What new knowledge and insights have emerged? What are the experiences of people themselves and their families and carers, and of health and social care staff, of service managers and policy makers? What are current practices, issues and challenges? These are the questions asked by this timely and important book.

We need to learn from what is happening in practice: how people die, where they die, what they die of, what support they receive and what their experiences are. It is good to see studies emerging that look at these questions, not only on a micro-level but also, increasingly, on a meso- and macro-level. We need numbers that set out the scope of the issues, as well as stories that bring understanding and insights. It is also good to see so many people, teams and organisations working hard to get end-of-life care right for people with intellectual and developmental disabilities, developing useful and practical resources, and sharing them. This book provides both a wider overview and a close-up look at some of the underlying issues. The book also provides a platform for the sharing of real-life personal stories, good practice and resources, that others can be inspired by, learn from, put into practice and build on.

One of the areas that has emerged as a focus for specific attention is the question of inclusion, truth telling, shared decision-making and advance care planning. There is growing recognition of the importance of listening to families and carers, but it is still unusual to hear the voices of people with intellectual and developmental disabilities themselves when it comes to end-of-life care. There are, however, tentative signs that people want the "Nothing" in "Nothing about us without us" to include "Nothing about our lives *and our deaths* without us". There are several chapters that focus on the latest research around these issues.

Inclusion means including *everyone*, but some people have been harder to include than others. The needs of those with severe, profound and multiple disabilities have perhaps been the most underresearched. It is good to see that creative approaches are emerging that enable us to learn more about this group of people. Other hard-to-reach groups are those in countries that are economically poor; a concerted effort is needed to address this gap.

There is much to learn too about specific groups of people. This book includes wide-ranging chapters about suicide and people with autism; the needs of people with dementia; and those of children with intellectual and developmental disabilities who require palliative care. Specific issues and new approaches include a look at inclusive funerals and the use of end-of-life doulas.

It is difficult to over-estimate the importance of learning how we can best support people with intellectual and developmental disabilities at the end of life. Caregivers will need help and support in meeting needs that are likely to be unfamiliar. Collaboration with other services, including cross-professional training is particularly important. One example in this book is the description of the process of dying, which will be utterly familiar to palliative care staff, but alarmingly unfamiliar to most disability staff and families. At an organisational level, services must be prepared to care for people with intellectual and developmental disabilities by making reasonable adjustments to their usual service delivery.

The importance of this book and the issues it addresses go beyond the obvious need to ensure that vulnerable people are provided with the high-quality care everyone deserves. People with intellectual and developmental disabilities need a very high level of personalisation in care provision. The systems, skills and policies needed to provide such individualised care are highly transferrable. If a hospital or hospice is consistently good at providing excellent support for patients who have intellectual and developmental disabilities, then I would be confident that they will provide good care for *all* of their patients. Similarly, if the day-to-day service providers for people with intellectual and developmental disabilities are able to support their clients well at the end of life, then I would put money on them being able to provide good support throughout the entire life-span. My mantra is that *getting it right* for people with intellectual and developmental disabilities who are dying could be a yardstick for the quality of health and social care services generally. Conducting research, developing policies, sharing stories and improving practice in this area are not marginal activities, but present opportunities for achieving all-round excellence.

Irene Tuffrey-Wijne
Professor of Intellectual Disability
and Palliative Care
Kingston University & St George's
University of London
London, UK

Reference

Tuffrey-Wijne, I., McLaughlin, D., Curfs, L., Dusart, A., Hoenger, C., McEnhill, L., Read, S., Ryan, K., Satgé, D., Straβer, B., Westergård, B., & Oliver, D. (2016). Defining consensus norms for palliative care of people with intellectual disabilities in Europe, using Delphi methods: A White Paper from the European Association of Palliative Care (EAPC). *Palliative Medicine, 30*(5), 446–455. https://doi.org/10.1177/0269216315600993

Irene Tuffrey-Wijne is a nurse with extensive clinical experience in both intellectual disability and palliative care services in the UK. Since 2001, she has led a programme of research around intellectual disability, bereavement and palliative care. She chairs the *European Association for Palliative Care* Reference Group on Intellectual Disabilities.

Preface

In offering this volume, we acknowledge that the content may be sensitive. We trust that the reader will take care.

End of life is a challenging topic and until recent decades was little considered by researchers and practitioners in the intellectual and developmental disability field. Fortunately, there is increasing international attention, with notable growth in both published research and available resources for practice. This volume is a reflection of that growing interest. We use the term *end of life* deliberately to encompass the dying phase, death and the period after death, and to include one's own death as well as the death of others. Much of this book was written during the COVID-19 pandemic, an experience that brought dying and death to the forefront and highlighted the harsh inequities encountered by people with disability, reminding us all too starkly about the importance of these issues.

As editors, we are delighted to have a truly international collection of expert authors from countries in Europe, Australia and North America. We thank them for their knowledge, wisdom and compassion. In keeping with the applied research and practice focus, the authors include both researchers and practitioners. Our one regret is that despite our best efforts, we ultimately were unable to include a chapter with people with intellectual and developmental disability as authors. A planned chapter became unworkable due to the constraints of the pandemic.

This volume presents the best contemporary research and practice. Research is important, but the most critical issue is how people with intellectual and developmental disability are included and supported to respond to the inevitability of dying and death. Therefore, this volume has a strong practical focus, with most chapters containing personal stories to illustrate the real-world effects of key issues. Many chapters conclude with realistic scenarios and reflection questions to prompt readers to consider translation into practice in their own situation. To support practice, chapters also present a list of resources, many of which are designed specifically for people with intellectual and developmental disability. We recognise that some readers, such as instructional faculty, will welcome the opportunity to assign to students specific chapters complete with the related resources. Others will be pleased to see all resources located in one chapter at the close of the book. We have done both to meet these differing needs.

In planning this publication, we sought to achieve several goals. Firstly, we intended to focus more on the experience of people with intellectual and developmental disability and less on the situation for family, caregivers and professionals. Secondly, we wanted to encompass intellectual disability and other types of developmental disability. In some cases, that aspiration has been constrained by the dominance of the intellectual disability literature. In chapters where intellectual disability research predominates, this fact is signalled in the chapter's title or in the introductory remarks. We also recognise that in some jurisdictions the terms intellectual disability and developmental disability are used interchangeably and/or together and in others, terms such as learning disability may be preferred. Thirdly, we sought to engage with new and challenging topics such as suicide, advance care planning, the COVID pandemic, accessible funerals, death doulas, and do not resuscitate orders. Finally, in a number of chapters, we addressed end-of-life care when people with intellectual and developmental disabilities are supported primarily by family which we saw as an important next step because traditionally the literature has tended to concentrate on care out of the family home.

We hope you find the book provocative, motivating and even inspiring. The importance of the issues is self-evident. As we conclude at the end of the last chapter, "For each individual, we don't have a second chance to enable a good death, so we need to do the very best we can the first, and only, time".

Sydney, Australia — Roger J. Stancliffe
Penrith, Australia — Michele Y. Wiese
Philadelphia, USA — Philip McCallion
Dublin, Ireland — Mary McCarron

About This Book

This book brings together two challenging topics, end of life and intellectual and developmental disability. It guides the reader about how to respectfully and sensitively involve people with intellectual and developmental disability and their caregivers (family and staff) in all aspects of mortality, dying and death.

With 16 chapters and 37 authors from 10 countries in Europe, Australia and North America, this volume presents both contemporary international research and its translation into practice. There is a particular focus on challenging and new topics, including suicide, do not resuscitate orders, children, people with profound and multiple disabilities, death doulas, accessible funerals and the COVID-19 pandemic.

The book identifies barriers and care challenges and offers practical responses. Chapters include real-world personal stories to ground the issues in lived experience and realistic scenarios with reflection questions to support translation into practice. Lists of resources, many designed specifically for people with intellectual and developmental disability, are provided to support practical implementation.

This publication is appropriate for researchers, practitioners and families across the disability and health sectors, who support people with disability in understanding and planning for end of life, have caring roles for the person who is dying or assist with post-death ritual and remembrance.

Praise for *End of Life and People with Intellectual and Developmental Disability*

"This book covers an often neglected area of life, practice and research. To have the issues and experiences discussed by informed leaders who have thought deeply about the complexity of the area is critically important for people with intellectual and developmental disability at this time. A very timely and important book."

—Nick Lennox, *Professor at the University of Queensland (Hon), Formerly Director of the Queensland Centre for Intellectual and Developmental Disability (QCIDD), Mater Hospitals, MRI/UQ, The University of Queensland (UQ), Australia*

"Thankfully, people with intellectual disabilities are living longer and better lives, but this makes it all the more important to consider advance directives, do not resuscitate orders, accessible funerals and many other issues which this book explores. I am thrilled to see this wonderful book, which is critically needed."

—Tom Shakespeare, *Professor of Disability Research, London School of Hygiene and Tropical Medicine, University of London, United Kingdom*

"Making a clear distinction between palliative and end of life care, this book enters new territory to highlight critical points of difference at end of life for people with intellectual and developmental disability. From advance care planning, do not resuscitate orders and suicide, to accessible funerals, distress among caregivers during COVID-19 and end of life issues for people with profound intellectual and multiple disability, no topic is left unexplored. The book concludes with a much-needed focus on future end of life research, policy and practice, reinforcing the key messages of self-determination, inclusion and equity. Written by experts in the field the book has a multidisciplinary appeal with much to offer health, allied health and social care professionals. It takes the reader to new, and at times challenging, places whilst remaining sensitive to the perspectives of people with intellectual and developmental disability."

—Karen Watchman, *Senior Lecturer, Faculty of Health Sciences and Sport, University of Stirling, Scotland, United Kingdom*

"A life course approach helps to provide appropriate and sensitive support at key times in the lives of people with intellectual and developmental disabilities. This important book contains a wealth of scholarship and practical information about end of life for people with intellectual and developmental difficulties. The range and depth of topics covered, from palliative care training to the particular needs of terminally ill children with intellectual disabilities, makes this text a vital resource for practitioners, policy makers, students and academics. Set against the background of the COVID pandemic, the challenges of which are addressed by many of the authors, this timely book highlights that the value we place on more vulnerable members of society is revealed by how they are treated in the final chapter of their lives."

—Andrew Jahoda, *Professor of Learning Disabilities, University of Glasgow, Glasgow, Scotland*

"This book provides an international flavour around the holistic needs of people with an intellectual disability at the end of life. It integrates theoretical perspectives with practical considerations in a truly meaningful way, making it an important book for researchers, clinicians and practitioners alike. A 'must have' book."

—Sue Read, *Emerita Professor of Learning Disability Nursing, School of Nursing and Midwifery, Faculty of Medicine and Health Sciences, Keele University, Stoke-on-Trent, United Kingdom*

"Death and dying for people with intellectual and developmental disabilities is an essential discussion that needs to be part of our everyday practice and conversation. Each chapter of this books provides an important insight to how we can start to open the dialogue to support what is simply part of living well."

—Angus Buchanan, *Head of School, Curtin School of Allied Health, Faculty of Health Sciences, Curtin University, Western Australia, Australia*

"This book is a 'must' for everyone working in palliative care and others in health and social care services linked to intellectual and developmental disability. It provides an excellent introduction to contemporary challenges but is also detailed enough to help guide the provision of more equitable and compassionate care. Hopefully it will make a real difference to people's lives who are either receiving care near the end of life or are supporting the delivery of care."

—Sheila Payne, *Emeritus Professor, International Observatory on End of Life Care, Division of Health Research, Faculty of Health and Medicine, Lancaster University, United Kingdom*

"This expansive and long-overdue book explores end-of-life in all its complexity. It confronts both the taboo that surrounds death and dying, and extends our understanding of how broader contemporary issues in this space impact on the lives of people with intellectual and developmental disabilities. Importantly, people with profound intellectual and multiple disability and children receive specific attention. As a collection of current research and practice, this book challenges us to include and promote the individual decisions made by people with intellectual and developmental disabilities about end-of-life issues."

—Brigit Mirfin-Veitch, *Director Donald Beasley Institute Dunedin, New Zealand*

"This is a very timely book on death, dying and end-of life care for people with intellectual and developmental disabilities, including autism as it addresses the latest developments in supported decision-making, self-determination strategies, technology, and palliative care practices. By bringing in the vast work that has been done with the general community and applying it to the unique needs of individuals with intellectual and developmental disability, this book is an important resource for researchers, carers, people with disabilities, and professionals who are tackling end-of-life issues."

—Tamar Heller, *Distinguished Professor/Head, Department of Disability and Human Development, University of Illinois Chicago, USA*

"I highly commend this publication which comes at a pivotal time due to the significantly increasing need for palliative and end-of-life care services for people with intellectual disability. The publication highlights the growing body of important end-of-life care research in a wide range of areas including advance care planning, dementia, COVID experiences, shared decision making, and children with intellectual disabilities requiring palliative care."

—Karen Charnley, *Director of All Ireland Institute of Hospice and Palliative Care, Dublin, Ireland*

"This is a thoughtful and compassionate book which examines every possible aspect of end of life and palliative care for people with intellectual and developmental disabilities. I can't recommend it too highly."

—Baroness Sheila Hollins, *Emeritus Professor of Learning Disability St. Georges, University of London, United Kingdom*

Contents

Notes on Contributors

Andrew P. Allen, Ph.D. is a Researcher interested in dementia in people with intellectual disability and the impact of COVID-19 on people living with intellectual disability.

Jane Bernal, MB, Ch.B., FRCPsych has researched death and dying as it affects people with intellectual disability for 20 years. For much of this time, she was Consultant in the Psychiatry of Intellectual Disabilities in London, later in Cornwall, but has now retired from clinical work. Her research interests include communication and less expected dying.

Melissa J. Bloomer, Ph.D. is a registered nurse with a strong profile of research examining end-of-life care in critical care and other inpatient settings. Her research focuses on the end-of-life care needs of vulnerable populations including the critically ill, older people, people with cognitive impairment and people from culturally and linguistically diverse backgrounds, and their families.

Dr. Éilish Burke, Ph.D. is the Ussher Assistant Professor in Ageing and Intellectual Disability, Trinity College Dublin, and the Associate Director of the Trinity Centre for Ageing and Intellectual Disability.

Her research interests include bone health and osteoporosis, inclusionary methods of healthcare screening and healthcare engagement, accessible health promotion and promoting research involvement of people with intellectual disability.

Alison Clarke, M.Sc. is a Research Assistant on the suicide prevention research team at Orygen, in Melbourne, Australia. She has worked across numerous youth suicide prevention projects, playing a leading or coordinating role in systematic reviews, Cochrane reviews, Delphi studies and intervention studies.

Luc Deliens, Ph.D. is Professor of Palliative Care Research at the VUB Brussels & Ghent University and Founding Director of the End-of-Life Care Research Group, Belgium. He has published over 500 papers in peer-reviewed journals, over 50 book chapters and successfully supervised over 50 PhDs in palliative care or end-of-life decision-making.

Carmel Doyle, Ph.D. has a specific interest in children with intellectual disability and complex care needs. She has completed several funded research projects in the field and has collaborated with key stakeholders. She has developed a strong network with disability service providers, has several publications and received funding nationally and internationally.

Pamela Dunne, M.Sc. works in a nurse-led specialist memory clinic for people with an intellectual disability providing comprehensive assessment and post-diagnostic support across the continuum of dementia, including support to peers, staff and family caregivers.

Caroline Ellison, Ph.D. is Associate Professor focused on Ageing and Disability at the University of South Australia. She works as an end-of-life doula and developmental educator with extensive multidisciplinary practice, teaching and research experience both independently and collaboratively across disability and human services in Australia and South-East Asia.

Lisa A. Ferretti, MSW conducts work focused on the development, implementation and dissemination of evidence-based programmes statewide and in large cities, and on translational research efforts advancing self-management in engaging and activating people in their health and self-care including end-of-life decisions while managing multiple chronic health conditions.

Rachel Forrester-Jones, Ph.D. is Professor and Director of the School of Health Studies at Western University, London, Ontario, Canada, and Honorary Professor of Social Policy at the University of Bath, UK. Her research interests scan the life course of people with intellectual and developmental disabilities from early transitions to death and bereavement, the focus being to further advance their social inclusion through social support networks. As a qualified Barrister, she also researches legal issues in relation to vulnerable groups.

Margaret Haigh, MNursRes is Project Manager of the Intellectual Disability Supplement to the Irish Longitudinal Study on Ageing (IDS-TILDA). She has a nursing background and has co-authored a number of publications focusing on the health and well-being of marginalised groups, as well as on health workforce issues.

Susan M. Hayward, Ph.D. is a Researcher with over 10 years of experience in a range of clinical and community service roles in both government and not-for-profit organisations. Her research interests include strengthening the psychological health and well-being of neurodivergent individuals, as well as the employment and social inclusion of these people.

Darren Hedley, Ph.D. has a background in psychology and has been working with autistic people for 20 years. His research concerns mental health and well-being in autism, focusing on understanding the mechanisms underpinning the high suicide risk associated with autism as well as translating research findings into effective suicide prevention strategies.

Kathyan Kelly, M.Sc. is a Social Researcher with a specific interest in social inclusion. She has worked extensively with vulnerable populations in the fields of mental health, homelessness, addiction, intellectual disability and at-risk families.

Martina Leigh, M.Sc. is an Advanced Nurse Practitioner in Intellectual Disability and Dementia and has extensive experience in supporting people with intellectual disability, their staff, peers and family caregivers across the continuum of dementia.

Philip McCallion, Ph.D., ACSW, FAASWSW, FIASSIDD, FGSA is Director/Professor, School of Social Work, Temple University focuses on caregiver support, dementia, palliative care and service systems. He is co-founder/co-principal investigator/co-applicant of the Intellectual Disability Supplement to the Irish Longitudinal Study on Aging and National Consultant on Intellectual Disabilities and Dementia, US National Alzheimer's and Dementia Resource Center.

Mary McCarron, Ph.D., M.A., BNS, RNID, RGN, FTCD is Professor of Ageing and Intellectual Disability at Trinity College University of Dublin, Ireland, and Executive Director of the Trinity Centre for Ageing and Intellectual Disability and the National Intellectual Disability Memory Clinic. She is Principal Investigator of the Intellectual Disability to the Irish Longitudinal Study on Ageing, IDS-TILDA, and has led extensive longitudinal research in Ireland and internationally on ageing, dementia and end-of-life issues for people with an intellectual disability.

Cathy McCarthy, M.Sc. has extensive clinical and managerial experience in supporting people with advanced dementia across a variety of care settings.

Jacqueline McGinley, Ph.D. has a research interest in improving care for people with intellectual and developmental disabilities who are nearing life's end and their families. She aims to understand how policy and practice inform advance care planning and decision-making for and with people who have intellectual and developmental disabilities.

Niamh Mulryan, MD has extensive experience in supporting and addressing the complex health concerns across the continuum of dementia and has been actively involved in numerous longitudinal studies on ageing and intellectual disability.

Louise O'Reilly, M.Sc. has extensive experience in providing end-of-life care to older people living with intellectual disability and dementia.

Claire Quinn, M.Sc. has a broad clinical and academic background and led the design and delivery of Ireland's only Masters programme for those caring for children with life-limiting illness at National University of Ireland, Galway.

Evelyn Reilly, PGDip is an Advanced Nurse Practitioner in Intellectual Disability and Dementia leading a specialist memory service for people with an intellectual disability providing comprehensive assessment and post-diagnostic support across the continuum of dementia, including support to peers, staff and family caregivers.

Karen Ryan, MB, BCH, BAO, BMedSc is a UCD Clinical Professor and Palliative Medicine Consultant at St Francis Hospice and the Mater Hospital, Dublin. She was Clinical Lead of the National Palliative Care Programme until 2018. Her research focuses on health services and palliative care for older people and people with intellectual disabilities.

Sabine Schäper, Dr. theol. has worked as a Lecturer for Inclusive Education since 2006, after leading services for people with intellectual disability for many years. Currently, she is conducting a research project on palliative care for people with intellectual disability in Germany.

Roger J. Stancliffe, Ph.D., FAAIDD, FIASSIDD is Professor Emeritus, University of Sydney, Australia. His research aims to make a difference in the lives of people with intellectual disability. He co-edited a special journal issue on end of life, is past editor of the *Journal of Intellectual & Developmental Disability* and received the AAIDD Research Award.

Mark A. Stokes, Ph.D. is a Psychologist, Researcher, Vice President of Kidsafe Victoria and former President of the Australasian Society for Autism Research. His research interests include how to best support autistic people transitioning to adulthood, sex and gender differences and their contribution to underdiagnosis in women, well-being and mental health.

Stuart Todd, Ph.D. has been a Researcher in Intellectual Disability for over 3 three decades. His interests have focused on death and dying, and the impact of family caregiving.

Mirko Uljarević, Ph.D. has a background in medicine, developmental psychopathology and big data science. His research is focused on improving clarity around the structure of key clinical domains such as core autism symptoms, anxiety and depression to improve assessments and understanding of underpinning mechanisms.

Hille Voss, Ph.D. has completed her doctoral research on advance care planning for people with intellectual disability and, as a Postdoc Researcher, leads research projects focusing on end-of-life care for people with intellectual disability.

Deborah Waldrop, Ph.D. conducts research centred on end-of-life care and decision-making for people who are nearing death and their families. She aims to focus on the disease and condition-specific needs and concerns of people with serious illnesses and disabilities.

Joanne Watson, Ph.D. conducts research informed by thirty years of practice as a Speech Pathologist. She has an extensive national and international profile, centred around supported decision-making, specifically as it relates to people with intellectual disability and complex communication support needs.

Michele Y. Wiese, Ph.D., MAPS, FASID is Adjunct Fellow, School of Psychology, Western Sydney University, Australia. She is also Honorary Senior Research Fellow at the University of Sydney and Fellow of the Australasian Society of Intellectual Disability. She is consulting editor to three international peer-reviewed journals.

Andrew Wormald, Ph.D. is a Research Fellow at Trinity College Dublin working on the IDS-TILDA project and a member of the Trinity Centre for Ageing and Intellectual Disabilities. His research interests include end of life and loneliness for people with an intellectual disability and their carers, and the experiences of living with Lewy Body Dementia.

List of Figures

List of Tables

1

Current and New Developments in Death, Dying and End-of-Life Care Policies and Practices

Luc Deliens

Introduction

A large majority of countries in the world face similar socio-demographic challenges; their populations are ageing, and they should expect rising numbers of deaths and of deaths in older age (Cohen & Deliens, 2012). This phenomenon is happening fast, in some countries faster than in others, and it requires not only an end-of-life policy response regarding the organisation of health services, but also the development of new public health approaches towards death and dying and towards palliative care. Guaranteeing a good end of life for all people with serious

L. Deliens (✉)
End-of-Life Care Research Group, VUB and Ghent University, Ghent, Belgium
e-mail: luc.deliens@vub.be

Australian Centre for Health Law Research, Queensland University of Technology, Brisbane, QLD, Australia

R. J. Stancliffe et al. (eds.), *End of Life and People with Intellectual and Developmental Disability*, https://doi.org/10.1007/978-3-030-98697-1_1

illnesses in our ageing societies requires appropriate policies, public health responses and access to end-of-life care services wherever people are being cared for. People die in all settings in our societies, in hospitals, in nursing homes and at home. In order to aim for a good end of life for all in need, the understanding of end-of-life care research, policy and practice needs to be based on a broad societal approach alongside the more traditional health services approach.

Palliative care is a multidisciplinary care approach and service-oriented answer to delivering better end-of-life care for people with cancer or other serious illnesses. In some countries, such as UK and USA, palliative care has now been part of the health care system for over 50 years. In 2018, we commemorated the 100th anniversary of the birth of Dame Cicely Saunders, who wrote her first medical paper on care for the dying in 1957, and by the summer of 1967 had initiated hospice care as an interdisciplinary concept of care at St. Christopher's Hospice in London (Clark, 2018). She is now acknowledged worldwide as one of the founders of the palliative care and hospice movement. Being confronted, as a nurse and then a doctor, with poor care for dying cancer patients in hospital, she aimed to develop and implement care interventions to prevent suffering for people with terminal illnesses and to improve quality of life for these patients and their families.

Ever since, the medical and nursing domain of palliative care and related research has developed rapidly across the world, and the scientific evidence for its effectiveness is steadily growing, as is the list of countries with a national policy on palliative care. The number of scientific journals concerning palliative care, end-of-life care, policy and practice is also growing rapidly. However, death, dying and palliative care is not yet recognised as a separate domain on the Web of Science.

Death, dying and end-of-life care will be experienced by everyone. Palliative care as a professional response of the needs of seriously ill people has been applied mainly in people with cancer, and the standard health care approaches are mostly conceived for people with cancer. However, people with intellectual and developmental disabilities have special needs, and for sure, the palliative care approaches need to be adapted for them. This chapter provides a context on end of life within which the situation for people with intellectual and developmental disabilities examined in other chapters of the book may be better understood.

In this chapter, a short introduction to end-of-life care and palliative careand end-of-life policy is outlined. The chapter will address issues such as place of death, palliative care, palliative care in the community, public health approaches to end-of-life care, advance care planning, end-of-life decision making, assisted dying (euthanasia in Belgium and the Netherlands, voluntary-assisted dying—VAD in Australia, medical assistance in dying—MAID in Canada, etc.) and palliative sedation. But first we will picture death, dying and place of death in the twenty-first century.

Death, Dying and Place of Death in the Twenty-First Century

Mortality has changed profoundly since the late nineteenth century, as a result of several societal developments (Cohen & Deliens, 2012). Improvements in standards of living, health behaviour, hygiene and nutrition, along with other public health measures such as public sanitation and immunisation, have resulted in an epidemiological transition. Mortality at a young age has dropped dramatically, average life expectancy has risen, and the main causes of death have shifted from infectious diseases to chronic and degenerative diseases such as cancer. Life expectancy at birth has risen resulting in ageing populations worldwide.

Most people whose death would not be unexpected tend to follow one of three different trajectories of functional decline (Murray et al., 2005):

1. Those who maintain comfort and functioning for a substantial period, followed by a short period of evident decline—mostly cancer patients,
2. Those who have long-term limitations with intermittent exacerbations—mostly organ system failure patients, or

3. Those with 'prolonged dwindling'—mostly frail older patients and those with a dementia-related disease.

In the twentieth century, the hospital setting has become a common place of end-of-life care and death (Reyniers, 2015). Throughout the nineteenth and twentieth centuries, medical knowledge and care increasingly became consolidated in hospitals. Moreover, a growing desire to postpone death and a belief in the ability of hospital medicine to do this, coupled with a reduced availability of home care for chronic and terminally ill patients, has led to an increase in the number of hospital deaths. Consequently, the proportion of deaths occurring in hospital rose steadily throughout the twentieth century. As a result, death has become institutionalised and has been removed from public view, so people increasingly are no longer exposed to it.

As there is a decline in home deaths, families might increasingly be unprepared for providing end-of-life care at home and have become inadequately supported to do so, eventually resulting in more hospital deaths (Seale, 2000). The late twentieth century is characterised by changes in household structures, such as smaller family size or more two-earner households, which have also made it more difficult to provide informal care at home. Moreover, there is a growing propensity for older people to live apart from their children and many of them live alone—mostly widows—as a result of gender differences in longevity, marriage patterns and an increase in divorces. As a result, there is a reduced availability of informal caregivers, caregivers being considered crucial in enabling people to die at home. At the same time, in recent decades, there has been a growing number of studies showing a consistent preference by people with serious illness for dying at home (Gomes et al., 2013a).

Given that research has shown that a majority prefer to die at home, national health care policies in several high-income countries aim to support end-of-life care and death at home for those residing at home, or in a nursing home for nursing home residents (Reyniers, 2015). Furthermore, the setting is considered to be important for the quality of death, as it entails important aspects of care, such as the type and intensity of care that can be delivered and the relationship of the caregivers to the dying patient. It is therefore assumed that death at home contributes to the overall quality of life of patients and families (Reyniers, 2015).

The proportion of home deaths is seen as a quality indicator for palliative care, in the sense that a higher proportion of home deaths reflects a better quality of end-of-life care provision (De Roo, 2015). Hence, an end-of-life care policy aimed at supporting patients to die at home can improve quality of life of patients at the end of life. Moreover, it is assumed that facilitating home deaths might reduce health care expenditure; inpatient deaths are shown to be more expensive than home deaths and shifting end-of-life care from the hospital to home might therefore reduce end-of-life care spending (Gomes et al., 2013b).

In recent decades, the proportion of people dying in hospital has decreased slightly in several countries. For example, in Belgium, it fell from 55.1 to 45.3% between 1998 and 2017. Despite countries such as the Netherlands or New Zealand reporting relatively low proportions of hospital deaths, the acute hospital setting consistently remains a frequent place of death in many high-income countries. Moreover, as end-of-life care provision within this setting is considered to be suboptimal, the need to further reduce the number of hospital deaths remains.

In the last decades of the twentieth century, the development of hospice care and palliative care has been one of the policy responses to the challenges of death and dying, improvement of care for the dying and the reduction of admissions to hospital late in a disease trajectory and at the end of life.

What Is Palliative Care?

Palliative care originates from the UK-instigated hospice movement and the principles of hospice care, but the term 'palliative care' was introduced in Canada by Dr. Balfourt Mount, a Montreal-based surgeon working at the Royal Victoria Hospital. He became aware that the word 'hospice' in the French language was associated with institutional care and that hospice care principles should be applied in all health care settings (Clark, 2018). He also knew that 'palliative treatment' was used in medicine for treatments when the disease could not be reversed. Dr.

Balfourt Mount wanted to apply the principles of hospice care in his hospital and he hit on the term 'palliative care' for a specially designated unit that he started in the mid-seventies in the Royal Victoria Hospital in Montreal.

Palliative medicine has been a recognised specialty in the UK since 1987 with the definition "the study and management of patients with active, progressive, far advanced disease, for whom the prognosis is limited and the focus of care is quality of life" (Saunders et al., 1995, p. 2). The World Health Organization (WHO, 1990) defined palliative care for the first time as "the active total care of patients whose disease is not responsive to curative treatment. Control of pain, of other symptoms and psychological, social and spiritual problems is paramount" (p. 11).

Most of the focus of palliative care in the earlier decades has been upon the total care of the person with very advanced cancer, but the approach was clearly also relevant for people with all kinds of advanced diseases, such as organ failure (COPD, heart failure, kidney failure, etc.), neurological diseases (ALS, MS, Parkinson, etc.) or frail older people with multiple pathologies and often with dementia. Soon after the WHO definition was created, it began to be criticised on the grounds that it focused on 'terminal care' while it may also be beneficial to people earlier in the illness trajectory and was limited to people with cancer, though may benefit people with other pathologies.

In 2002, the WHO redefined palliative care as "an approach that improves the quality of life of patients and their families facing the problems associated with life-threatening illness, through the prevention and relief of suffering by means of early identification and impeccable assessment and treatment of pain and other problems, physical, psychosocial, and spiritual" (World Health Organization, 2002, p. xv). The WHO definition further states that palliative care:

- Provides relief from pain and other distressing symptoms,
- Affirms life and regards dying as a normal process,
- Intends neither to hasten nor postpone death,
- Integrates the psychological and spiritual aspects of patient care,
- Offers a support system to help patients live as actively as possible until death,

- Offers a support system to help the family cope during the patient's illness and in their own bereavement,
- Uses a team approach to address the needs of patients and their families, including bereavement counselling, if indicated,
- Will enhance quality of life, and may also positively influence the course of illness,
- Is applicable early in the course of illness, in conjunction with other therapies that are intended to prolong life such as chemotherapy or radiation therapy, and
- Includes those investigations needed to better understand and manage distressing clinical complications (Sepúlveda et al., 2002, p. 94).

Through this 2002 definition, the WHO has expanded palliative care away from the dying phase and defined it as a care continuum (Fig. 1.1). The palliative approach may be relevant long before a patient is near the end of life. Attention to a person with a persistent serious illness, to the family whose life is disrupted by illness, as well as to control of any distressing symptoms, should be given much earlier, while active treatment is still pursued (Saunders et al., 1995). Hence, palliative care should also improve care upstream in the treatment trajectory to earlier stages of serious illness. Palliative care is not only terminal care and should not exclusively be given at the end of life or when curative or life-prolonging specialist treatments are not effective anymore and death is becoming more imminent or will happen in a foreseeable future.

At the same time, palliative care is not exclusively for cancer patients and can be applied to anyone with a serious illness confronted with pain or other distressing symptoms. Suffering is also broadly defined and can be caused by physical symptoms as well as by psychological, social or spiritual/existential problems. Such holistic care is complex, frequently multidisciplinary and is ideally initiated at the time of diagnosis of serious or life-threatening illness, independently of prognosis, and is delivered in concert with curative or life-prolonging therapies provided these are beneficial to the patient.

Furthermore, this complex care needs to be coordinated by a key caregiver; in some countries or health care systems, this is often a general practitioner. In nursing homes, it can be a palliative care physician, a

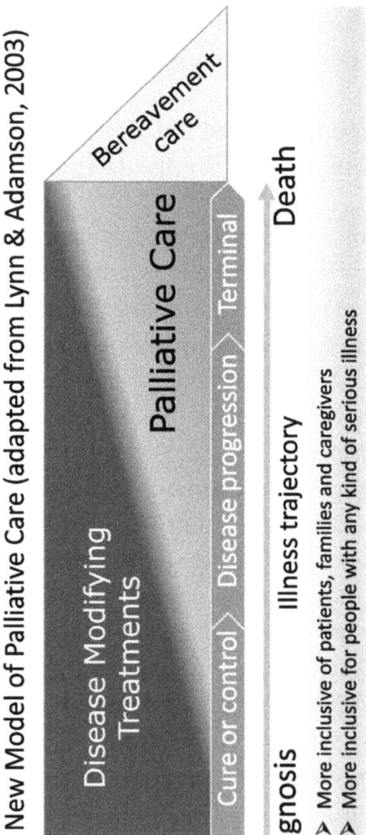

Fig. 1.1 Model of palliative care (adapted from Lynn & Adamson, 2003)

palliative care nurse or a district nurse (Cohen & Deliens, 2012). Hence, good palliative care requires a wide multidisciplinary clinical approach and coordinated care-giving at the end of life. While in most parts of the world this kind of high-quality end-of-life care is not yet available, there is also not one country in the world where high-quality palliative care is accessible for all in need, independent of their disease, age, gender, socioeconomic or ethnic background. There are pervasive cultural, attitudinal, structural and financial barriers to the optimal accessibility of palliative or end-of-life care, and researchers must study and identify these barriers and come up with solutions to overcome them.

The 2002 WHO definition of palliative care applies not only to individual caregiver–patient–family relationships but also applies to public health systems that need to monitor the problems people with serious illnesses are experiencing and create circumstances favourable to tackling them in the best possible ways. While the WHO has recognised palliative care as a major public health priority for all countries (Davies & Higginson, 2004), it remains a theme only sporadically dealt with within public health. However, in most countries, the history of palliative care can be written along the lines of the development of specialist palliative care clinical services. Most of these services are managed by doctors or nurses. From the point of view of the patients, palliative care should be a key element of any health care system and access to good palliative care is a basic human right (Human Rights Watch, 2009).

This WHO definition of palliative care has been criticised in recent decades by scientists and by several organisations (Krakauer, 2019). While it has been interpreted as limiting its purview to people facing life-threatening illness, this definition is not inclusive of people suffering unbearably without a life-threatening illness and makes little sense in low- and middle-income countries. Some limitations in the definition are noted, including lack of clarity on four points (Conner, 2019): first that palliative care is needed in chronic as well as in life-threatening conditions; second, that there should be no time or prognostic limit on the delivery of palliative care; third, that palliative care is needed at all levels of care including primary and secondary as well as tertiary care; and fourth, that palliative care is needed in all settings of care.

Palliative Care in the Community

Within a decade of the initiation of palliative care as developed for inpatients in St. Christopher's Hospice in London, people started applying the principles of palliative care in other contexts, in specialist inpatient wards in hospitals, but also in home care in the community (Clark, 2018). In most developed countries, palliative care has been developed and recognised by the public as a specialist health service in hospitals, in nursing homes or in primary care—for example, a hospice, a palliative care unit, a palliative day-care centre and a multidisciplinary palliative home care team. Research has repeatedly demonstrated that these specialist palliative care services improve the end-of-life care of patients and families suffering from serious illness (Temel et al., 2010; Vanbutsele et al., 2018, Zimmermann et al., 2014).

For instance, in Belgium, in 2002, the federal law on palliative care was approved, alongside the law legalising euthanasia. The ensuing royal decrees regulated minimum service provision in different palliative care settings and the funding of specialist palliative care services. In Belgium, five types of specialist multidisciplinary palliative care teams can be differentiated (Beernaert, 2015):

- At home, a multidisciplinary home care team of primary health care professionals with specialist training in palliative care work together with the regular care givers, as a second line consulting team. After consultation, they can also take over some aspects of the care;
- In nursing homes, similar to the home care setting, the provision of palliative care usually relies on the regular nursing home staff, although they can be supported in this by the coordinating and advisory physician of the nursing home and a nurse who specialises in palliative care (the reference nurse);
- Mobile multidisciplinary support teams can operate throughout the different wards of a hospital. They can support regular staff in the hospital by consulting the patient in the ward he or she is admitted to. The team is composed of at least three halftime members: a physician-specialist, a nurse and a psychologist. These teams are compulsory in every hospital in Belgium;

- Many hospitals also have a specialist inpatient palliative care unit with palliative beds reserved for patients in need of palliative care that they cannot receive at home or in their usual place of residence. This unit, usually consisting of six to 12 beds spread over several hospital units, or united in an independent unit, has a multidisciplinary team that provides specialised palliative care coordinated by a palliative care physician. There are 379 such beds available throughout Belgium;
- And finally, there are five palliative day-care centres that provide ambulatory palliative support to home care. These centres have been created as a complementary measure to home care to support informal caregivers by providing specialised palliative care that would not otherwise be possible at home.

As mentioned in Fig. 1.2, palliative care refers to three different approaches in our health care systems and our societies: specialist palliative care, generalist palliative care and community palliative care. The

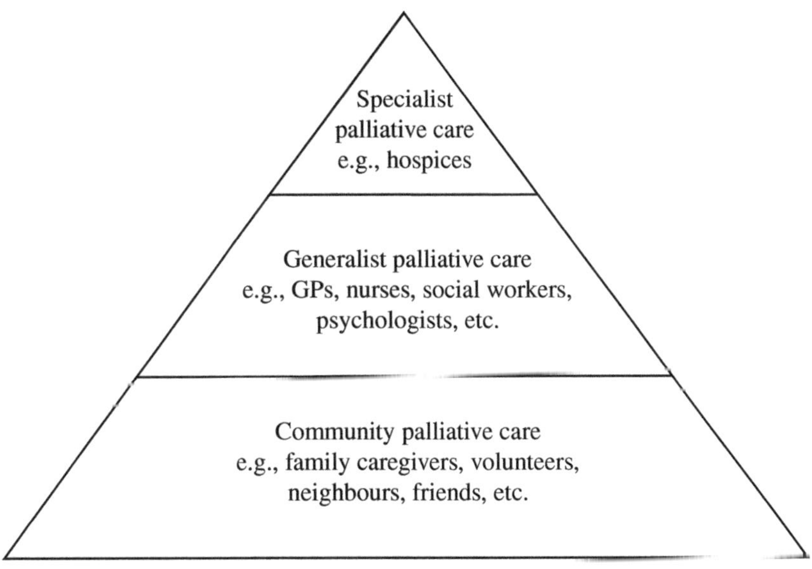

Fig. 1.2 Palliative care is a societal wide responsibility

latter is the least developed in high-income countries and the first is the least developed in low- and middle-income countries.

Generalist palliative care is provided by non-specialists or regular medical professionals involved in day-to-day care and support of patients with serious illness and their caregivers in their care setting (home, hospital or nursing home), but for whom palliative care is not the main focus of their clinical practice (e.g., primary care givers, oncologists, geriatricians, nurses) (Beernaert, 2015). Regular medical professionals should, in principle, be able to assess and meet the patient's palliative care needs under ordinary circumstances or seek advice from or refer to specialist palliative care services in more complex and challenging situations. Specialist palliative care is intended to be provided by professionals working solely in the field of palliative care and whose main activity is dealing with complex palliative care problems requiring specialist skills and competencies. However, palliative care is often misunderstood as only referring to the activities of specialist palliative care teams or services.

It is important to acknowledge that the specialist palliative care services should also be working with the patient's own doctors through consultation and education, and through these consulting and educational activities, palliative care can reach far more people than just those who are able to attend the specialist health services for palliative care. The introduction of general palliative care approaches among all kinds of health care providers is called the generalist palliative care approach (Fig. 1.2). In this way, palliative care becomes a skill and an attitude adopted by any kind of doctor, nurse, psychologist, chaplain or social worker who cares for people with serious illness.

By integrating basic skills in palliative care into the general health care professions through post-academic training, the capacity of palliative care can grow substantially and some countries, such as the Netherlands, have adopted this path as their primary nationwide palliative care policy. Educating general practitioners, nurses, social workers and psychologists in the community in generalist palliative care may be a more important development than the growth of specialist palliative care services themselves. Only by this approach can palliative care be fully integrated into all of our health care services.

Palliative care also includes a third approach, namely community palliative care (Fig. 1.2). Policy makers should recognise the fact that caring for seriously ill persons is, most of the time, taking place in the community and not being done by health care professionals but by family caregivers, neighbours, friends or volunteers. However, in people's minds, community palliative care is the least associated with palliative care, although capacity in the community is vastly larger than in professional health services (Kellehear, 2005).

Public Health Approaches to End-of-Life Care

Whatever the availability of quality end-of-life care services in a country, in order to know which patients and which people have access to them, societies must monitor not only the quality of palliative care provision but also its accessibility (Cohen & Deliens, 2012). The accessibility of end-of-life or palliative care services varies substantially according to differences in age, disease, culture and social characteristics and conditions. In order to develop a policy for proper end-of-life care for all, we need a population-wide perspective and not a single-service perspective to take into account different disease trajectories (not just cancer trajectories), different age groups and different cultures. Policy-making must be multi-sectoral: public health at the end of life integrates demographics, sociology, epidemiology, policy-making skills, organisational skills and economic issues, together with clinical palliative care knowledge and skills. In order to achieve good end-of-life care for all, it is not sufficient to have good nurses and good physicians; we must also have good monitoring systems in a context of advanced policies with a population approach.

Perhaps one of the greatest aids to the development of a public health policy for palliative care is the WHO World Health Assembly (WHA) resolution on palliative care of 2014 "Strengthening of palliative care as a component of comprehensive care throughout the life course" (Conner, 2019, p. 32). The resolution calls for a series of actions by all countries and by the WHO itself to ensure that palliative care is part of every country's health care system with an emphasis on community-based primary health care. National governments are called upon to:

- Develop and strengthen policies to integrate palliative care into health systems,
- Ensure adequate domestic funding and human resources,
- Support families, volunteers and caregivers,
- Integrate palliative care in health professional curricula at all levels,
- Assess domestic palliative care needs including for essential medicines,
- Review and revise drugs control legislation and regulation and add palliative care medicines to national essential medicine lists,
- Support partnerships between governments and civic society and
- Integrate palliative care into plans for management and control of non-communicable diseases.

In the last two decades, new public health approaches to end-of-life care have been developed (Sallnow & Paul, 2019). Questions have been raised regarding the appropriateness of building further health services for palliative care, as opposed to new perspectives on care, which see communities and professionals working in partnership. These perspectives are collectively known as the new public health perspective. This perspective endorses aspects of classic public health approaches such as the role of policy in improving health and the importance of prevention and health promotion, but also represents a departure from classical public health through its radical reframing of health as everyone's responsibility.

The new public health approaches to end-of-life care policy argue that the hospice and palliative care movements have become predominantly a professional and service-based response to the needs of seriously ill people and their families, and this has led to over-professionalisation of care and the disempowerment of communities in relation to death, dying, loss and bereavement (Kellehear, 2005). Two common terms to describe the new public health approach to end-of-life care are health promoting palliative care and compassionate communities (see also Chapter 16). Both approaches move beyond the existing structures and services for palliative care and focus more on the role of local communities, neighbourhoods, schools, businesses or councils in end-of-life care. In some Commonwealth countries, several models such as compassionate neighbours, compassionate universities, compassionate schools and compassionate cities have been implemented in recent years, but as yet there is little research evaluating these models.

Advance Care Planning

In the past, the paternalistic desire of doctors and other health care professionals to protect patients from bad news, such as the diagnosis or prognosis of a life-threatening disease, was culturally more accepted. But there has been a change in public attitudes and medical ethics towards a growing openness about such issues and the involvement of patients in health care decisions, although this does not yet apply to all medical professionals (De Vleminck & Deliens, 2017).

Advance care planning (ACP) initially began in the USA as a movement to establish advance directives (ADs) to refuse treatment, also known as living wills (De Vleminck & Deliens, 2017) (see also Chapters 4 and 15). Emphasis was mainly put on the documentation of an agreement regarding medical care between the physician and the patient. It was considered necessary to jointly prepare and complete a document regarding the patient's wishes and to record this document, for example in the patient's medical file (Wilkinson, 2011). As a result of developments in the USA, ADs have gradually been introduced in other countries and for the first time have allowed individuals to specify their choices regarding medical treatment before the loss of capacity. Yet, despite the legal and pragmatic utility of these documents, their uptake remains limited. Evidence also shows that there is no guarantee that ADs in isolation improve end-of-life care or correspond with future care preferences (Patel et al., 2004).

In recent decades, there has been growing recognition that the completion of documents alone is insufficient to ensure that patients receive care that is consistent with their values and preferences. In 2007, the RAND Corporation presented the results of a systematic review of the evidence published between 1990 and 2007 regarding the effectiveness of ADs for improving end-of-life outcomes in the USA (Wilkinson et al., 2007). One important conclusion of this report is that most Americans do not complete an AD and when they do, these documents often do not affect care because they are narrow and legalistic. ADs are too narrowly focused on the patient's right to refuse unwanted life-sustaining treatment, are often physically unavailable when needed, are too vague to be useful in decision-making or at odds with the patient's current clinical circumstances.

Following the evidence regarding the limitations of ADs, advance care planning (ACP) was developed as a new model in which emphasis is placed on the potential for discussions instead of documents. While the traditional model of ACP was mainly focused on preparing for the incapacity of patients by having specific treatment choices on paper, the developing model of ACP also focuses on *empowering competent patients* to have their say about their care and treatments. In more recent years, we have seen a shift to more emphasis being put on the process of communication and interaction rather than on a legal document. Completing an AD form should be seen as a means to achieving desired goals, not an end in itself. They can be the important result of an ACP process, but good communication underlying or preceding the formalisation into a document is a prerequisite for high-quality care.

The rationale behind ACP is that it provides information to health care professionals about a person's wishes and preferences in case a time comes when they are unable to express themselves (De Vleminck & Deliens, 2017). ACP provides a way of increasing the congruence between a patient's wishes, values and needs and their future care and is also a possible means by which people can exert greater control over their treatment, often but not exclusively relating to end-of-life issues. ACP entails discussions with patients and their representatives about the goals and desired direction of care, particularly end-of-life care, in the event that the patient loses capacity and is unable to make decisions about their care.

ACP is increasingly implemented in oncology and beyond, and a consensus definition of ACP was published only in 2017 in Lancet Oncology (Rietjens et al., 2017). Following a formal Delphi consensus process, this study developed a definition of ACP and provided recommendations for its application. Of the 109 experts (82 from Europe, 16 from North America and 11 from Australia) who rated the ACP definitions and its 41 recommendations, agreement for each definition or recommendation was between 68 and 100%. In this paper, ACP was defined in a short and in a long version (Rietjens et al., 2017, p. e546):

Extended Definition of Advance Care Planning (ACP): Advance care planning enables individuals who have decisional capacity to identify their values, to reflect upon the meanings and consequences of serious illness scenarios, to define goals and preferences for future medical treatment and care, and to discuss these with family and health care providers. ACP addresses individuals' concerns across the physical, psychological, social and spiritual domains. It encourages individuals to identify a personal representative and to record and regularly review any preferences, so that their preferences can be taken into account should they, at some point, be unable to make their own decisions.

Brief Definition of Advance Care Planning (ACP): Advance care planning enables individuals to define goals and preferences for future medical treatment and care, to discuss these goals and preferences with family and health care providers and to record and review these preferences if appropriate.

Recommendations from this consensus paper included the adaptation of ACP based on the 'readiness of the individual', targeting ACP content as the individual's health condition worsens and using trained non-physician facilitators to support the ACP process (Rietjens et al., 2017, p. e548). The following recommendations received the strongest agreement and very strong consensus among the experts:

1. *Exploration of leeway*: ACP should include an exploration of the extent to which the individual allows their personal representative or proxy to consider their current clinical context in addition to their previously stated preferences.
2. *Appropriate health care providers* are needed for clinical elements of ACP, such as discussing diagnosis, prognosis, treatment and care options, exploring the extent to which goals and preferences for

future medical treatment and care are realistic, and documenting the discussion in the medical file of the patient.

3. *Advance care directives* need both a structured format to enable easy identification of specific goals and preferences in emergency situations, and an open-text format so individuals can describe their values, goals and preferences.
4. And two recommended constructs to be assessed in ACP studies, namely 'self-efficacy to engage in ACP' (rated by individuals, family and health care professionals) and the type of health care interventions desired.

ACP discussions can play a major role in facilitating adaptation to illness realities by providing patients with information about diagnosis and prognosis, by leading to appropriate decision-making, by alleviating anxiety and by improving quality of life throughout the trajectory of the illness. Previous intervention studies have shown that ACP interventions stimulate discussions about goals of care between patients and their care providers (Clayton et al., 2007), improve concordance between a patient's preferences and the end-of-life care they receive (Detering et al., 2010) and improve the quality of care at the end of life (Bischoff et al., 2013). They are also associated with positive family outcomes such as improved satisfaction with care and reduced stress and anxiety (Detering et al., 2010). Effective ACP discussions support not only end-of-life care but quality of life throughout the illness trajectory, including the period before death is imminent.

Moreover, some studies indicated that many patients see ACP as having other important roles besides enhancing the patient's autonomy in health care decisions, such as lessening the burden on families and increasing trust and understanding between family members and with health professionals (Seymour et al., 2004). Health professionals seem to have similar views, seeing the value of ACP not only in terms of allowing a patient's wishes to be honoured, but also in giving peace of mind to patients and encouraging meaningful communication between family members about end-of-life issues. The process outcomes of ACP are very important; for example, how ACP may be a means of communication about the end of life through enhancing mutual understanding and openness, enabling discussion of concerns, relieving fears about the burden of decision-making for surrogate decision-makers and strengthening family ties.

End-of-Life Decision-Making

In the second half of the twentieth century, improvements in clinical management have increased survival for many types of cancer and the risk of cardiac death has significantly decreased due to effective drug management, technical devices and sophisticated intensive care (van der Heide & Rietjens, 2012). However, the growing number of life-prolonging options also complicates decision-making at the end of life. People increasingly wish to have some control over their last phase of life and end-of-life care frequently involves questions about when the application of burdensome or costly life-prolonging interventions is appropriate, when forgoing treatment is appropriate, and when the need to alleviate suffering is more important than the risk of hastening death.

Despite the availability of palliative care services, inappropriate treatment at the end of life is still widespread. Appropriate end-of-life care has been defined as "treatment and/or medication for which the expected health benefit (i.e., improved quality of life, increased life expectancy, pain relief, improvement of functional status) exceeds possible negative outcomes (i.e., mortality, decrease in quality of life, pain, symptom burden)" (De Schreye, 2020, p. 13). Inappropriate end-of-life care, then, is the opposite, where the benefits do not exceed possible negative outcomes.

Examples of inappropriate end-of-life care could include unwanted transfer to hospital close to death, people not dying in their preferred place, aggressive life-prolonging treatments that have a negative effect on the patient's quality of life and receiving medication close to death that should only be used for long-term prevention, such as statins. Some treatments like blood transfusion or chemotherapy have considerable burdensome side-effects and are very stressful for the patients and those close to them. Such an aggressive approach to treatment, especially with patients near the end of their lives, could be considered inappropriate care. Admission to hospital often brings aggressive treatment, even in patients near the end of their life. Of all people eligible for palliative care in 2008 in Belgium ($N = 44{,}229$; i.e., 43% of all deaths), 51% died in hospital, 25% at home and 24% in long-term care settings (Houttekier et al., 2014). Of those officially living at home at the time of death, hospital death occurred in more than 60%; of those living in long-term care settings, this was 16%.

Advance care planning was promoted as one of the tools to avoid inappropriate treatment near the end of life, to enhance shared decision-making at the end of life and to prepare doctors, patients and families for end-of-life decisions (Sudore & Fried, 2010). The defining elements in categorising an end-of-life decision are whether the physician committed or omitted an act, the potentially life ending intention of the physician and the request of the patient (Van den Block et al., 2011). The first dimension, the medico-technical dimension, involves what the physician actually does. Two kinds of actions can be distinguished: the withholding or withdrawing of potentially life-prolonging treatment, and the use of potentially life-shortening drugs. The physician's intention and the patient's involvement in the decision-making process belong to the second, medico-ethical dimension. The physician can make a decision with the knowledge that it may hasten death but without the intention of hastening death, with the co-intention of hastening death or with the explicit intention of hastening death. The patient may explicitly request the physician to make, or not to make, a specific type of decision.

The most common end-of-life decisions with their description are listed in Table 1.1.

Empirical data about end-of-life decision making practices are scarce, but the few countries in which the incidences of these end-of-life decisions have been measured have shown that they are very common in modern medical practice; end-of-life decision making preceding death takes place in 40–50% of dying people (Chambaere et al., 2015; van der Heide et al., 2017).

Table 1.1 Classification of most common end-of-life decisions (Deliens et al., 2000)

Non-treatment decisions	These are decisions to withhold or withdraw potentially life-prolonging treatment. Examples of such decisions include forgoing radiation or chemotherapy, artificial respiration, resuscitation, antibiotics treatment, artificial administration of food and fluid, etc.	
Intensified alleviation of pain and symptoms	These decisions concern the administration of drugs for pain and/or symptom relief in doses that may also have a life-shortening effect. Life shortening can in these cases be taken into account	
Assisted dying	Encompasses the administration, supply or prescription of drugs with the explicit intention of hastening the patient's death. Depending on the absence or presence of the patient's explicit request and the person administrating the drugs, the following subdivision can be made	
	Euthanasia	The administration of drugs by someone other than the patient with the explicit intention of hastening the patient's death, at the patient's explicit request
	Physician-assisted suicide	The supply or prescription of drugs to be taken by the patient him/herself, with the explicit intention of hastening the patient's death, at the patient's explicit request
	Life-ending drug use without the patient's explicit request	Illegal in all countries is the administration of drugs with the explicit intention of hastening the patient's death, without explicit request from the patient
Continuous deep sedation until death	The use of drugs to keep the patient continuously and deeply sedated or in a coma until death	
Voluntarily stopping eating and drinking	The conscious and voluntary decision of the patient to stop all food and fluids intake with the explicit aim of hastening death	

Assisted Dying and Palliative Sedation

There is growing impetus for people with serious illness to take more control over their end of life. In modern societies, increased value is being placed on personal autonomy—exercising one's own choices and the right to self-determination—and these values are also increasingly being recognised at the end of life (Singer et al., 1999). In 18 jurisdictions in the world, this has led to the development of legislation that allows clinicians (doctors or nurses) under strict legal criteria to assist people in dying, all using different labels: voluntary-assisted dying or VAD in Australia (legal in some states and territories), euthanasia in Belgium, medical assistance in dying or MAID in Canada, euthanasia and physician-assisted suicide in the Netherlands and physician-assisted dying in the USA. Over 200 million people in the world have access to euthanasia and/or physician-assisted suicide, and new legislation is being crafted or considered in more countries (Mroz et al., 2021).

Assisted dying is receiving increasing media and research attention, and legalisation is expanding internationally (Diamond et al., 2019). The potential legitimation of this practice is laden with medical, legal and ethical considerations. Regardless of the legality or willingness to participate, clinicians working with people with advanced and serious illness will be confronted with requests to hasten death and must be able to respond to enquiries about this topic, whether the patient's aim is information gathering or is a formal request to hasten death.

Assisted dying practice characteristics differ and there is also considerable variation in the terminology and labels used for assisted dying, which can add to the confusion and controversy around the practices (Mroz et al., 2021). Frequency of use also varies greatly by jurisdiction between 0.1% (California, USA) and 4% (the Netherlands) of all deaths, though a consistent increase has been seen in European countries including Belgium, the Netherlands and Switzerland as well as some jurisdictions with long-standing physician-assisted dying laws, such as Oregon and Washington.

All assisted dying legislation includes *substantive and procedural requirements*, such as minimum age, waiting period, health condition, physician consultation and reporting procedure; however, some are extensive and detailed while others are more limited (Mroz et al., 2021). As access to assisted dying expands in new and existing jurisdictions, research must also expand to examine the impact on patients, specifically among vulnerable populations, as well as on health care practitioners, health care systems and communities.

The only end-of-life decision of last resort that is legal in most countries in the world is *palliative sedation*. When practised until death, it is also described as continuous deep sedation until death, i.e., "*the practice whereby a clinician administers sedative drugs resulting in the continuous reduction or taking away of a patient's consciousness until death follows*" (Robijn, 2020, p. 16). When nutrition and hydration are forgone, this practice is often called *terminal sedation*. Continuous sedation should, all guidelines agree, generally be administered proportionally to the severity of a patient's symptoms.

The incidence of continuous deep sedation until death was estimated in Flanders, Belgium, at 12% of all dying people in 2013 (Chambaere et al., 2015) and at 16% in 2008 in the UK (Seale, 2009). Hence, we can assume that this is a very common practice in modern medicine at the end of life.

The status of continuous deep sedation with respect to its possible life-shortening effect is unclear and much debated. Research has shown that, in most cases, death is expected, but not explicitly intended.

Conclusion

In recent years, there has been a rapid increase in end-of-life care research and this has resulted in important developments concerning end-of-life care policies and practices. The importance of palliative care has been brought sharply into focus by the experiences of the COVID-19 pandemic in many countries; dying alone without the chance to prepare, without contact with those close to you, perhaps while undergoing invasive and aggressive treatment, maybe in chaotic circumstances, this is the very antithesis of the values, and practice, of palliative care and it is to be hoped that having gone through this experience as a global community we will become even more aware of the need for good palliative care.

References

Beernaert, K. (2015) *Identifying and addressing palliative care needs throughout the illness trajectory* (Unpublished doctoral dissertation). Vrije Universiteit Brussel Belgium & Ghent University Amsterdam Netherlands.

Bischoff, K. E., Sudore, R., Miao, Y., Boscardin, W. J., & Smith, A. K. (2013). Advance care planning and the quality of end-of-life care in older adults. *Journal of the American Geriatrics Society, 61*, 209–214. https://doi.org/10.1111/jgs.12105

Chambaere, K., Vander Stichele. R., Mortier, F., Cohen, J., & Deliens, L. (2015). Recent trends in euthanasia and other end-of-life practices in Belgium. *New England Journal of Medicine, 372*, 1179–1181. https://doi.org/10.1056/NEJMc1414527

Clark, D. (2018). Cicely Saunders: A life and legacy. *Oxford University Press.* https://doi.org/10.1093/oso/9780190637934.001.0001

Clayton, J. M., Butow, P. N., Tattersall, M. H., Devine, R. J., Simpson, J. M., Aggarwal, G., Clark, K. J., Currow, D. C., Elliott, L. M., Lacey, J., Lee, P. G., & Noel, M. A. (2007). Randomized controlled trial of a prompt list to help advanced cancer patients and their caregivers to ask questions about prognosis and end-of-life care. *Journal of Clinical Oncology, 25*(6), 715–723. https://doi.org/10.1200/JCO.2006.06.7827

Cohen, J., & Deliens, L. (Eds.) (2012). *A public health perspective on end of life care*. Oxford University Press. https://doi.org/10.1093/acprof:oso/9780199599400.001.0001

Conner, S. R. (2019). Global aspects of palliative care. In R. D. MacLeod & L. Van den Block (Eds.), *Textbook of palliative care* (pp. 29–36). Springer. https://doi.org/10.1007/978-3-319-31738-0_4-1

Davies, E., & Higginson, I. J. (2004). *Palliative care: The solid facts*. World Health Organization Europe. ISBN 9289010916

Deliens, L., Mortier, F., Bilsen, J. Cosyns, M., Vander Stichele, R., Johan Vanoverloop, J., & Ingels, K. (2000). End-of-life decisions in medical practice in Flanders, Belgium: A nationwide survey. *Lancet, 356*(9244), 1806–1811. https://doi.org/10.1016/s0140-6736(00)03233-5

De Roo, M. (2015). *Quality indicators for palliative care* (Unpublished doctoral dissertation). VU University Medical Centre. ISBN 978-94-6108-896-3.

De Schreye, R. (2020). *Population-level evaluation of the appropriateness of end-of-life care in Belgium* (Unpublished doctoral dissertation). Vrije Universiteit Brussel. ISBN 9789461170385.

Detering, K. M., Hancock, A. D., Reade, M. C., & Silvester, W. (2010). The impact of advance care planning on end of life care in elderly patients: Randomised controlled trial. *BMJ,* 340, c1345. https://doi.org/10.1136/bmj.c1345

De Vleminck, A., & Deliens, L. (2017). Advance care planning for older people. In J. P. Michel, B. L. Beattie, F. C. Martin, & J. D. Walston (Eds.), *Oxford textbook of geriatric medicine* (3rd ed., pp. 1235–1240). Oxford University Press. https://doi.org/10.1093/med/9780198701590.003.0160

Diamond, R. E., Khurana, P., & Quill, T. (2019). Request for assisted suicide. In R.D. MacLeod & L. Van den Block (Eds.), *Textbook of palliative care* (pp. 1639–1647). Springer. https://doi.org/10.1007/978-3-319-77740-5_90

Gomes, B., Calanzani, N., Gysels, M., Hall, S., & Higginson I. J. (2013a). Heterogeneity and changes in preferences for dying at home: A systematic review. *BMC Palliative Care, 12*(1), 7. https://doi.org/10.1186/1472-684X-12-7

Gomes, B., Calanzani, N., Curiale, V., McCrone, P., & Higginson, I. J. (2013b). Effectiveness and cost-effectiveness of home palliative care services for adults with advanced illness and their caregivers. *Cochrane Database of Systematic Reviews,* (6), CD007760. https://doi.org/10.1002/14651858.CD007760

Houttekier, D., Cohen, J., Pepersack, T., & Deliens, L. (2014). Dying in hospital: A study of incidence and factors related to hospital death using death certificate data. *European Journal of Public Health, 24*(5), 751–756. https://doi.org/10.1093/eurpub/ckt209

Human Rights Watch. (2009). *Please, do not make us suffer anymore. Access to pain treatment as a human right*. Human Rights Watch. https://www.hrw.org/sites/default/files/reports/health0309webwcover_1.pdf

Kellehear, A. (2005). *Compassionate cities: Public health and end-of-life care.* Routledge. ISBN 9780415367738.

Krakauer, E. L. (2019). Palliative care, towards a more responsive definition. In R. D. MacLeod & L. Van den Block (Eds.), *Textbook of palliative care* (pp. 105–109). Springer. https://doi.org/10.1007/978-3-319-31738-0_1-1

Lynn, J., & Adamson, D. M. (2003). *Living well at the end of life: Adapting health care to serious chronic illness in old age*. RAND. https://doi.org/10.7249/WP137

Mroz, S., Dierickx, S., Deliens, L., Cohen, J., & Chambaere, K. (2021). Assisted dying around the world: A status quaestionis. *Annals of Palliative Medicine, 10*(3), 3540–3553. https://doi.org/10.21037/apm-20-637

Murray, S. A., Kendall, M., Boyd, K., & Sheikh, A. (2005). Illness trajectories and palliative care. *BMJ, 330*, 1007–1011. https://doi.org/10.1136/bmj.330.7498.1007

Patel, R. V., Sinuff, T., & Cook, D. J. (2004). Influencing advance directive completion rates in non-terminally ill patients: A systematic review. *Journal of Critical Care, 19*(1), 1–9. https://doi.org/10.1016/j.jcrc.2004.02.002

Reyniers, Th. (2015). *Hospital admissions at the end of life. Reasons, appropriateness and avoidability* (Unpublished doctoral dissertation). Vrije Universiteit Brussel Belgium & Ghent University Amsterdam Netherlands. ISBN 9789057184239.

Rietjens, J. A. C., Sudore, R. L., Connolly, M., van Delden, J. J., Drickamer, M. A., Droger, M., van der Heide, A., Heyland, D. K., Houttekier, D., Janssen, D. J. A., Orsi, L., Payne, S., Seymour, J., Jox, R. J., & Korfage, I. J., on behalf of the European Association for Palliative Care. (2017). Definition and recommendations for advance care planning: An international consensus supported by the European Association for Palliative Care. *Lancet Oncology, 18*(9), e543–e551. https://doi.org/10.1016/S1470-2045(17)30582-X

Robijn, L. (2020). *Towards the improvement of continuous sedation until death in nursing homes* (Unpublished doctoral dissertation). Vrije Universiteit Brussel Belgium & Ghent University Amsterdam Netherlands. ISBN 9789461170484.

Sallnow, L., & Paul, S. (2019). New public health approaches to end-of-life care. In R. D. MacLeod & L. Van den Block (Eds.), *Textbook of palliative care* (pp. 1739–1748). Springer. https://doi.org/10.1007/978-3-319-31738-0_97-1

Saunders, C., Baines, M., & Dunlop, R. (1995). *Living with dying: A guide to palliative care* (3rd Ed.). Oxford University Press. https://doi.org/10.1093/acprof:oso/9780192625144.001.0001

Seale, C. (2000). Changing patterns of death and dying. *Social Science & Medicine,* 51(6), 917–930. https://doi.org/10.1016/s0277-9536(00)00071-x

Seale, C. (2009). End-of-life decisions in the UK involving medical practitioners. *Palliative Medicine, 23*(3), 198–204. https://doi.org/10.1177/0269216308102042

Sepúlveda, M., Marlin, A., Yoshida, T., & Ullrich, A. (2002). Palliative care: The World Health Organization's global perspective. *Journal of Pain and Symptom Management, 24*(2), 91–96. https://doi.org/10.1016/S0885-3924(02)00440-2

Seymour, J., Gott, M., Bellamy, G., Ahmedzai, S. H., & Clark, D. (2004). Planning for the end of life: The views of older people about advance care statements. *Social Science & Medicine, 59*(1), 57–68. https://doi.org/10.1016/j.socscimed.2003.10.005

Singer, P. A., Martin, D. K., & Kelner, M. (1999). Quality end-of-life care; Patients' perspectives. *Journal of the American Medical Association, 281*(2), 163–168. https://doi.org/10.1001/jama.281.2.163

Sudore, R. L., & Fried, T. R. (2010). Redefining the "planning" in advance care planning: Preparing for end-of-life decision making. *Annals of Internal Medicine, 153*(4), 256–261. https://doi.org/10.7326/0003-4819-153-4-201008170-00008

Temel, J. S., Greer, J. A., Muzikansky, A., Gallagher, E. R., Admane, S., Jackson, V. A., Dahlin, C. M., Blinderman, C. D., Jacobsen, J., Pirl, W. F., Billings, J. A., & Lynch, T. J. (2010). Early palliative care for patients with metastatic non-small-cell lung cancer. *New England Journal of Medicine, 363*(8), 733–742. https://doi.org/10.1056/NEJMoa1000678

Vanbutsele, G., Pardon, K., Van Belle, S., Surmont, V., De Laat, M., Colman, R., Eecloo, K., Cocquyt, V., Geboes, K., & Deliens, L. (2018). Effect of early and systematic integration of palliative care in patients with advanced cancer: A randomised controlled trial. *Lancet Oncology, 19*(3), 394–404. https://doi.org/10.1016/S1470-2045(18)30060-3

Van den Block, L., van der Heide, A., & Deliens, L. (2011). End-of-life decisions in cancer patients. In E. Azoulay (Ed.), *Pulmonary involvement in patients with hematological malignancies* (pp. 693–700). Springer Verlag. ISBN 9783642157424 – 9783642157417.

van der Heide, A., Onwuteaka-Philipsen, B. D., & van Delden, J. J. M. (2017). End-of-life decisions in the Netherlands over 25 years. *New England Journal of Medicine, 377*(5), 492–494. https://doi.org/10.1056/NEJMc1705630

van der Heide, A., & Rietjens J. A. C. (2012). End-of-life decisions. In J. Cohen & L. Deliens (Eds.), *A public health perspective on end of life care* (pp. 50–59). Oxford University Press. https://doi.org/10.1093/acprof:oso/9780199599400.001.0001

Wilkinson, A. M. (2011). Advance directives and advance care planning: the US experience. In K. Thomas & B. Lobo (Eds.), *Advance care planning in end of life care*. Oxford University Press. https://doi.org/10.1093/acprof:oso/9780199561636.003.0017

Wilkinson, A. M., Wenger, N., & Shugarman, L. R. (2007). *Literature review on advance directives.* U.S. Department of Health and Human Services, Assistant Secretary for Planning and Evaluation, Office of Disability, Aging and Long-Term Care Policy. https://aspe.hhs.gov/reports/literature-review-advance-directives-0

World Health Organization. (1990). WHO expert committee on cancer pain relief and active supportive care & World Health Organization: Cancer pain relief and palliative care: Report of a WHO expert committee [meeting held in Geneva from 3 to 10 July 1989]. https://apps.who.int/iris/handle/10665/39524

World Health Organization (WHO) (2002). *National cancer control programmes: Policies and managerial guidelines.* World Health Organization. ISBN 9241545577. https://apps.who.int/iris/bitstream/handle/10665/42494/9241545577.pdf?sequence=1&isAllowed=y

Zimmermann, C., Swami, N., Krzyzanowska, M., Hannon, B., Leighl, N., Oza, A., Moore, M., Rydall, A., Rodin, G., Tannock, I., Donner, A., & Lo, C. (2014). Early palliative care for patients with advanced cancer: A cluster-randomised controlled trial. *Lancet, 383*(9930), 1721–1730. https://doi.org/10.1016/S0140-6736(13)62416-2

2

Experience of End-of-Life Issues by People with Intellectual Disability

Roger J. Stancliffe, Michele Y. Wiese, Philip McCallion, and Mary McCarron

Introduction

This chapter aims to examine available end-of-life research involving people with intellectual disability, with a focus on their experiences, rather than the views of others such as caregivers and clinicians. In part, it is intended to complement Chapter 1, which dealt with general

R. J. Stancliffe
Centre for Disability Research and Policy, University of Sydney, Sydney, NSW, Australia
e-mail: roger.stancliffe@sydney.edu.au

M. Y. Wiese (✉)
School of Psychology, Western Sydney University, Penrith, NSW, Australia
e-mail: m.wiese@westernsydney.edu.au

P. McCallion
School of Social Work, Temple University, Philadelphia, PA, USA
e-mail: philip.mccallion@temple.edu

R. J. Stancliffe et al. (eds.), *End of Life and People with Intellectual and Developmental Disability*,
https://doi.org/10.1007/978-3-030-98697-1_2

community experiences. This chapter provides an overview of the current participation and experience of people with intellectual disability in end-of-life matters. Many ensuing chapters offer further detail and these are referred to throughout.

We planned to include studies concerning people with all types of developmental disability, but have been constrained by the available research. The strong and growing body of end-of-life research concerning people with intellectual disability, together with a relative paucity of research examining the experiences of individuals without intellectual disability but with autism spectrum disorder (ASD), cerebral palsy and other early onset disabilities, means that most of the studies we discuss concern people with intellectual disability, a reality reflected in our chapter's title. Because of the disproportionate focus on this group, our chapter employs a convention that where the nature of the developmental disability is not stated, readers should assume that people with intellectual disability are being referred to, whereas when discussing people with autism spectrum disorder or cerebral palsy for example, we will explicitly state this. One exception to this research imbalance involves the growing research attention to suicide among people with ASD (see Chapter 3), which will receive limited attention here.

Some previously published research reviews examine specific issues in greater depth than is possible here, or cover topics not considered in the current chapter. To assist readers to locate reviews relevant to their interests, Table 2.1 lists such publications by the review's main topic. All reviews listed are of research involving people with intellectual disability.

M. McCarron
Trinity Centre for Ageing and Intellectual Disability, Trinity College Dublin, Dublin, Ireland
e-mail: mccarrm@tcd.ie

Table 2.1 End-of-life research literature reviews involving the participation and experiences of people with intellectual disability by main topic

Main topic	Reviews
Bereavement	Dodd et al. (2005)
Decision making at end of life	Kirkendall et al. (2017)
End-of-life care	Moro et al. (2017)
Palliative care	Adam et al. (2020), Dunkley and Sales (2014) and Tuffrey-Wijne et al. (2007)
Quality of life of elderly people	Schepens et al. (2019)
Staff readiness to provide support	Lord et al. (2017) and McCallion et al. (2012)
Suicide	Dodd et al. (2016) and Mollison et al. (2014)

The Experiences of People with Intellectual Disability

What People with Intellectual Disability Say, Feel and Do About End-of-Life Issues

There is a range of evidence that end-of-life issues are important to people with intellectual disability. For example:

- Individuals explained that end of life was very important to them and recounted poignant examples of how dying and death had affected them (McEvoy et al., 2012; McRitchie et al., 2014; Read & Carr, 2014).
- Adults with intellectual disability experienced grief (Dodd et al., 2005, 2021; McRitchie et al., 2014; Ryan et al., 2011b).
- Adults with intellectual disability indicated clearly that determining and honouring the person's wishes is crucial for good end-of-life care (Bekkema et al., 2016).
- Adults with intellectual disability who knew that they were dying actively participated in their own advance care planning (McKenzie et al., 2017, p. 1094) and all asked that support people (staff and family) "talk straight" and provide clear information so they could understand.

- Many adults regularly *initiate* discussion or activities about end of life or a specific dead person, indicating that there are issues they want to know about and are motivated to discuss (Hollins & Esterhuyzen, 1997; Stancliffe et al., 2021; Tuffrey-Wijne et al., 2020).
- When asked, almost all adults with intellectual disability who could speak for themselves mentioned the death of someone they knew, signifying that this was a salient experience (McEvoy et al., 2012; Stancliffe et al., 2016).
- Adolescents and young adults with intellectual disability worry more about the possible death of a loved one than age peers without intellectual disability (Forte et al., 2011; Young et al., 2016). Adults with intellectual disability reported greater fear of death than adults without intellectual disability, but with wide variability between individuals (Stancliffe et al., 2016).
- People with intellectual disability say they feel a responsibility to attend funerals to show respect for and say farewell to the deceased and report finding the ceremony worthwhile (Forrester-Jones, 2013; McRitchie et al., 2014).

This wide range of evidence consistently shows that end of life is very important to people with intellectual disability. Individuals with other developmental disabilities presumably also consider end of life to be important, but direct research evidence about their views is scarce. Dark et al. (2011) reported the clear sense of loss and grief experienced by adults with cerebral palsy related to the death of someone close. However, the literature is silent regarding the other issues listed above as they are experienced by people with cerebral palsy or autism spectrum disorder and without intellectual disability.

Exclusion from End-of-Life Discussions and Activities

Even though end of life is important to people with intellectual disability, studies regularly show that caregivers and professionals avoid involving people with intellectual disability in a wide range of end-of-life situations, such as:

- funerals (Dowling et al., 2006; Forrester-Jones, 2013; McRitchie et al., 2014; Chapter 10).
- informing people about their own dying (see *Your Own Death* below).
- involving them in their own end-of-life decisions or advance care planning (see *Your Own Death* below and Chapter 4).
- disclosing the terminal diagnosis and/or death of friends/family members (see *The Death of Others* below), and sometimes even deliberately not responding to end-of-life concerns raised by people with intellectual disability (Ryan et al., 2011b; Wiese et al., 2013).

There are exceptions to this pattern of protection and exclusion for people with intellectual disability (Forrester-Jones, 2013; McKenzie et al., 2017; McMaugh et al., 2017; McRitchie et al., 2014), but those exceptions would be seen as unremarkable, even typical involvement in end of life for the general population.

The Death of Others

In the first systematic study of bereavement experienced by adults with intellectual disability, Hollins and Esterhuyzen (1997) found widespread exclusion before, at and after death, of adults who attended UK day centres and had experienced parental bereavement in the previous two years. Two-thirds of the parents did not die suddenly, but only 17% of those who did not live with the deceased parent regularly visited their dying family member. While 54% were reported to have attended the funeral, and 68% had mementos of the dead parent, only 16% had taken part in a pre-bereavement discussion about the possibility that their parent would die. Moreover, a mere 16% of these adults had an opportunity to visit the grave, or to receive any bereavement counselling. These findings contrast with research involving the general community reporting, for example, much greater disclosure of expected parental death to offspring. Sheehan et al.'s (2014) study of hospice patients found that *all* families discussed the dying parent's illness and expected death with their adolescent children without disability.

Multiple studies have confirmed that people with intellectual disability are often not told that a relative or friend is dying (Cithambaram et al.,

2019; Forrester-Jones, 2013; Ryan et al., 2011b). One important effect of this practice is that, even in cases of what would otherwise be *expected* death of others, people with intellectual disability experience sudden and unexpected deaths, having been denied the opportunity to go through anticipatory grief and to say goodbye.

By contrast, Ryan et al. (2011b) noted disability staff reports of the positive benefits of involving peers before the expected death of a fellow service user. Likewise, in a study of bereaved adults with intellectual disability who had experienced a family member's death, Dowling et al. (2006) showed that counselling from volunteer bereavement counsellors resulted in improved mental health and behaviour of the bereaved, as well as comfort and willingness to talk about the dead person.

Your Own Death

People with intellectual disability are rarely told that they are dying (Todd, 2004, 2013; Tuffrey-Wijne et al., 2020). There is consistent evidence that people with intellectual disability frequently are not involved in their own end-of-life decision-making or advance care planning (Flynn et al., 2016; Kirkendall et al., 2017; Moro et al., 2017; Voss et al., 2017, 2019). Such exclusion may be unavoidable in the final days/weeks of life when people are acutely ill and unable to speak for themselves, but this widespread disenfranchisement goes much further to encompass times when the person is able to express their views and exercise self-determination. This pattern has been reported many times, but there are only a handful of counter examples confirming what is possible; that when a person with intellectual disability who is dying is given clear information and opportunities to take part in their own end-of-life planning and decision-making, more favourable outcomes are achieved (Flynn et al., 2016; McKenzie et al., 2017). The story below illustrates some of the benefits of involving a man with intellectual disability in planning for his remaining life and his funeral, knowing that he is dying.

An indigenous man's story

Cultures vary in their practices and beliefs about end of life and funerals. This story involves an Australian indigenous man with intellectual disability and in accordance with indigenous custom, his name will not be used. He had become separated from his birth family as a child, and spent most of his life living with a non-indigenous family. At the time he developed a life-limiting illness, he was being supported by an intellectual disability service. He was aware he was dying, and with the support of an indigenous staff member, was able to learn about relevant indigenous end-of-life ceremonies and customs, and to reconnect with his birth family. Through a series of conversations with this worker, he expressed his wishes about the rest of his life and what he wanted at his funeral. Because he was able to have these conversations *before* he died, he had the opportunity to meet two of his brothers, and to have his funeral wishes honoured.

An online video *Honouring Cultural Wishes* from the *Talking End of Life* (TEL) website presents more details about this man's story: https://vimeo.com/260009282 or https://www.caresearch.com.au/tel/tabid/4659/Default.aspx

Disclosure and Non-disclosure

In the largest such study to date, Tuffrey-Wijne et al. (2020) found that people may be told that they are ill, but most (82%) are not told that they are going to die (see also Ryan et al., 2011a). As noted previously, people with intellectual disability may not be told about the terminal condition or even the death of people they are close to. This tendency towards non-disclosure can be evident even when people with intellectual disability ask questions; Wiese et al. (2013) and Ryan et al. (2011b) reported examples of disability staff deliberately withholding information and avoiding the topic.

Reasons for Non-disclosure

Often described as *protection and avoidance* (Wiese et al., 2015), reasons for non-disclosure and unwillingness to discuss end of life encompass factors related to people with intellectual disability, to caregivers and to disability service settings. The reasons include (a) concern about upsetting the person or causing emotional/psychological harm, (b) problems of understanding and communication, (c) caregivers being uncomfortable with the topic or unsure what to say, (d) problems with the knowledge, skills, training or resources available to caregivers and (e) disability service policies on end of life and/or confidentiality that are absent/unclear (Cithambaram et al., 2019; Lord et al., 2017; Ryan et al., 2011a, 2011b; Todd, 2013; Tuffrey-Wijne & Rose, 2017; Tuffrey-Wijne et al., 2013; Voss et al., 2019; Wiese et al., 2012, 2013, 2015). Tuffrey-Wijne et al.'s (2020) UK national survey found that dying people with intellectual disability with less understanding of the concept of death and those with poorer communication skills were significantly less likely to be told they were going to die. Ryan et al. (2011a, p. 201) found that disability staff "were concerned that their lack of skill and experience in the area would cause harm if they engaged in open conversations" about death and dying.

Worry about upsetting people with intellectual disability and causing psychological harm has been reported as a particularly widespread concern constraining caregivers from disclosure and from involving people with intellectual disability in end-of-life activities, discussions and planning (Lord et al., 2017). Yet there is growing evidence that such concerns are overstated. Multiple studies now exist in which adults with intellectual disability have been asked about potentially sensitive end-of-life issues without any evident distress, even among those who have experienced recent parental bereavement (Dodd et al., 2021; McRitchie et al., 2014; O'Keeffe et al., 2019). In a controlled, longitudinal intervention study, Stancliffe et al. (2021) found no evidence of psychological harm and only what would be considered typical (mostly low) discomfort levels, when people with intellectual disability talked about end of life (see *How do People with Intellectual Disability Respond Emotionally to Discussing End of Life?* below).

Dealing with Disclosure Issues

Unless the person with intellectual disability has no or almost no contact with the dying individual (which does happen), it is difficult to envisage how that individual's eventual death could be hidden. Therefore, the issue is when, how and how much to tell the person with intellectual disability. Also, in the case of expected deaths, there is a question if people should also be offered opportunities (and any needed support) to spend time with the dying person, to say goodbye and to prepare themselves for the eventual death. Research shows benefits to people with intellectual disability spending time with the dying person (Ryan et al., 2011b).

Research also indicates that some people, with or without disability, are uncomfortable with these issues and should have the opportunity to make an informed choice about the information they receive and their degree of participation. For example, one of McEvoy et al.'s (2012, p. 198) participants said, "I didn't attend the funeral. I wanted to go but couldn't bring myself to go" (see also Chapter 10).

While the end of life is inevitable, engaging with it is challenging, especially if the person has learning and communication needs. Available research supports the notion that ideally, suitable training and resources should be available to assist caregivers to hold these conversations in an individualised and sensitive manner (Tuffrey-Wijne et al., 2020). To help with these delicate judgements, we draw readers' attention to the intellectual-disability-specific research, resources and guidelines about breaking bad news (Tuffrey-Wijne, 2013a, 2013b; Tuffrey-Wijne et al., 2010, 2013). Tuffrey-Wijne's (2013a) guidelines emphasise honouring the person's wishes about disclosure, if the person has capacity. If the person lacks capacity, best-interest decisions are guided by possible harm of disclosure or of non-disclosure, together with close consideration of the person's wish to know or not. The issue of capacity and its consequences for decision-making is increasingly contested (see also Chapter 5) and there is no agreed gold standard for capacity assessment. It can be difficult to predict how the person may react. For example, disclosure of terminal cancer to a person with intellectual disability resulted in distress in one case, but in apparent acceptance without evident upset in another (Tuffrey-Wijne et al., 2010). Conversely, a woman whose terminal cancer was not disclosed was regularly distressed by her unexplained and deteriorating health problems (Tuffrey-Wijne et al., 2010).

Staff Readiness to Provide Support

Often there is not good communication between family and staff caregivers of persons with intellectual disability and hospice and palliative care providers (Fahey-McCarthy et al., 2009). On a system level, specialist palliative care staff need training to address their communication difficulties with persons with intellectual disability, and to redesign approaches to achieve better integration of services structures in intellectual disability and specialist palliative care (McCallion et al., 2012). On a day-to-day operational level, Todd (2004) highlighted that when staff working in intellectual disability care are equipped to report important changes, this has proven critical to good symptom management in terminal care. The need for caregivers to support people with intellectual disability to take part in advance care planning, discussions about death and dying and the experience of death rituals may also require training for caregivers. There are reports on evidence-informed training interventions (Fahey-McCarthy et al., 2009; Hahn & Cadogan, 2011; Voss et al., 2021) which have been shown to change palliative care and advanced dementia care practices. However, beyond grief and bereavement and advance care planning (Dowling et al., 2006; McKenzie et al., 2017; Voss et al., 2021), there is an absence of evaluated staff training interventions to improve inclusion of the person with intellectual disability.

Active Participation in End-of-Life Research

Referring primarily to intellectual-disability-specific research and active participation (e.g., by self-reporting) by people with disability in that research, Stancliffe, Wiese et al. (2017, p. 977) noted "the apparent underrepresentation of people with intellectual disability in end-of-life research". Kirkendall et al. (2017) reviewed 12 papers on end-of-life decision-making, but only one involved people with intellectual disability as active participants. Likewise, Voss et al. (2017) examined 14 studies on advance care planning for people with intellectual disability and found no examples of publications reporting their own perspectives. Instead, studies obtained the views of professionals, family or disability service staff, or were based on medical-records data. We know of no end-of-life research where children with intellectual disability were active participants, whereas participation is more common by children without disability, even children as young as 4 years (Slaughter & Griffiths, 2007).

Research participation by those with lived experience of disability has improved over time, but varies markedly by topic. For example, there is somewhat greater representation of adults with intellectual disability as active participants in research on more subjective topics such as grief (e.g., Dodd et al., 2021; McEvoy et al., 2012; McRitchie et al., 2014). However, even in grief research, the tendency is to focus on caregiver views, observation or assessment of behavioural or psychiatric symptoms, often with proxies as informants (e.g., Hollins & Esterhuyzen, 1997). We are not aware of any published end-of-life studies that involve people with intellectual disability as co-researchers, although some fine examples exist of jointly constructed personal stories (e.g., Read & Carr, 2014).

Stancliffe, Wiese et al. (2017, p. 977) also asked "whether the widely recognized issue of avoidance and protection by families and disability staff … also extends to researchers, ethics committees and those who gate-keep access to research by people with intellectual disability". Undoubtedly, many groups can contribute to understanding end of life, but the voices of people with intellectual disability are central, as a matter of both social justice and research validity. On the latter point, Dodd et al., (2005, pp. 541–542) noted that separate interviews with carers and bereaved individuals with intellectual disability:

> have again exposed the once common perception amongst carers, that people with intellectual disability have no response or merely a limited response to a death. However, this was broadly contradicted by the

bereaved people themselves, who clearly expressed their sadness, distress and anxiety at their loss.

This clear divergence is an example of the larger issue of important differences between proxy and self-report data (Dodd et al., 2021), where proxies, unable to report accurately about unobservable emotional responses, instead rely on their *interpretation* of the person's observable behaviour, resulting in data which does not accurately reflect the views of people with intellectual disability themselves. Reliance on proxy data or medical records in end-of-life research perhaps suggests that researchers have been unwilling or unable to directly involve people with intellectual disability. Possible reasons for exclusion include:

- disability researchers' own overprotective attitudes
- gatekeepers protecting people with intellectual disability from opportunities to participate in what those gatekeepers see as potentially risky research
- participants with intellectual disability not meeting inclusion criteria, such as awareness of their own terminal condition, which in turn may result from overprotection by family or professionals (McKenzie et al., 2016, 2017)
- perceived or actual problems with ethics committees and their willingness to approve end-of-life research involving vulnerable participants, and/or difficulties with consent processes (Savage et al., 2015)
- the relative lack of validated self-report end-of-life measurement scales adapted for people with intellectual disability (see *Measurement Instruments Designed or Adapted Specifically for People with Intellectual Disability below, Resources* at the end of this chapter, and Chapters 3 and 17).

Moro et al. (2017, p. 1052) commented that "encouraging the active participation of adults with intellectual disabilities in end-of-life research is critical". Stancliffe et al. (2021) (see *How do People with Intellectual Disability Respond Emotionally to Discussing End of Life?* below) found no apparent psychological or emotional harm from discussing end of life with adults with intellectual disability. Yet, McKenzie et al. (2016) highlighted a particular recruitment difficulty for research on collaborative advance care planning; over 60% of potential participants with intellectual disability were ineligible because they had not been told they were dying. Apart from this study, we know of no other direct research on the inclusion of people with intellectual disability in end-of-life research. The relative absence of the voices of people with intellectual disability in this literature suggests researchers may need to reflect on their own attitudes, adapt instruments to the needs/capacities of people with intellectual disability and educate gatekeepers in disability services and ethics committees.

Savage et al. (2015) discussed approaches to better address sensitive involvement of people with intellectual disability in end-of-life research, such as a multi-faceted approach to identify and respond to any participant discomfort or distress. Even so, Savage and her colleagues were unable to fully overcome problems of communication and understanding, and like most end-of-life research involving people with intellectual disability, only included participants who could verbally communicate their views and experiences.

What People with Intellectual Disability Understand About End of Life?

People with intellectual disability reportedly have a limited understanding of the concept of death. Not unexpectedly, those with milder intellectual disability have a better comprehension than individuals with more significant impairment, but level of understanding of death varies greatly (McEvoy et al., 2012, 2017; Stancliffe et al., 2016). Further, lack of basic biological knowledge of bodily functioning (McEvoy et al., 2017), and limited everyday opportunities to learn about death (Wiese et al., 2015) likely constrain the opportunities available to people with intellectual disability to improve their understanding.

The inevitability of one's own death is often poorly understood (McEvoy et al., 2012; Stancliffe et al., 2016). Nevertheless, there are reported individual examples of people with intellectual disability understanding that they were going to die (McKenzie et al., 2017; Tuffrey-Wijne et al., 2010). Ryan et al. (2011b, p. 264) noted that people with intellectual disability who "had a poor understanding of the concept of death appeared to be especially vulnerable to high levels of distress when either they or their friends became unwell. They worried that they would be next to die".

How Do People with Intellectual Disability Respond Emotionally to Discussing End of Life?

There is consistent anecdotal research evidence showing that, in studies excluding participants with recent bereavement or themselves approaching death, adults with mild/moderate intellectual disability and verbal communication are willing and able to talk about end-of-life without evident ill effects (Forrester-Jones, 2013; McEvoy et al., 2012, 2017; Stancliffe et al., 2016; Stancliffe, Wiese, Read et al., 2017). This is also true of individuals who experienced bereavement in recent years (Dodd et al., 2021; McRitchie et al., 2014; O'Keeffe et al., 2019; Savage et al., 2015). In all studies, no or very few participants experienced transient emotional discomfort. In the handful of studies involving discussion with individuals of *their own* terminal condition, not surprisingly, emotional discomfort was more common (McKenzie et al., 2017; Tuffrey-Wijne et al., 2010).

Moving beyond anecdotal and observational evidence, Stancliffe et al.'s (2021) study was the first to directly assess mental health and emotional effects on adults with intellectual disability of engaging in conversations and activities (encounters) about end of life. Key findings following the six-month intervention were that:

- over two-thirds of encounters were initiated by the individual with intellectual disability, a clear indication that the person wanted to engage with end-of-life issues
- there was no pre-post change in self-reported depression or fear of death, while anxiety improved

- people with intellectual disability were comfortable in most encounters. Of the 8% of encounters that were emotionally very uncomfortable, these mostly concerned the death of a known person or pet and were participant-initiated, yet revealed a clear desire to talk about the issue despite feeling emotional.

Death can be a sensitive topic and it is not surprising that in a few cases people can become upset. However, Stancliffe et al. (2021) reported that this discomfort was transient and responded to reassurance from staff. Other studies have also found that emotional discomfort sometimes occurs but typically is short-lived (Cithambaram et al., 2019) and helped by emotional support from others (McKenzie et al., 2017). McKenzie et al. (2017) stated that planned support strategies were beneficial when clients were upset, and could include both immediate and longer-term strategies, such as pre-programmed phones, so the person with intellectual disability could easily call key support people, and/or scheduling extra visits from staff or friends. Taken together, these studies suggest that caregivers should expect occasional emotional discomfort and should plan ahead with helpful supports in case they are needed. Overall, the research clearly shows that talking about end of life is not harmful and importantly, if approached sensitively, can be beneficial.

Stancliffe et al.'s (2021) data showed many people with intellectual disability themselves initiated conversations about end of life, but others may not have the words, or may not know how to start these discussions (Tuffrey-Wijne et al., 2020). This issue may be compounded by caregivers' lack of recognition of the person's emotional response and need for support: "Caregivers regularly interpreted the lack of emotional expression as an indicator that the patients were coping well, and had not needed to seek emotional support, with most neglecting to ask questions about emotions" (Flynn et al., 2016, p. 1202). Fortunately, end-of-life resources designed specifically for people with intellectual disability are available to help guide these discussions and to help people with intellectual disability to understand (see *Resources* at the end of this chapter and Chapter 17).

Measurement Instruments Designed or Adapted Specifically for People with Intellectual Disability

Reliable and valid end-of-life assessment instruments are one important resource for clinicians, practitioners and researchers. Without these tools, the subjective experiences and knowledge of people with intellectual disability about end of life cannot be measured appropriately. Until recently, end-of-life instruments tailored to the cognitive and communication needs of people with intellectual disability have not existed. Instead, proxy reporting using available mainstream instruments has been relied upon, meaning that self-reports by people themselves have often not been elicited. This is problematic because subjective issues such as grief or suicidality risk, and cognitive matters including understanding death, are best evaluated by asking the person directly. Fortunately, in the last decade or so several robust and useful intellectual disability-specific self-report instruments have begun to appear.

These purpose-designed scales include helpful features to assist comprehension and responding, such as simple question wording, pictures, vignettes (brief stories) to make the issue more concrete, simple response formats and pictorial response scales to enable non-verbal responding (e.g., pointing). The availability of these instruments means that people with intellectual disability can better respond to various end-of-life topics, including their knowledge, plans, feelings and fears.

In Table 2.2, we offer a list of the known available instruments. The table summarises both self-report as well as proxy instruments, all designed for people with intellectual disability. More detailed information about the self-report instruments is given in the Resources list at the end of this chapter and in Chapter 17.

Though these instruments represent a promising start, their availability and applicability are currently limited. Prospective users need to contact the relevant researchers for access, the instruments' clinical utility has not been explored, their psychometric properties mostly have not been evaluated independently and training to use the instruments is not

Table 2.2 End-of-life assessment instruments for people with intellectual disability by topic and respondent

Topic	Respondent: Person with intellectual disability (self-report)	Respondent: Proxy
Complicated grief	Dodd et al. (2021) and O'Keeffe et al. (2019)—*Complicated Grief Questionnaire for People with Intellectual Disabilities (CGQ-intellectual disability)*	Guerin et al. (2009). *Complicated Grief Questionnaire for People with Intellectual Disabilities (CGQ-intellectual disability)*
Understanding the concept of death	McEvoy et al. (2012)—*Concept of Death Questionnaire (CODQ)*	
Palliative and end-of-life care		Hunt et al. (2019). *VOICES-SF*—retrospective mortality follow-back survey of palliative and end-of-life care. This instrument was designed for the general community but has been used successfully (with minor modifications) in intellectual disability research (e.g., Todd et al., 2020)
End-of-life planning	Stancliffe, Wiese, Read et al. (2017). *End-of-Life planning scale (EOLPS)*	
Fear of death	Stancliffe, Wiese, Read et al. (2017). *Fear of death scale 3.0-intellectual disability (FODS3.0-intellectual disability)*	

yet available. Further, critical instruments have not been developed. For example, Dodd et al. (2016) noted the absence of a valid intellectual disability-specific scale to measure suicidality.

Commentary

The available research has highlighted many critical issues including the need for more education and support for staff caregivers and for people with intellectual disability themselves, and better relationships, understanding and communication between staff in intellectual disability services and those in hospice and palliative care services (see also Chapter 8). The largely absent voices of people with intellectual disability themselves (as well as their peers and their family members) poses a challenge to researchers to ensure a more inclusive perspective. Families have been largely absent in intellectual disability service provision and their views about the inclusion of people with intellectual disability have also been filtered through staff perceptions and reports. The same has been true in the past with respect to the views of people with intellectual disability themselves. The changing landscape of more families continuing to provide care, person-centred planning and advance care planning capturing the wishes of the person with intellectual disability and then becoming the driving force for services and greater weight given to self-determination mean that the focus of research has been changing and needs to change more.

Most end-of-life research to date has involved identifying *problems and inequities* experienced by people with intellectual disability as perceived by others, rather than direct data collection and the evaluation of approaches to overcome these issues. There are a small number of published intervention studies (Dowling et al., 2006; McKenzie et al., 2017). Dowling et al. (2006) compared two approaches to bereavement support of adults with intellectual disability, and McKenzie et al. (2017) described the process and effects of supported participation by adults with intellectual disability in their own advance care planning. As noted by Adam et al. (2020, p. 1006) in their systematic review of palliative care, "there is a lack of research into strategies to improve practice". All of these issues need to be more fully addressed and as noted previously, will require the development of new measures to better capture a range of voices. Where there has been participation by people with intellectual disability, it has been by people with milder intellectual disability. We know less about the experiences and views of people with severe and profound intellectual disability (see also Chapter 13). Likewise, overwhelmingly the available research has involved adults with intellectual disability, with little known about children. These are important areas to begin expanding involvement of people with intellectual disability, both as participants and co-researchers.

Another issue is that, while there has been significant research related to people with intellectual disability and end of life, apart from mortality research (e.g., Landes et al., 2019), there has been far less research attention to these issues for people with other types of developmental disability such as cerebral palsy and autism spectrum disorder. The exception to this is one peer-reviewed study about communication issues related to grief and loss experienced by people with cerebral palsy (Dark et al., 2011, 2016) and another offering a narrative review and clinical advice about persons with autism coping with complex social emotional issues including grief (Ahlers et al., 2017). There is, however, a significant and growing literature concerning autism and suicide (see also Chapter 3), and a number of non-peer-reviewed publications that provide advice around how to support people with autism to cope with grief and loss (e.g., Forrester-Jones & Broadhurst, 2007; Lipsky, 2013). The widespread avoidance and protection by caregivers that has been reported for people with intellectual disability should also be explored with persons with cerebral palsy and autism. The numbers of persons with autism, in particular, are growing and researchers must respond in an inclusive manner.

In all of these endeavours to expand and report on research knowledge, a continued absence of agreed instruments and the reliance on a limited range of reporters will restrain our building of that knowledge and our ability to provide evidence for practice.

Reflection

The following story captures a number of issues discussed in this chapter. Gina's story clearly reflects exclusion: about the death of others, non-disclosure, active participation in the end of life and the individual's emotional responses to it. This story is unfortunately not an unusual one for people with intellectual disability and their experience of end of life.

Gina's* story
Gina is 43. She lives in supported accommodation, with drop-in support every day. Five years ago her mother, who lived about 300 kilometres away, died after a long illness. Gina had visited her mother during the illness and was aware she was dying. Following her mother's death, Gina's father and sister told Gina she "wasn't allowed" to attend the funeral. The reasons given were that it was too far away, that her family members would not have time to drive the long distance to collect Gina, and that it was "better all around" if Gina stayed at home. She reluctantly complied.

Gina made her own memory box of beloved objects and photos commemorating her mother. About a year after the death, she began to ask the drop-in staff about her mother and where she was buried. This became, in the staff member's words, "almost obsessive, like a broken record". When her father phoned occasionally, Gina's attempts to ask him where mum was buried were responded to vaguely. By the second year, staff began trying to use distraction techniques. Gina escalated to shouting and crying inconsolably about her mother. The cycle continued. Gina was said to have developed challenging behaviour.

* Assumed names are used in this story

Reflection Questions

1. How could Gina have been supported to have a different experience that was more inclusive? In answering this question, think about the various issues mentioned in this chapter, such as the death of others, non-disclosure about dying and death, active participation in the end of life, understanding what death is and the individual's emotional responses to it.
2. Caregivers and family clearly wish to protect people with intellectual disability from end-of-life discussions. The research is clear that

talking about the end of life is not harmful for the individual with intellectual disability, so how can caregivers be encouraged to protect less and more fully involve the person in these discussions?

Resources

Self-Report Instruments

1. *Complicated Grief Questionnaire for People with Intellectual Disabilities (CGQ-intellectual disability)* (Dodd et al., 2021; O'Keeffe et al., 2019). This 23-item self-report scale is administered by interview. It complements the existing proxy version of the CGQ-intellectual disability (Guerin et al., 2009). The authors propose that the scale is suitable for individuals with mild-moderate intellectual disabilities. Questions and response options are both supported with visual aids to enhance comprehension and to enable non-verbal responding (pointing). The CGQ-intellectual disability includes a practice pre-test intended to aid comprehension and minimise response bias. Enquiries should be directed to the authors.
2. *Concept of Death Questionnaire (CODQ)*—developed by McEvoy et al. (2012). This 13-item scale evaluates understanding of the five components of death (Causality, Finality, Non-functionality, Universality and Inevitability). Items relate to three simply worded vignettes concerning a person's death. The interviewer reads each vignette aloud, then asks the questions about the components of death. CODQ internal consistency and inter-rater agreement are satisfactory (McEvoy et al., 2012). The CODQ's utility and psychometrics have also been evaluated by an independent team and found to be sound (see Stancliffe et al., 2016; Stancliffe, Wiese, Read et al., 2017). Contact the CODQ authors for more information about availability.
3. *End-of-Life Planning Scale (EOLPS)*—developed by Stancliffe, Wiese, Read et al. (2017). This scale is designed to measure a person with intellectual disability's understanding and self-determination about end-of-life planning. It is not a tool to do such planning. The EOLPS has seven subscales: (a) Bequeathing, (b) Preferred place of care, (c)

Funeral wishes, (d) Preferred carers, (e) Advance care planning, (f) Things to take with me and (g) Organ donation. Each subscale has a simple vignette that the interviewer reads aloud before asking questions related to who decides, and what the person can do now (i.e., while alive) so that their wish is known. Each vignette is supported by one or two colour photos. Comprehensibility, reliability and validity data were mostly satisfactory or better (Stancliffe, Wiese, Read et al., 2017). A copy of the EOLPS is available from the authors.

4. *Fear of Death Scale 3.0-intellectual disability (FODS3.0-intellectual disability).* This scale was adapted by Stancliffe, Wiese, Read et al. (2017) from an existing fear of death scale for the general population (Lester & Abdel-Khalek, 2003). Adaptation involved simplifying the wording and response options. The FODS 3.0-intellectual disability has 28 items in four 7-item factors: (a) others' dying, (b) others' death, (c) your own dying and (d) your own death. The 3-point response scale ranges from 1 (not frightened), to 3 (very frightened). It is supported by a 3-point pictorial scale to enable pointing. Comprehensibility, reliability and validity were sound (Stancliffe, Wiese, Read et al., 2017). A copy of the FODS3.0-intellectual disability is available from the authors.

References

Adam, E., Sleeman, K. E., Brearley, S., Hunt, K., & Tuffrey-Wijne, I. (2020). The palliative care needs of adults with intellectual disabilities and their access to palliative care services: A systematic review. *Palliative Medicine, 34*(8), 1006–1018. https://doi.org/10.1177/0269216320932774

Ahlers, K. P., Gabrielsen, T. P., Lewis, D., Brady, A. M., & Litchford, A. (2017). Supporting individuals with autism spectrum disorder in understanding and coping with complex social emotional issues. *School Psychology International, 38*(6), 586–607. https://doi.org/10.1177/0143034317719942

Bekkema, N., de Veer, A. J., Hertogh, C. M., & Francke, A. L. (2016). Perspectives of people with mild intellectual disabilities on care relationships at the end of life: A group interview study. *Palliative Medicine, 30*(7), 625–633. https://doi.org/10.1177/0269216316640421

Cithambaram, K., Duffy, M., & Courtney, E. (2019). End-of-life care research on people with intellectual disabilities: Challenges for proactive inclusion in an Irish context. *British Journal of Learning Disabilities, 47*, 70–76. https://doi.org/10.1111/bld.12260

Dark, L., Balandin, S., & Clemson, L. (2011). Communicating about loss: Experiences of older Australian adults with cerebral palsy and complex communication needs. *Communication Disorders Quarterly, 32*(3), 176–189. https://doi.org/10.1177/1525740109353936

Dark, L. J., Clemson, L., & Balandin, S. (2016). Communication changes experienced by adults with cerebral palsy as they age. *International Journal of Speech-Language Pathology, 18*(6), 521–532. https://doi.org/10.3109/17549507.2016.1143976

Dodd, P., Doherty, A., & Guerin, S. (2016). A systematic review of suicidality in people with intellectual disabilities. *Harvard Review of Psychiatry, 24*(3), 202–213. https://doi.org/10.1097/HRP.0000000000000095

Dodd, P., Dowling, S., & Hollins, S. (2005). A review of the emotional, psychiatric and behavioural responses to bereavement in people with intellectual disabilities. *Journal of Intellectual Disability Research, 49*, 537–543. https://doi.org/10.1111/j.1365-2788.2005.00702.x

Dodd, P., McEvoy, J., Lockhart, K., Burke, T., O'Keeffe, L., & Guerin, S. (2021). An exploratory study of self-reported complicated grief symptoms in parentally bereaved adults with intellectual disability. *Journal of Intellectual Disability Research, 65*, 297–305. https://doi.org/10.1111/jir.12812

Dowling, S., Hubert, J., White, S., & Hollins, S. (2006). Bereaved adults with intellectual disabilities: A combined randomized controlled trial and qualitative study of two community-based interventions. *Journal of Intellectual Disability Research, 50*(4), 277–287. https://doi.org/10.1111/j.1365-2788.2005.00759.x

Dunkley, S., & Sales, R. (2014). The challenges of providing palliative care for people with intellectual disabilities: A literature review. *International Journal of Palliative Nursing, 20*(6), 279–284. https://doi.org/10.12968/ijpn.2014.20.6.279

Fahey-McCarthy, E., McCarron, M., Connaire, K., & McCallion, P. (2009). Developing an education intervention for staff supporting persons with intellectual disability and advanced dementia. *Journal of Policy and Practice in Intellectual Disabilities, 6*(4), 267–275. https://doi.org/10.1111/j.1741-1130.2009.00231.x

Flynn, S., Hulbert-Williams, N. J., Hulbert-Williams, L., & Bramwell, R. (2016). "You don't know what's wrong with you": An exploration of cancer-related experiences in people with an intellectual disability. *Psycho-Oncology, 25*(10), 1198–1205. https://doi.org/10.1002/pon.4211

Forrester-Jones R. (2013). The road barely taken: Funerals, and people with intellectual disabilities. *Journal of Applied Research in Intellectual Disabilities, 26,* 243–256. https://doi.org/10.1111/jar.12022

Forrester-Jones, R., & Broadhurst, S. (2007). *Autism and loss.* Jessica Kingsley Publishers.

Forte, M., Jahoda, A., & Dagnan, D. (2011). An anxious time? Exploring the nature of worries experienced by young people with a mild to moderate intellectual disability as they make the transition to adulthood. *British Journal of Clinical Psychology, 50*(4), 398–411. https://doi.org/10.1111/j.2044-8260.2010.02002.x

Guerin, S., Dodd, P., Tyrell, J., McEvoy, J., Buckley, S., & Hillery, J. (2009). An initial assessment of the psychometric properties of the complicated grief questionnaire for people with intellectual disabilities (CGQ-intellectual disability). *Research in Developmental Disabilities, 30*(6), 1258–1267. https://doi.org/10.1016/j.ridd.2009.05.002

Hahn, J. E., & Cadogan, M. P. (2011). Development and evaluation of a staff training program on palliative care for persons with intellectual and developmental disabilities. *Journal of Policy and Practice in Intellectual Disabilities, 8*(1), 42–52. https://doi.org/10.1111/j.1741-1130.2011.00288.x

Hollins, S., & Esterhuyzen, A. (1997). Bereavement and grief in adults with learning disabilities. *British Journal of Psychiatry, 170*, 497–502. https://doi.org/10.1192/bjp.170.6.497

Hunt, K., Richardson, A., Darlington, A., & Addington-Hall, J. (2019). Developing the methods and questionnaire (VOICES-SF) for a national retrospective mortality follow-back survey of palliative and end-of-life care in England. *BMJ Supportive & Palliative Care*, *9*, e5. https://doi.org/10.1136/bmjspcare-2016-001288

Kirkendall, A., Linton, K., & Farris, S. (2017). Intellectual disabilities and decision making at end of life: A literature review. *Journal of Applied Research in Intellectual Disabilities, 30*(6), 982–994. https://doi.org/10.1111/jar.12270

Landes, S. D., Stevens, J. D., & Turk, M. A. (2019). Heterogeneity in age at death for adults with developmental disability. *Journal of Intellectual Disability Research, 63*(12), 1482–1487. https://doi.org/10.1111/jir.12672

Lester, D., & Abdel-Khalek, A. (2003). The Collett-Lester fear of death scale: A correction. *Death Studies, 27*(1), 81–85. https://doi.org/10.1080/07481180302873

Lipsky, D. (2013). *How people with autism grieve and how to help*. Jessica Kingsley Publishers.

Lord, A. J., Field, S., & Smith, I. C. (2017). The experiences of staff who support people with intellectual disability on issues about death, dying and bereavement: A metasynthesis. *Journal of Applied Research in Intellectual Disabilities, 30*(6), 1007–1021. https://doi.org/10.1111/jar.12376

McCallion, P., McCarron, M., Fahey-McCarthy, E., & Connaire, K. (2012). Meeting the end of life needs of older adults with intellectual disabilities. In E. Chang & A. Johnson (Eds.). *Contemporary and innovative practice in palliative care* (pp. 255–267). Intech. https://www.intechopen.com/books/contemporary-and-innovative-practice-in-palliative-care/meeting-the-end-of-life-needs-of-older-adults-with-intellectual-disabilities

McEvoy, J., MacHale, R., & Tierney, E. (2012). Concept of death and perceptions of bereavement in adults with intellectual disabilities. *Journal of Intellectual Disability Research, 56*, 191–203. https://doi.org/10.1111/j.1365-2788.2011.01456.x

McEvoy, J., Treacy, B. & Quigley, J. (2017). A matter of life and death: knowledge about the body and concept of death in adults with intellectual disabilities. *Journal of Intellectual Disability Research, 61*(1), 89–98. https://doi.org/10.1111/jir.12347

McKenzie, N., Brandford, S., Conder, J., & Mirfin-Vietch, B. (2016). Why won't you tell me I'm dying? *Journal of Intellectual Disability Research, 60*(7–8), 707.

McKenzie, N., Mirfin-Veitch, B., Conder, J., & Brandford, S. (2017). "I'm still here": Exploring what matters to people with intellectual disability during advance care planning. *Journal of Applied Research in Intellectual Disabilities, 30*(6), 1089–1098. https://doi.org/10.1111/jar.12355

McMaugh, P. J., Wiese, M. Y., & Stancliffe, R. J. (2017). The experiences of parents in supporting their son or daughter with intellectual disability to learn about dying and death. *Journal of Intellectual & Developmental Disability, 42*(3), 285–294. https://doi.org/10.3109/13668250.2016.1236367

McRitchie, R., McKenzie, K., Quayle, E., Harlin, M., & Neumann, K. (2014). How adults with an intellectual disability experience bereavement and grief: A qualitative exploration. *Death Studies, 38*(3), 179–185. https://doi.org/10.1080/07481187.2012.738772

Mollison, E., Chaplin, E., Underwood, L., & McCarthy, J. (2014). A review of risk factors associated with suicide in adults with intellectual disability. *Advances in Mental Health and Intellectual Disabilities, 8*(5), 302–308. https://doi.org/10.1108/AMHintellectual.disability-05-2014-0021

Moro, T. T., Savage, T. A., & Gehlert, S. (2017). Agency, social and health-care supports for adults with intellectual disability at the end of life in out-of-home, non-institutional community residences in Western nations: A literature review. *Journal of Applied Research in Intellectual Disabilities, 31*, 484–489. https://doi.org/10.1111/jar.12374

O'Keeffe, L., Guerin, S., McEvoy, J., Lockhart, K., & Dodd, P. (2019). The process of developing self-report measures in intellectual disability: A case study of a complicated grief scale. *British Journal of Learning Disabilities, 47*(2), 134–144. https://doi.org/10.1111/bld.12261

Read, S., & Carr, M. (2014). Living with loss. In S. Read (Ed.), *Supporting people with intellectual disabilities experiencing loss and bereavement: Theory and compassionate practice* (pp. 36–45). Jessica Kingsley Publishers.

Ryan, K., Guerin, S., Dodd, P., & McEvoy, J. (2011a). Communication contexts about illness, death and dying for people with intellectual disabilities and life-limiting illness. *Palliative and Supportive Care, 9*, 201–208. https://doi.org/10.1017/S1478951511000137

Ryan, K., Guerin, S., Dodd, P., & McEvoy, J. (2011b). Exploring the experiences of people with intellectual disabilities when service users die. *British Journal of Learning Disabilities, 39*, 259–265. https://doi.org/10.1111/j.1468-3156.2010.00662.x

Savage, T. A., Moro, T. T., Boyden, J. Y., Brown, A. A., & Kavanaugh, K. L. (2015). Implementation challenges in end-of-life research with adults with intellectual and developmental disabilities. *Applied Nursing Research, 28*, 202–205. https://doi.org/10.1016/j.apnr.2014.10.002

Schepens, H., van Puyenbroeck, J., & Maes, B. (2019). How to improve the quality of life of elderly people with intellectual disability: A systematic literature review of support strategies. *Journal of Applied Research in Intellectual Disabilities, 32*, 483–521. https://doi.org/10.1111/jar.12559

Sheehan, D. K., Draucker, C. B., Christ, G. H., Mayo, M. M., Heim, K., & Parish, S. (2014). Telling adolescents a parent is dying. *Journal of Palliative Medicine, 17*(5), 512–520. https://doi.org/10.1089/jpm.2013.0344

Slaughter, V., & Griffiths, M. (2007). Death understanding and fear of death in young children. *Clinical Child Psychology and Psychiatry, 12*(4), 525–535. https://doi.org/10.1177/1359104507080980

Stancliffe, R. J., Wiese, M. Y., & Read, S. (2017). Editorial: End of life and people with intellectual disability. *Journal of Applied Research in Intellectual Disabilities, 30*(6), 977–981. https://doi.org/10.1111/jar.12383

Stancliffe, R. J., Wiese, M. Y., Read, S., Jeltes, G., & Clayton, J. M. (2016). Knowing, planning for and fearing death: Do adults with intellectual disability and disability staff differ? *Research in Developmental Disabilities, 49–50*, 47–59. https://doi.org/10.1016/j.ridd.2015.11.016

Stancliffe, R. J., Wiese, M. Y., Read, S., Jeltes, G., & Clayton, J. M. (2017b). Assessing knowledge and attitudes about end of life: Evaluation of three instruments designed for adults with intellectual disability. *Journal of Applied Research in Intellectual Disabilities, 30*(6), 1076–1088. https://doi.org/10.1111/jar.12358

Stancliffe, R. J., Wiese, M. Y., Read, S., Jeltes, G., Barton, R., & Clayton, J. M. (2021). Does talking about end of life with adults with intellectual disability cause emotional discomfort or psychological harm? *Journal of Applied Research in Intellectual Disabilities, 34*, 659–669. https://doi.org/10.1111/jar.12835

Todd, S. (2004). Death counts: The challenge of death and dying in learning disability services. *Learning Disability Practice, 7*(10), 12–15. https://doi.org/10.7748/ldp2004.12.7.10.12.c1551

Todd, S. (2013). "Being there": The experiences of staff in dealing with matters of dying and death in services for people with intellectual disabilities. *Journal of Applied Research in Intellectual Disabilities, 26*(3), 215–230. https://doi.org/10.1111/jar.12024

Todd., S., Bernal, J., Shearn, J., Worth, R., Jones, E., Lowe, K., Madden, P., Barr, O., Forrester Jones, R., Jarvis, P., Kroll., T., McCarron, M., Read, S., & Hunt, K. (2020). Last months of life of people with intellectual disabilities: A UK population-based study of death and dying in intellectual disability community services. *Journal of Applied Research in Intellectual Disabilities, 33*(6), 1245–1258. https://doi.org/10.1111/jar.12744

Tuffrey-Wijne, I. (2013a). *How to break bad news to people with intellectual disabilities: A guide for carers and professionals*. Jessica Kingsley.

Tuffrey-Wijne, I. (2013b). A new model for breaking bad news to people with intellectual disabilities. *Palliative Medicine, 27*(1), 5–12. https://doi.org/10.1177/0269216311433476

Tuffrey-Wijne, I., Bernal, J., & Hollins, S. (2010). Disclosure and understanding of cancer diagnosis and prognosis for people with intellectual disabilities: Findings from an ethnographic study. *European Journal of Cancer Nursing, 14*, 224–230. https://doi.org/10.1016/j.ejon.2010.01.021

Tuffrey-Wijne, I., Hogg, J., & Curfs, L. (2007). End-of-life and palliative care for people with intellectual disabilities who have cancer or other life-limiting illness: A review of the literature and available resources. *Journal of Applied Research in Intellectual Disabilities, 20*, 331–344. https://doi.org/10.1111/j.1468-3148.2006.00350.x

Tuffrey-Wijne, I., Giatras, N., Butler, G., Creswell, A., Manners, P., & Bernal, J. (2013). Developing guidelines for disclosure or non-disclosure of bad news around life-limiting illness and death to people with intellectual disabilities. *Journal of Applied Research in Intellectual Disabilities, 26*, 231–242. https://doi.org/10.1111/jar.12026

Tuffrey-Wijne, I., Finlayson, J., Bernal, J., Taggart, L., Lam, C. K. K., & Todd, S. (2020). Communicating about death and dying with adults with intellectual disabilities who are terminally ill or bereaved: A UK-wide survey of intellectual disability support staff. *Journal of Applied Research in Intellectual Disabilities, 33*(5), 927–938. https://doi.org/10.1111/jar.12714

Tuffrey-Wijne, I., & Rose, T. (2017). Investigating the factors that affect the communication of death-related bad news to people with intellectual disabilities by staff in residential and supported living services: An interview study. *Journal of Intellectual Disability Research, 61*(1), 727–736. https://doi.org/10.1111/jir.12375

Voss, H., Vogel, A., Wagemans, A. M., Francke, A. L., Metsemakers, J. F., Courtens, A. M., & de Veer, A. J. E. (2017). Advance care planning in palliative care for people with intellectual disabilities: A systematic review. *Journal of Pain and Symptom Management, 54*, 938–960. https://doi.org/10.1016/j.jpainsymman.2017.04.016

Voss, H., Vogel, A., Wagemans, A. M. A., Francke, A. L., Metsemakers, J. F. M., Courtens, A. M., & de Veer, A. J. E. (2019). Advance care planning in the palliative phase of people with intellectual disabilities: Analysis of medical files and interviews. *Journal of Intellectual Disability Research, 63*(10), 1262–1272. https://doi.org/10.1111/jir.12664

Voss, H., Vogel, A., Wagemans, A. M. A., Francke, A. L., Metsemakers, J. F. M., Courtens, A. M., & de Veer, A. J. E. (2021). Development, implementation, and evaluation of an advance care planning program for professionals in palliative care of people with intellectual disability. *Intellectual and Developmental Disabilities, 59*(1), 39–54. https://doi.org/10.1352/1934-9556-59.1.39

Wiese, M., Dew, A., Stancliffe, R. J., Howarth, G., & Balandin, S. (2013). 'If and when?': The beliefs and experiences of community living staff in supporting older people with intellectual disability to know about dying.

Journal of Intellectual Disability Research, 57, 980–992. https://doi.org/10.1111/j.1365-2788.01593.x

Wiese, M., Stancliffe, R. J., Balandin, S., Howarth, G., & Dew, A. (2012). End-of-life care and dying: Issues raised by staff supporting older people with intellectual disability in community living services. *Journal of Applied Research in Intellectual Disabilities, 25*(6), 571–583. https://doi.org/10.1111/jar.12000

Wiese, M., Stancliffe, R. J., Read, S., Jeltes, G., & Clayton, J. (2015). Learning about dying, death and end-of-life planning: Current issues informing future actions. *Journal of Intellectual & Developmental Disability, 40*(2), 230–235. https://doi.org/10.3109/13668250.2014.998183

Young, R., Dagnan, D., & Jahoda, A. (2016). Leaving school: A comparison of the worries held by adolescents with and without intellectual disabilities. *Journal of Intellectual Disability Research, 60*(1), 9–21. https://doi.org/10.1111/jir.12223

3

Suicide and Autism: A Lifespan Perspective

Darren Hedley, Susan M. Hayward, Alison Clarke, Mirko Uljarević, and Mark A. Stokes

Suicide accounts for 1.4% of deaths worldwide, is ranked as the 14th leading cause of death across all ages, and the second leading cause of death amongst 15 to 29-year-olds (Casey et al., 2008; Roth et al., 2018).

The original version of this chapter was revised: Co-author's middle initial has been corrected. The correction to this chapter is available at https://doi.org/10.1007/978-3-030-98697-1_18

D. Hedley (✉)
La Trobe University, Melbourne, VIC, Australia
e-mail: D.Hedley@latrobe.edu.au

S. M. Hayward · A. Clarke · M. Uljarević
The University of Melbourne, Melbourne, VIC, Australia
e-mail: susan.hayward@deakin.edu.au

M. Uljarević
e-mail: mirko.uljarevic@unimelb.edu.au

A. Clarke
Orygen, Melbourne, VIC, Australia
e-mail: alison.clarke@orygen.org.au

M. A. Stokes
Deakin University, Melbourne, VIC, Australia
e-mail: mark.stokes@deakin.edu.au

R. J. Stancliffe et al. (eds.), *End of Life and People with Intellectual and Developmental Disability*, https://doi.org/10.1007/978-3-030-98697-1_3

Suicide has a significant impact on individuals, families, and society; in Australia suicide and the associated mental health impacts cost the economy in excess of AUD $51 billion annually (Productivity Commission, 2019). Compared to people in the general population, people with a diagnosis of Autism Spectrum Disorder (ASD; henceforth "autism")[1] are at increased risk of suicide behaviour, including thoughts of suicide (ideation), planning for suicide, non-fatal suicide attempt, and death by suicide (Hedley & Uljarević, 2018; Hirvikoski et al., 2016; Kirby et al., 2019). In this chapter, we provide a detailed overview of the current knowledge concerning suicide behaviour in autistic people, supported by examples from people with lived experience.

"It [the thought of suicide] isn't always present, thanks largely to medication and other factors such as exercise and therapy, but it is always at the back of my mind, lurking, like a burglar waiting for his opportunity to steal my levity".

Will Attwood, autistic author.[2]

Autism

The Centers for Disease Control and Prevention (CDC) estimates 1 in 54 children are diagnosed with autism in the United States (Maenner et al., 2020). Autism is a lifelong neurodevelopmental condition that is more prevalent in males than females (i.e., 2:1 to 4:1); at least 30% of autistic individuals have a co-occurring intellectual developmental disorder (Maenner et al., 2020). Autism is hallmarked by varying degrees of impairment in social interaction and communication, restricted, repetitive behaviours and interests (RRBI), and can

[1] Consistent with the most recent edition of the Publication Manual of the American Psychological Association (2019), and the preferences of individuals with lived experience of autism (Bury et al., 2020), we utilise identity first language.

[2] Permission to be identified and quoted in this chapter provided by W.A. via personal communication with author D.H., June 28, 2020.

include hypo- or hyper-sensory sensitivities (American Psychiatric Association, 2013). Cognitive profiles are often uneven, reflecting relative areas of strength and weakness (Oliveras-Rentas et al., 2012; Rabiee et al., 2019). Specific cognitive domains may also be impacted (e.g., cognitive control, emotional regulation, executive functioning; Geurts et al., 2004; Karalunas et al., 2018).

Autism is marked by significant heterogeneity; however, symptoms are often pervasive across all aspects of development, affecting treatment options and response (Masi et al., 2017). Treatment response, as well as health and well-being, are particularly confounded by a high number of co-occurring neuropsychiatric conditions (Lai et al., 2019; Rodriguez-Seijas et al., 2020). This chapter specifically considers depression, anxiety, sleep disorders, substance use, and psychosis, due primarily to their association with suicidal behaviour in the general population (Joiner et al., 2005; Kessler et al., 2005). We also review the evidence concerning suicidal behaviour in intellectual and developmental disability due to its high prevalence in autism.

Risk and Prevalence of Suicide Behaviour in Autism

There is robust evidence of increased risk of suicidal behaviour in the autistic population, with suicide rates ranging from 0.17% to 0.4% (Hirvikoski et al., 2016; Kirby et al., 2019; Kõlves et al., 2021). In a large population-based study from Sweden, Hirvikoski et al. (2016) reported a sevenfold increased risk of premature death by suicide amongst autistic people compared to the general population. In a large cohort study from Denmark, Kõlves et al. (2021) reported a threefold higher rate of both suicide attempt and suicide in the autistic sample. Risk may be impacted by degree of intellectual impairment, with a lower relative risk in autistic people with co-occurring intellectual and developmental disability (Hirvikoski et al., 2016). Kirby et al. (2019) reported 73% of suicides by autistic people were violent (e.g., hanging/strangulation, blunt force injury; non-violent means included asphyxiants or intoxication, including drug overdose), a rate similar to the non-autistic comparison group. However, autistic people were less likely than non-autistic people to use firearms. Autistic males and females did not differ significantly on the method used; in the general population, males are more likely to use violent means than females (Ajdacic-Gross et al., 2008).

Lifespan Perspective. The average age of those who die by suicide may be lower in autistic groups than in the general population (e.g., 32 vs. 41 years; Kirby et al., 2019). There is also a higher incidence of suicide amongst younger autistic (0.16%) than non-autistic (0.07%) people (Kirby et al., 2019). Suicide risk may be greater when diagnosis is deferred until later in life, possibly as individuals diagnosed during early childhood may be more likely to access supports and interventions throughout critical developmental periods. For example, of 374 people diagnosed with autism in adulthood, 66% reported suicidal ideation and 35% reported having made suicide plans or attempts (Cassidy et al., 2014). When diagnosis is delayed until adulthood, there may also be a history of diagnostic overshadowing (Reiss et al., 1982), leading to other mental health conditions being treated without considering autism (Andrea et al., 2014). Finally, there is a growing portion of people diagnosed with autism in their fifties and beyond, with this group commonly reporting other psychiatric diagnoses (e.g., depression, anxiety), feeling isolated, and "alien" throughout their lives (Stagg & Belcher, 2019).

Below we present a story about "Max" as a way of illustrating the early age that suicidal behaviour can present, difficulties navigating the healthcare system, and the complex interrelationship between autism symptoms, mental health, social relationships, and suicide attempts.

Max's Story* Max is an Indigenous Australian who was diagnosed with autism at 13 years following a series of suicide attempts. Max is verbal but struggles with social/pragmatic communication and expressive and receptive language. Max experiences difficulties with changes and surprises, attention, sensory regulation, and becomes easily overwhelmed by auditory information. Max was diagnosed with severe anxiety, childhood depression, and Post Traumatic Stress Disorder (PTSD) prior to receiving their autism diagnosis. Max faces significant mental health challenges and sleep difficulties. These were unsuccessfully treated with medication and affect access to services. Max is currently 21 years of age and presents with a primary diagnosis of autism and several additional co-occurring diagnoses.

As a child, Max was active in sports but was regularly excluded from group activities. Max struggled throughout childcare and school, was in conflict with teachers, and was the victim of bullying. Max was unsuccessful in attempts to make friends and experienced social isolation. By the sixth grade Max had attended numerous schools, struggled with exclusion from school programmes, and displayed high levels of non-attendance. Max's parents returned home one day to find that Max had attempted suicide by strangulation.

Despite seeking professional assistance, Max's parents state that "no one took the suicide attempt seriously; they labelled it as attention seeking". Max's suicide attempts continued, resulting in an attempt by strangulation at a public location. The family sought help from a local hospital where they were instructed to go home and initiate a 24-hours suicide watch. About six months after hospital presentation, Max was diagnosed with autism. Max's suicide attempts continued, including at school. The school developed a plan to manage Max's behaviour, however, the family reported neither they nor Max were consulted in this process. Following Max's 14th birthday, the family were informed the school could no longer support Max. Max's mother reported feeling dismissed by professionals and unprepared to provide the high level of care that Max required.

During visits to the local mental health service, Max struggled with the noise of the waiting room, which led to sensory overload and escape attempts. Max also had difficulty with changes to staff and long wait times to see specialists. Max's anxiety increased with each visit. Max's mother describes a cycle of being admitted to a service, often receiving additional diagnoses, followed by referral elsewhere due to Max's complex needs. They failed to develop an effective solution to Max's challenges; at times Max was engaging in suicide behaviour on a weekly basis. Attempts were violent and unplanned. Max increasingly withdrew from the outside world deeming it to be "unsafe". Max slept more and began engaging in non-suicidal self-injurious behaviour, including severe head banging. At 16 years of age, Max had completely withdrawn from the outside world.[3]

* Assumed names are used in this story.

Subthreshold Symptomatology. Little is understood or reported about suicidal behaviour in people who are autistic but remain undiagnosed, or who exhibit high autistic traits but whose symptoms do not reach the diagnostic threshold. Although speculative, these people may face similar challenges to those who receive a formal diagnosis. They may also face added challenges due to not understanding or knowing the underlying cause of their difficulties or failing to access treatments that consider the influence of autistic traits.

Gender. Autistic females are at a significantly higher risk of suicide than non-autistic females (Hirvikoski et al., 2016; Kirby et al., 2019; Kõlves et al., 2021). In the general population, being female is a protective factor for suicide (Centers for Disease Control & Prevention, 2020). Compared to autistic males, autistic females report a relatively higher number of psychiatric diagnoses (Hedley et al., 2018; Lai et al., 2019; Sedgewick et al., 2021). Females may be more likely to be diagnosed with autism in adulthood (Kirby et al., 2019) or may not receive a formal autism diagnosis (Lai et al., 2015). Females may also be at greater risk of traumatic experiences, including sexual abuse (Kirby et al., 2019), and could experience identity conflict (Bargiela et al., 2016; Cassidy et al., 2018). Concealing autistic traits ("camouflaging") is more common in autistic females and has been reported to account for 3.5% of the variance in suicidal behaviour (Cassidy et al., 2018). A relatively high portion of autistic females report non-binary gender preference or minority sexual orientation (George & Stokes, 2018b). Incongruence between sex and

[3] At the time of writing, Max is supported by a multidisciplinary team. Team members have experience with autism and are willing to learn about Max's needs and wishes for the future. Max has an Assistance Dog and is beginning to venture outside again. Whilst suicide attempts have now stopped, Max continues to experience suicidal ideation to varying degrees.

gender can have subsequent impacts on mental health and well-being (George & Stokes, 2018a).

Intellectual Disability/intellectual Developmental Disorder and Suicide Risk

Risk of suicide may be greater in autistic individuals without co-occurring intellectual and developmental disability (Hirvikoski et al., 2016); however, co-occurring intellectual and developmental disability has been found to be associated with greater risk of suicide attempt or self-inflicted injury (e.g., intentional suffocation, drowning, firearm wounds, poisoning, injuries from objects, and other mechanisms of self-injury; Hand et al., 2019). Overall, there is a very limited amount of research concerning suicide risk in autistic people with intellectual and developmental disability. Similarly, suicide in people with intellectual and developmental disability in the absence of autism remains under-explored and poorly characterised (Ludi et al., 2012; Wark et al., 2018), with similar (Erlangsen et al., 2020) or lower (Patja et al., 2001) rates of suicide reported than in the general population. Frequent risk factors for suicidal behaviour include concurrent mental health difficulties and *severity* of intellectual and developmental disability, with those with milder intellectual and developmental disability at greater risk (Dodd et al., 2016). However, in adolescents with borderline intellectual impairment, a recent study reported no clear association between suicidal thoughts and behaviours and level of intellectual and developmental disability (King et al., 2019).

Reported methods of suicide by people with mild or moderate intellectual and developmental disability, who were mostly in psychiatric institutions, included hanging, drowning, intoxication by eating cigarettes or medication, and lying on a train track or jumping under a train (Patja et al., 2001). Males in this study were at a lower risk of suicide compared to males in the general population, but risk for females was similar. People with borderline or mild intellectual functioning may be at increased risk compared to those with moderate to profound intellectual and developmental disability (Weinheimer, 2018). However, suicidal intent can be difficult to assess in people with profound levels of intellectual impairment. Other factors associated with suicidal behaviour in intellectual and developmental disability include younger age, history of abuse or self-harm, trauma, and familial psychopathology, a prior visit to an emergency department, a history of psychiatric hospitalisation, and co-occurring physical disability (Lunsky et al., 2012; Weinheimer, 2018). Female sex may be a risk factor in those who attempt suicide (Lunsky et al., 2012). People with intellectual and developmental disability also report low levels of social support, high levels of loneliness, stress, anxiety, and depression, which may present as additional risk factors for suicide in this population (Lunsky, 2004).

Correlates, Risk and Protective Factors

Suicide rates are elevated across a wide range of psychiatric disorders and are compounded by the presence of multiple conditions (Joiner et al., 2005; Kessler et al., 2005). In autism, there is a higher prevalence of co-occurring mental health diagnoses than in the general population (Hudson et al., 2018; Lai et al., 2019). The association between autism and suicide deaths, as well as higher prevalence of self-harm, may be explained by the presence of co-occurring psychiatric conditions (Jokiranta-Olkoniemi et al., 2021), with one study reporting that over 90% of autistic people who died by suicide had at least one other co-occurring condition (Kõlves et al., 2021). Due to their relevance to suicide we focus on depression, anxiety, and sleep disorders, substance use and psychosis.

Depression. Lifetime rates of depressive disorders are high in the autistic population, with estimates ranging from 28 to 49% (Uljarević, Hedley, Cai et al., 2020a, 2020b; Uljarević, Hedley, Foley et al., 2020a, 2020b), about four times that observed in the general population (11% to 17%; Hudson et al., 2018; Lai et al., 2019). This is notable given a strong positive association between depression and suicidal behaviour in autistic people (Dell'Osso et al., 2019; Hand et al., 2019). Risk of depression may be higher in individuals without co-occurring intellectual and developmental disability; however, depression may present differently in people with intellectual and developmental disability (Hudson et al., 2018; Rai, Culpin, et al., 2018; Rai, Heuvelman, et al., 2018). Psychosocial risk factors that may be symptomatic of the social communication difficulties experienced in autism, such as loneliness and reduced social support, predict depressive scores (Mazurek, 2014) and suicidal ideation (Hedley et al., 2018). Bullying is also prevalent and a risk factor for depression in autistic children (Rai et al., 2018; Rai, Heuvelman, et al., 2018).

Anxiety. Anxiety has been shown to increase suicide risk in non-autistic populations (De La Vega et al., 2018; Kanwar et al., 2013). Anxiety disorders are more prevalent and may have greater symptom severity in autistic compared to clinical and non-clinical populations (Uljarević, Hedley, Cai et al., 2020a, 2020b). Although the prevalence of anxiety in autism varies considerably across studies, the majority of large-scale studies and meta-analytic reviews indicate that at least 40% of children and adolescents (van Steensel et al., 2011) and up to 60% of adults meet the diagnostic threshold for anxiety disorders (Lever & Geurts, 2016; Uljarević, Hedley, Foley et al., 2020a, 2020b). However, it is important to emphasise that anxiety can present atypically in autism (Kerns et al., 2014; Lau et al., 2019). Thus, it can be missed by standardised screening and diagnostic instruments suggesting likely underestimation of prevalence and impact.

Research to date has mainly focused on exploring the effects of sex, age, cognitive functioning, and core autism symptoms on the presentation and severity of anxiety in the autistic population. There is some evidence that female sex (Hedley et al., 2018; Lai et al., 2019), higher IQ (Hallett et al., 2013), and severity of autism symptoms (Wigham et al., 2015) are associated with higher prevalence and severity of anxiety, although not all studies report significant results (Duvekot et al., 2017; Hollocks et al., 2016). In autism, relationships have been identified between anxiety and both emotion regulation (Cai et al., 2019) and intolerance of uncertainty (Cai et al., 2018; Maisel et al., 2016). These are consistent with findings in non-autistic samples (Aldao et al., 2010).

Sleep Disorders. Sleep problems are prevalent in autism and may include short sleep duration, low sleep quality and efficiency, insomnia, daytime sleepiness, and circadian sleep desynchronisation (Carmassi et al., 2019; Mazzone et al., 2018; Reynolds et al., 2019). In both general and non-autistic clinical populations, there is a strong support of an association between suicide behaviour and sleep complaints (Bernert & Joiner, 2007; Bernert et al., 2015; Malik et al., 2014). Sleep disorders are modifiable with treatment (Bernert et al., 2015; Lovato et al., 2016) and some sleep medications may reduce suicidal ideation (e.g., zolpidem-CR; McCall et al., 2019). Studies have yet to examine relationships between sleep problems and suicide in autistic people, although autistic traits (and shorter sleep duration) have been found to predict suicidal ideation in a non-autistic sample (Hochard et al., 2020).

Alcohol and Other Drugs. Relatively little research has examined substance use amongst autistic people, which is thought to affect from 0.7% to 36% of the autistic population (Arnevik & Helverschou, 2016; Lugo-Marín et al., 2019). It was once thought that substance use was rare amongst autistic people, or primarily associated with co-occurring attention deficit hyperactivity disorder (Butwicka et al., 2017). Currently, however, the risk of substance use-related problems in autistic people is thought to be double than that of the general population (Butwicka et al., 2017). Substance dependence significantly increases risk of death by suicide, with a high portion of suicide deaths involving alcohol and opioids (Esang & Ahmed, 2018)—risk may be more than doubled in the presence of substance use disorder (Poorolajal et al., 2016). Substance use clearly needs to be considered as a potential risk factor for suicide in autism in both research and applied settings.

Psychosis. Autism often co-occurs with psychosis and schizophrenia (Chisholm et al., 2015; Upthegrove et al., 2018), with evidence of shared genetic susceptibility (Ruzzo & Geschwind, 2016). Suicide is a significant risk for people with early onset first-episode psychosis (Sanchez-Gistau et al., 2013), and there is a high overlap between schizophrenia and suicide, returning a lifetime risk of suicide around 5% (Hor & Taylor, 2010). In patients with first episode psychosis, a high level of autistic traits and positive symptoms of schizophrenia have been found to be associated with depression, hopelessness, and suicidal behaviour (Upthegrove et al., 2018). Clinicians should therefore assess for psychosis or other positive symptoms of schizophrenia in addition to autism. Where psychotic symptoms are present in conjunction with high autistic traits or autism diagnosis, depression and suicidal behaviour needs to be seriously considered. However, it is also important to acknowledge that similar to autism itself, psychosis is phenotypically and etiologically complex, with a number of domains within both positive and negative symptom dimensions that are at least partially distinct in terms of correlates, outcomes, and underpinning mechanisms (Strauss et al., 2018). Research that combines dimensional frameworks with fine-grained latent variable approaches is required in order to understand how transdiagnostic interactions between specific, precisely defined symptom domains that occur across autism, psychosis, and the schizophrenia spectrum give rise to suicide risk.

There has been a recent and significant shift towards consideration of transdiagnostic or dimensional constructs that might underlie suicidal risk and behaviour (Glenn et al., 2017, 2018; O'Connor & Portzky, 2018). In the next section, we identify potential transdiagnostic factors associated with autism that might increase suicide risk in both autistic and non-autistic populations.

Dimensional Constructs

Dimensional constructs represent basic, biologically meaningful dimensions of functioning that span the full range of human behaviour, from normative to atypical functioning (Cuthbert & Insel, 2013). These dimensions represent building blocks of normative functioning, and if disrupted, can result in specific behaviours/traits/characteristics or symptoms seen across a range of disorders. Therefore, studying specific dimensional constructs, and cross-dimensional interactions, measured across different levels, from genes and neurocircuitry to observable behaviours, has been proposed as a better way of defining and understanding mental disorders than symptom-based classification systems (Cuthbert & Insel,

2013). In the context of this chapter, dimensional constructs are important as they offer a way to understand common risk factors across clinically diverse populations.

Autistic Traits. Autistic traits are normally distributed in the general population (Ruzich et al., 2015). There is emerging evidence that autistic traits are risk markers for suicide in non-autistic (Pelton & Cassidy, 2017; Stanley et al., 2021; Upthegrove et al., 2018) and autistic (Cassidy et al., 2014; Hedley et al., 2018) populations. Autistic trait severity has been found to directly (Cassidy et al., 2014) or indirectly (through mediator variables including loneliness, low perceived social support, burdensomeness, belonging; Hedley et al., 2018; Pelton & Cassidy, 2017) predict increased suicide risk. Thus, autistic traits may serve as a useful marker for suicide risk across clinically diverse populations.

Social Communication. Social communication difficulties contribute to poor social relationships and support, loneliness, and depression (Mazurek, 2014; Rai, Culpin, et al., 2018; Rai, Heuvelman, et al., 2018) and are identified as important transdiagnostic risk factors for suicide (Glenn et al., 2017, 2018). In non-autistic populations, impairments in different domains of social and communication functioning are associated with increased risk of suicidal ideation and behaviour (Hedley, Batterham, et al., 2021; Hedley, Uljarević, et al., 2021; Stanley et al., 2021). In autistic people, social communication challenges may lead to difficulties recognising and communicating thoughts or feelings about suicide and seeking help or support; for example, clearly articulating difficulties to health professionals. For autistic people who are non-verbal, or have limited verbal communication skills, there are likely to be additional challenges detecting and correctly diagnosing mental health problems.

Restricted, Repetitive Behaviours and Interests (RRBI). Cognitive rigidity and rumination are associated with suicidal behaviour in both general (Fazakas-DeHoog et al., 2017; Hedley, Batterham, et al., 2021; Hedley, Uljarević, et al., 2021; Smith et al., 2006) and autistic (Dell'Osso et al., 2019) populations. As it relates to RRBI, autistic people may have difficulties with broad cognitive domains including cognitive control and executive function (Geurts et al., 2004; Karalunas et al., 2018). These cognitive challenges likely underpin cognitive and behavioural rigidity and social communication difficulties (Bos et al., 2019). Emotional dysregulation is another transdiagnostic dimension characteristic of autism (Cai et al., 2018, 2019) that is strongly associated with RRBI (Samson et al., 2014). Emotion dysregulation is associated with both depressive symptoms and suicidal behaviour in non-autistic people (Crandall et al., 2018); however, it may also serve as a protective measure against lethal self-harm (Anestis et al., 2011). People who have difficulty with emotional regulation, low distress tolerance, and high negative urgency exhibit higher levels of suicidal *desire* (i.e., perceived burdensomeness, thwarted belongingness), but may also have a lower *capability* for suicide (Anestis et al., 2011).

Risk Assessment and Instruments

There is a general paucity of mental health and other instruments designed specifically for autistic people (see also Chapter 16); thus, instruments designed for the general population are typically substituted for this population (Uljarević et al., 2018). Cassidy, Bradley, et al. (2020), Cassidy, Nicolaidis, et al. (2020)) examined the psychometric properties of the Suicidal Behaviours Questionnaire-Revised (SBQ-R; Osman et al., 2001) in autistic adults. The SBQ-R is a four-item instrument assessing lifetime ideation and attempts, frequency of ideation over 12 months, threat of a suicide attempt, and likelihood of future suicidal behaviour. In the study, autistic participants scored higher on all items than participants from the general population. However, SBQ-R items loaded differently in the two populations suggesting the instrument may perform differently for each group. Autistic participants reported difficulty interpreting and responding to questions and cited response options as lacking relevance or not adequately capturing their experience; for example, some autistic participants reported difficulty understanding the concept of a suicide plan (Cassidy et al., 2020; Cassidy, Nicolaidis, et al., 2020). A modified version of the SBQ-R for use by autistic people has been developed for research use, returning improved psychometric characteristics in autistic people (although predominantly females diagnosed in adulthood) compared to the original version (Cassidy, Bradley, et al., 2021).

The Interpersonal Needs Questionnaire-10 (INQ-10; van Orden et al., 2012) measures thwarted belonging and perceived burden; important constructs in Joiner's (2005) interpersonal-psychological theory of suicidal behaviour. However, the INQ-10 may function differently in autistic than general populations, possibly due to inclusion of items requiring awareness of mental states of others, which may be a challenging concept for autistic respondents (Pelton et al., 2020).

The Acquired Capability for Suicide Scale–Fearlessness About Death (ACCS-FAD) is a scale that assesses suicidal capability (Ribeiro et al., 2014). Although the use of non-concrete language and negatively worded responses affected performance, there is some support for its psychometric properties in autistic populations with one item removed (Pelton et al., 2020).

Notably, these studies have, to date, excluded autistic participants with co-occurring intellectual and developmental disability who may have experienced additional difficulties understanding and responding to the questionnaire. Instruments designed for use by people with co-occurring intellectual and developmental disability may need to be adapted by simplifying question wording and providing less complex response options (Nicolaidis et al., 2020).

Given problems associated with instruments that are not specifically designed for use in autistic populations, it may be important to combine validated instruments with a clinical interview conducted by professionals experienced with both suicide risk assessment and autism. Two tools developed in Australia, the Suicidal Ideation Attributes Scales (SIDAS; van Spijker et al., 2014) and the Suicide Assessment Kit (SAK; Deady et al., 2015) are examples of instruments that can be used as part of a clinical risk assessment by trained and suitably qualified health practitioners. The chapter authors have developed and are currently evaluating modified versions of these two instruments for use in the autistic population (Hedley, Batterham, et al., 2021; Hedley, Uljarević, et al., 2021).

Non-Suicidal Self-Injury (NSSI). NSSI in autistic people contributes to reduced quality of life and difficulty accessing services (Steenfeldt-Kristensen et al., 2020). NSSI is highly prevalent in the autistic population, including in those with and without co-occurring intellectual and developmental disability, with an estimated prevalence around 42% (Steenfeldt-Kristensen et al., 2020). Hand-hitting, skin picking, and hitting self or objects are amongst the most common forms of self-injury; rubbing self on surfaces, self-pinching, and self-cutting are less common (Steenfeldt-Kristensen et al., 2020). Due to lack of research, it is difficult to evaluate the influence of intellectual impairment on behaviour, although one study found similar rates in adults with average cognitive ability (31%) and severe intellectual and developmental disability (32%) (Ballaban-Gil et al., 1996). NSSI is associated with suicide risk in the autistic population (Moseley et al., 2020), and may be more prevalent in autistic women than men (Maddox et al., 2017). In the general population, NSSI often precedes suicide attempts (Olfson et al., 2017).

Suicide attempts may be misdiagnosed as NSSI in the autistic population. Because NSSI can be predictive of suicide attempt, whether harming behaviour is a suicide attempt or NSSI is an important consideration for suicide prevention. It is therefore critically important that health professionals carefully evaluate autistic people who present with NSSI for suicide risk, including in the presence of intellectual and developmental disability.

Prevention and Service Access

Autistic people describe significant difficulties accessing mental health support with lack of appropriate treatment and support options as significant barriers to their health and well-being (Camm-Crosbie et al., 2019), reflecting a general paucity of specialised services for this population (Marrus et al., 2014). The problem is particularly common in older autistic adults (Vogan et al., 2017). Families of younger children also report difficulties accessing mental health services (Jackson et al., 2020), suggesting mental health challenges begin early in autism. It is therefore essential for policy makers and clinicians to improve their understanding of co-occurring mental health conditions and suicidal behaviour in autistic people in order to appropriately inform suicide management, treatment, and prevention (Camm-Crosbie et al., 2019; Hedley & Uljarević, 2018).

Currently, there is a dearth of quality research examining mental health and other interventions for autistic adolescents and adults (Foley & Trollor, 2015). Interventions for mental health and well-being with some efficacy include social skills programmes (Reichow et al., 2013), CBT (Wood et al., 2020), and peer-mediated approaches (Crane et al., 2021). Interventions that target key risk factors (e.g., social isolation, social support, sleep concerns, depression) may prove beneficial but require research (Hedley & Uljarević, 2018). Treatments that directly target suicide risk should be prioritised (Office of the Surgeon General and the National Action Alliance for Suicide Prevention, 2012).

Programmes informed by lived experiences of autistic people with suicidal behaviour may be important for improving service access (Maple et al., 2018). Co-design and development create better informed services that are more likely to meet the needs of consumers, increase utilisation, and reduce stigma around help-seeking (Maple et al., 2018). Given that differences in gender identity and sexual orientation are relatively common within the autistic population (George & Stokes, 2018a, 2018b), supports are needed for autistic youth in relation to their sexual identity and development (Vanbergeijk et al., 2008). Concerns or confusion regarding gender identity and sexual orientation may also negatively influence help-seeking behaviour (Pinder-Amaker, 2014).

Lack of knowledge and difficulty navigating services are identified as significant barriers to accessing appropriate mental health services by autistic people (Lake et al., 2014). Adjustments that may improve access include provision of quiet rooms, removal of potential sensory distractions, and allowance for additional time to articulate problems and concerns (Lunsky et al., 2018). Multi-disciplinary service models incorporating health professionals who have autism knowledge, who are willing to listen and patiently develop a therapeutic alliance with the autistic person, that are co-developed with autistic people, and that take a holistic and lifespan approach are recommended and urgently needed. Training healthcare providers in autism generally, and specifically in the management of mental health conditions in autistic people, will be essential in developing effective suicide prevention strategies in this population.

Coronavirus (COVID-19) Pandemic

There is no doubt that the current global coronavirus (COVID-19) pandemic is negatively impacting the health and well-being of people across the globe, particularly those with neuropsychiatric conditions (Fontenelle & Miguel, 2020). Australian researcher and autistic advocate, Dr Jaclyn den Houting (2020) describes this impact on autistic people associated with restrictions on activity, changes to work routine, disturbing media coverage, and uncertainty that may enhance pre-existing anxiety. Restrictions to movement and extended lockdowns may further affect access to regular health services and social supports (Cassidy, Bradley, et al., 2020; Cassidy, Nicolaidis, et al., 2020). These experiences may also be common amongst non-autistic people; the significant impact of the pandemic on the mental health of society is only beginning to be felt. The following excerpt was written by an autistic person describing the compounding impact of COVID-19 for her as news of the pandemic worsened and people were losing their jobs.

> I could feel the rush of adrenaline circulate throughout my body every time there was a stressor. It would just build up and build up and build up each day. I was finding it really hard to get to sleep at night and my sleep quality was terrible and every morning was the same. Every time something happened unexpectedly, or any time an additional 'pressure' was felt, my whole body would feel this rush. I persisted and persisted. I had to work. I had to parent. I had to domesticate. Then the crying started. I was the only one in my family working. It got to a point where my workplace suggested I take some time off. It was such a relief and I felt supported. Although I wasn't going to get paid for taking time off, I needed it. I spent two full days sleeping to start the recovery process.
> Anonymous

Summary

The prevailing evidence strongly suggests increased risk of suicidal behaviour amongst autistic people. There is a critical need for research that seeks to better understand the phenomena of suicide in autism, particularly amongst those with intellectual and developmental disability, to enable the development of effective prevention strategies. As we have outlined in this chapter, many of the risk and protective factors associated with suicidal behaviour are similar for autistic and non-autistic people (e.g., depression, anxiety, sleep, substance use, psychosis). However, there are also factors that may be uniquely characteristic to autism or that have yet to be identified. Taking a transdiagnostic and dimensional approach, autistic traits align and overlap with mental health symptomology and other factors that may heighten vulnerability to suicide. Service access remains a significant barrier. It is critical that healthcare professionals consider the possibility of suicide as well as the presence of other psychiatric conditions (e.g., depression, psychosis) in autistic people. Adopting a transdiagnostic approach may also benefit those who exhibit high autistic traits but are not formally diagnosed and may offer insights into suicide risk and behaviour in non-autistic populations. Finally, programmes adapted and informed by the lived experiences of autistic people with suicidal behaviour may be necessary for improving service accessibility, screening, intervention, and ultimately suicide prevention.

Reflection

Michael's Story* Michael is a 15-year old male who has been referred to you due to difficulties with schoolwork. Michael lives with his father on a small rural farm and attends a local school. Michael's father reports that as a child, Michael was generally happy but liked to play alone rather than with other children. Michael reports he currently does not have any close friends. In his spare time, he likes to watch the trucks that pass the farm. During the interview, Michael's father tells you that he has seen Michael lying down in the middle of the road. Although the road is not overly busy, he is concerned about the behaviour as it could be dangerous for Michael. He has not talked to Michael about it though, and is not sure how to bring it up. It is likely that Michael has a mild intellectual disability and, although he has not been assessed for it, it is possible that he may have autism.

*Assumed names are used in this story.

1. How might Michael's provisional diagnoses of mild intellectual disability and possibly autism influence the approach taken when supporting Michael and his family?
2. Would you have any concerns for suicidal behaviour based on the information provided here? Why/why not?
3. What additional information or questions might be helpful to determine whether Michael is at risk of suicide?
4. Reflect on how you might discuss Michael's behaviour with him and his family.
5. How might you help to build a comprehensive system of supports around Michael to a) reduce risk and b) increase quality of life experiences?

List of Quality Assessments

Presently, there are no clinically available suicide risk assessment instruments that have been developed specifically for use by autistic people. When using instruments developed for non-autistic populations, the interpretation of specific items should be verified through follow-up questions and a clinical interview. Importantly, if there is any clinical concern, it is important to ask directly about the presence of any suicidal behaviour or risk (e.g., thoughts or plans about suicide or self-harm) as well as to determine available supports. Interviews should ideally be conducted by a health professional with expertise in mental health and intellectual and developmental disabilities.

The list below is not intended to be extensive, but describes three instruments that have either been used in research involving autistic adults, or are in development for use in this population. It is recommended that all instruments listed below be administered by a healthcare professional or someone with training in the administration and interpretation of health questionnaires.

1. *Patient Health Questionnaire-9* (PHQ-9) includes a single question concerning potential for self-harm/suicidal ideation. In the absence of alternative instruments, the PHQ-9 provides a useful, brief screening tool for suicide risk in autistic people that has been used in research settings (Arnold et al., 2019). The PHQ-9 is widely available online.
2. *Suicide Behaviours Questionnaire-Revised (SBQ-R)* is a four question self-report instrument designed to assess risk factors for suicide. The instrument provides cut-off scores to indicate risk of suicidal behaviour in the general population and in people with clinical diagnoses. The original SBQ-R has been used in research settings with autistic people and a revised version for autistic people has been developed (SBQ-ASC). The revised version is only recommended for use in research settings and is not recommended for assessing risk of future suicide attempts or self-harm. The SBQ-ASC has not been tested in people with co-occurring intellectual disability. Readers interested in the SBQ-ASC are referred to Cassidy, Bradley, Cogger-Ward, and Rodgers (2020; please refer to the References).
3. *Suicidal Ideation Attributes Scale* (SIDAS) is a five question web-based measure of the severity of suicidal ideation and the *Suicide Assessment Kit* (SAK) is a comprehensive assessment and policy package designed to assess and manage suicide risk, which incorporates an 11-question suicide risk screen. Both instruments are included here as they have been modified for use by autistic people by the chapter authors in collaboration with the authors of the original instruments. The original versions of these instruments are available through the Australian National University (SIDAS; https://rsph.anu.edu.au/research/tools-resources/suicidal-ideation-attributes-scale-sidas) and the University of New South Wales (SAK; https://ndarc.med.unsw.edu.au/suicide-assessment-kit). Because SIDAS and SAK are currently being evaluated for use with autistic people, caution is recommended if they

are to be used in this population. At the time of writing, the modified instruments are not yet available for general use (enquiries can be directed to the lead author of this chapter).

Resources

1. *Autism Speaks—Eight Critical Measures to Counter Suicide.* A list of tips for talking about and preventing suicide in autistic people. www.autismspeaks.org/blog/8-critical-measures-counter-suicide
2. *International Association for Suicide Prevention (IASP)* provides a worldwide directory of resources and hotlines for suicide support, as well as a forum for academics, mental health professionals, crisis workers, and suicide survivors. www.iasp.info
3. *National Autistic Society—Suicidality in autism: risk and prevention.* Practical suggestions for assessment and prevention of suicide for autistic people. network.autism.org.uk/good-practice/evidence-base/suicidality-autism-risk-and-prevention.
4. *Suicide Prevention Australia* is the peak body for suicide prevention in Australia. Provides information on suicide prevention policy and advocacy, resources, webinars, research and publications. www.suicidepreventionaust.org
5. *Suicide Response Project* is an evidence-based free suicide prevention educational toolkit and resource that provides information about how to detect and respond to people at risk of suicide, including neurodivergent and sex/gender diverse people. www.suicideresponseproject.com

Acknowledgements We would like to thank Will Attwood and Lachlan Dummigan for their insights which helped shape the writing of this chapter, and Melissa Dummigan for reading and commenting on a draft version. We would especially like to thank the contributors of the case studies and quotes presented in this chapter for so freely sharing their very personal stories. D.H. and S.M.H. contributed equally to the writing of this chapter.

Funding The writing of this chapter was supported by a Suicide Prevention Australia National Suicide Prevention Research Fellowship awarded to D.H. M.U. is supported by the Australian Research Council Discovery Early Career Researcher Award (DE180100632). D.H. also receives research funding to examine suicide prevention in autism from Untapped Holdings Pty Ltd.

References

Ajdacic-Gross, V., Weiss, M. G., Ring, M., Hepp, U., Bopp, M., Gutzwiller, F., & Rössler, W. (2008). Methods of suicide: International suicide patterns derived from the WHO mortality database. *Bulletin of the World Health Organization, 86*, 657–736. https://doi.org/10.2471/BLT.07.043489

Aldao, A., Nolen-Hoeksema, S., & Schweizer, S. (2010). Emotion-regulation strategies across psychopathology: A meta-analytic review. *Clinical Psychology Review, 30*, 217–237. https://doi.org/10.1016/j.cpr.2009.11.004

American Psychiatric Association. (2013). *Diagnostic and statistical manual of mental disorders* (5th ed.). American Psychiatric Publishing. https://doi.org/10.1176/appi.books.9780890425596

American Psychological Association. (2019). *Publication manual of the American Psychological Association* (7th ed.). American Psychological Association. https://doi.org/10.1037/0000165-000

Andrea, T., Katharine, D., Nicole, K., Nicole, K., Thomas, O., & Susan, W. (2014). Under-identification of autism spectrum disorder in females: A case series illustrating the unique presentation of this disorder in young women. *Scandinavian Journal of Child and Adolescent Psychiatry and Psychology, 2*, 66–76. https://doi.org/10.21307/sjcapp-2014-010

Anestis, M. D., Bagge, C. L., Tull, M. T., & Joiner, T. E. (2011). Clarifying the role of emotion dysregulation in the interpersonal-psychological theory of suicidal behavior in an undergraduate sample. *Journal of Psychiatric Research, 45*, 603–611. https://doi.org/10.1016/j.jpsychires.2010.10.013

Arnevik, E. A., & Helverschou, S. B. (2016). Autism spectrum disorder and co-occurring substance use disorder: A systematic review. *Substance Abuse: Research and Treatment, 10*, 69–75. https://doi.org/10.4137/SART.S39921

Arnold, S., Uljarebić, M., Hwang, Y. I., Richdale, A. L., Trollor, J. N., & Lawson, L. P. (2019). Brief report; Psychometric properties of the Patient Health Questionnaire-9 (PHQ-9) in autistic adults. *Journal of Autism and*

Developmental Disorders, 50(6), 2217–2225. https://doi.org/10.1007/s10803-019-03947-9

Ballaban-Gil, K., Rapin, I., Tuchman, R., & Shinnar, S. (1996). Longitudinal examination of the behavioral, language, and social changes in a population of adolescents and young adults with autistic disorder. *Pediatric Neurology, 15*, 217–223. https://doi.org/10.1016/S0887-8994(96)00219-6

Bargiela, S., Steward, R., & Mandy, W. (2016). The experiences of late-diagnosed women with autism spectrum conditions: An investigation of the female autism phenotype. *Journal of Autism and Developmental Disorders, 46*, 3281–3294. https://doi.org/10.1007/s10803-016-2872-8

Bernert, R. A., & Joiner, T. E. (2007). Sleep disturbances and suicide risk: A review of the literature. *Neuropsychiatric Disease and Treatment, 3*, 735–743. https://doi.org/10.2147/NDT.S1248

Bernert, R. A., Kim, J. S., Iwata, N. G., & Perlis, M. L. (2015). Sleep disturbances as an evidence-based suicide risk factor. *Current Psychiatry Reports, 17*, Article e554. https://doi.org/10.1007/s11920-015-0554-4

Bos, D. J., Silverman, M. R., Ajodan, E. L., Martin, C., Silver, B. M., Brouwer, G. J., Di Martino, A., & Jones, R. M. (2019). Rigidity coincides with reduced cognitive control to affective cues in children with autism. *Journal of Abnormal Psychology, 128*, 431–441. https://doi.org/10.1037/abn0000423

Bury, S. M., Jellett, R., Spoor, J. R., & Hedley, D. (2020). "It defines who I am" or "it's something I have": What language do [autistic] Australian adults [on the autism spectrum] prefer? *Journal of Autism and Developmental Disorders*. Advance online publication. https://doi.org/10.1007/s10803-020-04425-3

Butwicka, A., Långström, N., Larsson, H., Lundström, S., Serlachius, E., Almqvist, C., Frisén, L., & Lichtenstein, P. (2017). Increased risk for substance use-related problems in autism spectrum disorders: A population-based cohort study. *Journal of Autism and Developmental Disorders, 47*, 80–89. https://doi.org/10.1007/s10803-016-2914-2

Cai, R. Y., Richdale, A. L., Dissanayake, C., Trollor, J., & Uljarević, M. (2019). Emotion regulation in autism: Reappraisal and suppression interactions. *Autism, 23*, 737–749. https://doi.org/10.1177/1362361318774558

Cai, R. Y., Richdale, A. L., Dissanayake, C., & Uljarević, M. (2018). Inter-relationship between emotion regulation, intolerance of uncertainty, anxiety, and depression in youth with autism spectrum disorder. *Journal of Autism and Developmental Disorders, 48*, 316–325. https://doi.org/10.1007/s10803-017-3318-7

Camm-Crosbie, L., Bradley, L., Shaw, R., Baron-Cohen, S., & Cassidy, S. (2019). "People like me don't get support": Autistic adults' experiences of support and treatment for mental health difficulties, self-injury and suicidality. *Autism, 23*, 1431–1441. https://doi.org/10.1177/1362361318816053

Carmassi, C., Palagini, L., Caruso, D., Masci, I., Nobili, L., Vita, A., & Dell'Osso, L. (2019). Systematic review of sleep disturbances and circadian sleep desynchronization in autism spectrum disorder: Toward an integrative model of a self-reinforcing loop. *Frontiers in Psychiatry, 10*, Article e366. https://doi.org/10.3389/fpsyt.2019.00366

Casey, P., Dunn, G., Kelly, B. D., Lehtinen, V., Dalgard, O. S., Dowrick, C., & Ayuso-Mateos, J. L. (2008). The prevalence of suicidal ideation in the general population: Results from the Outcome of Depression International Network (ODIN) study. *Social Psychiatry and Psychiatric Epidemiology, 43*, 299–304. https://doi.org/10.1007/s00127-008-0313-5

Cassidy, S., Bradley, L., Cogger-Ward, H., & Rodgers, J. (2021). Development and validation of the suicide behaviours questionnaire-Autism spectrum conditions in a community sample of autistic, possibly autistic and non-autistic adults. *Molecular Autism, 12,* Article e46. https://doi.org/10.1186/s13229-021-00449-3

Cassidy, S., Bradley, L., Cogger-Ward, H., Shaw, R., Bowen, E., Glod, M., Baron-Cohen, S., & Rodgers, J. (2020a). Measurement properties of the suicidal behaviour questionnaire-revised in autistic adults. *Journal of Autism and Developmental Disorders, 50*, 3477–3488. https://doi.org/10.1007/s10803-020-04431-5

Cassidy, S., Bradley, L., Shaw, R., & Baron-Cohen, S. (2018). Risk markers for suicidality in autistic adults. *Molecular Autism, 9*, Article e42. https://doi.org/10.1186/s13229-018-0226-4

Cassidy, S., Bradley, P., Robinson, J., Allison, C., McHugh, M., & Baron-Cohen, S. (2014). Suicidal ideation and suicide plans or attempts in adults with Asperger's syndrome attending a specialist diagnostic clinic: A clinical cohort study. *Lancet Psychiatry, 1*, 142–147. https://doi.org/10.1016/S2215-0366(14)70248-2

Cassidy, S., Nicolaidis, C., Davies, B., Rosa, S. D. R., Eisenman, D., Onaiwu, M. G., Kapp, S. K., Kripke, C. C., Rodgers, J., & Waisman, T. C. (2020b). An expert discussion on autism in the COVID-19 pandemic. *Autism in Adulthood, 2*, 106–117. https://doi.org/10.1089/aut.2020.29013.sjc

Centers for Disease Control and Prevention. (2020). *Web-based Injury Statistics Query and Reporting System (WISQARS) fatal injury data 1999–2018*. Centers for Disease Control and Prevention.

Chisholm, K., Lin, A., Abu-Akel, A., & Wood, S. J. (2015). The association between autism and schizophrenia spectrum disorders: A review of eight alternate models of co-occurrence. *Neuroscience and Biobehavioral Reviews, 55*, 173–183. https://doi.org/10.1016/j.neubiorev.2015.04.012

Crandall, A., Allsop, Y., & Hanson, C. L. (2018). The longitudinal association between cognitive control capacities, suicidality, and depression during late adolescence and young adulthood. *Journal of Adolescence, 65*, 167–176. https://doi.org/10.1016/j.adolescence.2018.03.009

Crane, L., Hearst, C., Ashworth, M., Davies, J., & Hill, E. L. (2021). Supporting newly identified or diagnosed autistic adults: An initial evaluation of an autistic-led programme. *Journal of Autism and Developmental Disorders, 51*, 892–905. https://doi.org/10.1007/s10803-020-04486-4

Cuthbert, B.N., & Insel, T.R. (2013). Toward the future of psychiatric diagnosis: The seven pillars of RDOC. *BMC Medicine, 11*, Article e126. https://doi.org/10.1186/1741-7015-11-126

de la Vega, D., Giner, L., & Courtet, P. (2018). Suicidality in subjects with anxiety or obsessive-compulsive and related disorders: Recent advances. *Current Psychiatry Reports, 20*, Article e26. https://doi.org/10.1007/s11920-018-0885-z

Deady, M., Ross, J., & Darke, S. (2015). *Suicide Assessment Kit (SAK): A comprehensive assessment and policy development package*. National Drug and Alcohol Research Centre.

Dell'Osso, L., Carpita, B., Muti, D., Morelli, V., Salarpi, G., Salerni, A., Scotto, J., Massimetti, G., Gesi, C., Ballerio, M., Signorelli, M. S., Luciano, M., Politi, P., Aguglia, E., Carmassi, C., & Maj, M. (2019). Mood symptoms and suicidality across the autism spectrum. *Comprehensive Psychiatry, 91*, 34–38. https://doi.org/10.1016/j.comppsych.2019.03.004

den Houting, J. (2020). Stepping out of isolation: Autistic people and COVID-19. *Autism in Adulthood, 2*, 103–105. https://doi.org/10.1089/aut.2020.29012.jdh

Dodd, P., Doherty, A., & Guerin, S. (2016). A systematic review of suicidality in people with intellectual disabilities. *Harvard Review of Psychiatry, 24*, 202–213. https://doi.org/10.1097/HRP.0000000000000095

Duvekot, J., van der Ende, J., Verhulst, F. C., Slappendel, G., van Daalen, E., Maras, A., & Greaves-Lord, K. (2017). Factors influencing the probability of a diagnosis of autism spectrum disorder in girls versus boys. *Autism, 21*, 646–658. https://doi.org/10.1177/1362361316672178

Erlangsen, A., Stenager, E., Conwell, Y., Andersen, P. K., Hawton, K., Benros, M. E., Nordentoft, M., & Stenager, E. (2020). Association between neurological disorders and death by suicide in Denmark. *JAMA, 323*, 444–454. https://doi.org/10.1001/jama.2019.21834

Esang, M., & Ahmed, S. (2018). A closer look at substance use and suicide. *American Journal of Psychiatry Residents' Journal, 13*, 6–8. https://doi.org/10.1176/appi.ajp-rj.2018.130603

Fazakas-DeHoog, L. L., Rnic, K., & Dozois, D. J. A. (2017). A cognitive distortions and deficits model of suicide ideation. *Europe's Journal of Psychology, 13*, 178–193. https://doi.org/10.5964/ejop.v13i2.1238

Foley, K., & Trollor, J. (2015). Management of mental ill health in people with autism spectrum disorder. *Australian Family Physician, 44*, 784–790. https://www.racgp.org.au/afp/2015/november/management-of-mental-ill-health-in-people-with-autism-spectrum-disorder

Fontenelle, L. F., & Miguel, E. C. (2020). The impact of coronavirus (COVID-19) in the diagnosis and treatment of obsessive-compulsive disorder. *Depression and Anxiety, 37*, 510–511. https://doi.org/10.1002/da.23037

George, R., & Stokes, M. A. (2018a). A quantitative analysis of mental health among sexual and gender minority groups in ASD. *Journal of Autism and Developmental Disorders, 48*, 2052–2063. https://doi.org/10.1007/s10803-018-3469-1

George, R., & Stokes, M. A. (2018b). Sexual orientation in autism spectrum disorder. *Autism Research, 11*, 133–141. https://doi.org/10.1002/aur.1892

Geurts, H. M., Verte, S., Oosterlaan, J., Roeyers, H., & Sergeant, J. A. (2004). How specific are executive functioning deficits in attention deficit hyperactivity disorder and autism? *Journal of Child Psychology and Psychiatry, 45*, 836–854. https://doi.org/10.1111/j.1469-7610.2004.00276.x

Glenn, C. R., Cha, C. B., Kleiman, E. M., & Nock, M. K. (2017). Understanding suicide risk within the Research Domain Criteria (RDoC) framework: Insights, challenges, and future research considerations. *Clinical Psychological Science, 5*, 568–592. https://doi.org/10.1177/2167702616686854

Glenn, C. R., Kleiman, E. M., Cha, C. B., Deming, C. A., Franklin, J. C., & Nock, M. K. (2018). Understanding suicide risk within the Research Domain Criteria (RDoC) framework: A meta-analytic review. *Depression and Anxiety, 35*, 65–88. https://doi.org/10.1002/da.22686

Hallett, V., Lecavalier, L., Sukhodolsky, D. G., Cipriano, N., Aman, M. G., McCracken, J. T., McDougle, C. J., Tierney, E., King, B. H., Hollander, E., Sikich, L., Bregman, J., Anagnostou, E., Donnelly, C., Katsovich, L., Dukes,

K., Vitiello, B., Gadow, K., & Scahill, L. (2013). Exploring the manifestations of anxiety in children with autism spectrum disorders. *Journal of Autism and Developmental Disorders, 43*, 2341–2352. https://doi.org/10.1007/s10803-013-1775-1

Hand, B. N., Benevides, T. W., & Carretta, H. J. (2019). Suicidal ideation and self-inflicted injury in Medicare enrolled autistic adults with and without co-occurring intellectual disability. *Journal of Autism and Developmental Disorders, 50*, 3489–3495. https://doi.org/10.1007/s10803-019-04345-x

Hedley, D., Batterham, P., Bury, S. M., Clapperton, A., Denney, K., Dissanayake, C., Fox, P., Gallagher, E., Hayward, S. M., Robinson, J., Sahin, E., Trollor, J., Uljarević, M., & Stokes, M. A. (2021). The Suicide Ideation Attributes Scale-Modified (SIDAS-M): Development and preliminary validation of a new scale for the measurement of suicidal ideation in autistic people. Manuscript submitted for publication.

Hedley, D., & Uljarević, M. (2018). Systematic review of suicide in autism spectrum disorder: Current trends and implications. *Current Developmental Disorders Reports, 5*, 65–76. https://doi.org/10.1007/s40474-018-0133-6

Hedley, D., Uljarević, M., Cai, R. Y., Bury, S. M., Stokes, M. A., & Evans, D. W. (2021b). Domains of the autism spectrum disorder phenotype, cognitive control, and rumination as transdiagnostic predictors of DSM-5 suicide risk. *PLoS ONE, 16*, Article e0245562. https://doi.org/10.1371/journal.pone.0245562

Hedley, D., Uljarević, M., Foley, K. R., Richdale, A., & Trollor, J. (2018). Risk and protective factors underlying suicidal ideation in autism spectrum disorder. *Depression and Anxiety, 35*, 648–657. https://doi.org/10.1002/da.22759

Hirvikoski, T., Mittendorfer-Rutz, E., Boman, M., Larsson, H., Lichtenstein, P., & Bölte, S. (2016). Premature mortality in autism spectrum disorder. *British Journal of Psychiatry, 208*, 232–238. https://doi.org/10.1192/bjp.bp.114.160192

Hochard, K. D., Pendrous, R., Mari, T., & Flynn, S. (2020). Examining the relationship between autism traits and sleep duration as predictors of suicidality. *Journal of Autism and Developmental Disorders, 50*, 3575–3584. https://doi.org/10.1007/s10803-020-04405-7

Hollocks, M. J., Pickles, A., Howlin, P., & Simonoff, E. (2016). Dual cognitive and biological correlates of anxiety in autism spectrum disorders. *Journal of Autism and Developmental Disorders, 46*, 3295–3307. https://doi.org/10.1007/s10803-016-2878-2

Hor, K., & Taylor, M. (2010). Suicide and schizophrenia: A systematic review of rates and risk factors. *Journal of Psychopharmacology, 24*, 81–90. https://doi.org/10.1177/1359786810385490

Hudson, C. C., Hall, L., & Harkness, K. L. (2018). Prevalence of depressive disorders in individuals with autism spectrum disorder: A meta-analysis. *Journal of Abnormal Child Psychology, 47*, 165–175. https://doi.org/10.1007/s10802-018-0402-1

Jackson, L., Keville, S., & Ludlow, A. K. (2020). Mothers' experiences of accessing mental health care for their child with an autism spectrum disorder. *Journal of Child and Family Studies, 29*, 534–545. https://doi.org/10.1007/s10826-019-01662-8

Joiner, T. E. (2005). *Why people die by suicide*. Harvard University Press.

Joiner, T. E., Brown, J. S., & Wingate, L. R. (2005). The psychology and neurobiology of suicidal behavior. *Annual Review of Psychology, 56*, 287–314. https://doi.org/10.1146/annurev.psych.56.091103.070320

Jokiranta-Olkoniemi, E., Gyllenberg, D., Sucksdorff, D., Suominen, A., Kronström, K., Chudal, R., & Sourander, A. (2021). Risk for premature mortality and intentional self-harm in autism spectrum disorders. *Journal of Autism and Developmental Disorders, 51*, 3098–3108. https://doi.org/10.1007/s10803-020-04768-x

Kanwar, A., Malik, S., Prokop, L. J., Sim, L. A., Feldstein, D., Wang, Z., & Murad, M. H. (2013). The association between anxiety disorders and suicidal behaviors: A systematic review and meta-analysis. *Depression and Anxiety, 30*, 917–929. https://doi.org/10.1002/da.22074

Karalunas, S. L., Hawkey, E., Gustafsson, H., Miller, M., Langhorst, M., Cordova, M., Fair, D., & Nigg, J. T. (2018). Overlapping and distinct cognitive impairments in attention-deficit/hyperactivity and autism spectrum disorder without intellectual disability. *Journal of Abnormal Child Psychology, 46*, 1705–1716. https://doi.org/10.1007/s10802-017-0394-2

Kerns, C. M., Kendall, P. C., Berry, L., Souders, M. C., Franklin, M. E., Schultz, R. T., Miller, J., & Herrington, J. (2014). Traditional and atypical presentations of anxiety in youth with autism spectrum disorder. *Journal of Autism and Developmental Disorders, 44*, 2851–2861. https://doi.org/10.1007/s10803-014-2141-7

Kessler, R. C., Berglund, P., Borges, G., Nock, M., & Wang, P. S. (2005). Trends in suicide ideation, plans, gestures, and attempts in the United States, 1990–1992 to 2001–2003. *JAMA, 293*, 2487–2495. https://doi.org/10.1001/jama.293.20.2487

King, T. L., Milner, A., Aitken, Z., Karahalios, A., Emerson, E., & Kavanagh, A. M. (2019). Mental health of adolescents: Variations by borderline intellectual functioning and disability. *European Child and Adolescent Psychiatry, 28*, 1231–1240. https://doi.org/10.1007/s00787-019-01278-9

Kirby, A. V., Bakian, A. V., Zhang, Y., Bilder, D. A., Keeshin, B. R., & Coon, H. (2019). A 20-year study of suicide death in a statewide autism population. *Autism Research, 12*, 658–666. https://doi.org/10.1002/aur.2076

Kõlves, K., Fitzgerald, C., Nordentoft, M., Wood, S. J., & Erlangsen, A. (2021). Assessment of suicidal behaviors among individuals with autism spectrum disorder in Denmark. *JAMA Network Open, 4*, Article e2033565. https://doi.org/10.1001/jamanetworkopen.2020.33565

Lai, M.-C., Kassee, C., Besney, R., Bonato, S., Hull, L., Mandy, W., Szatmari, P., & Ameis, S. H. (2019). Prevalence of co-occurring mental health diagnoses in the autism population: A systematic review and meta-analysis. *The Lancet Psychiatry, 6*, 819–829. https://doi.org/10.1016/S2215-0366(19)30289-5

Lai, M.-C., Lombardo, M. V., Auyeung, B., Chakrabarti, B., & Baron-Cohen, S. (2015). Sex/gender differences and autism: Setting the scene for future research. *Journal of the American Academy of Child and Adolescent Psychiatry, 54*, 11–24. https://doi.org/10.1016/j.jaac.2014.10.003

Lake, J.K., Perry, A., & Lunsky, Y. (2014). Mental health services for individuals with high functioning autism spectrum disorder. *Autism Research and Treatment, 2014*, Article e502420. https://doi.org/10.1155/2014/502420

Lau, W.K.W., Leung, M.-K., & Lau, B.W.M. (2019). Resting-state abnormalities in autism spectrum disorders: A meta-analysis. *Scientific Reports, 9*, Article e3892. https://doi.org/10.1038/s41598-019-40427-7

Lever, A. G., & Geurts, H. M. (2016). Psychiatric co-occurring symptoms and disorders in young, mintellectual and developmental disabilityle-aged, and older adults with autism spectrum disorder. *Journal of Autism and Developmental Disorders, 46*, 1916–1930. https://doi.org/10.1007/s10803-016-2722-8

Lovato, N., Lack, L., & Kennaway, D. J. (2016). Comparing and contrasting therapeutic effects of cognitive-behavior therapy for older adults suffering from insomnia with short and long objective sleep duration. *Sleep Medicine, 22*, 4–12. https://doi.org/10.1016/j.sleep.2016.04.001

Ludi, E., Ballard, E. D., Greenbaum, R., Pao, M., Bridge, J., Reynolds, W., & Horowitz, L. (2012). Suicide risk in youth with intellectual disabilities: The challenges of screening. *Journal of Developmental and Behavioral Pediatrics, 33*, 431–440. https://doi.org/10.1097/DBP.0b013e3182599295

Lugo-Marín, J., Magán-Maganto, M., Rivero-Santana, A., Cuellar-Pompa, L., Alviani, M., Jenaro-Rio, C., Díez, E., & Canal-Bedia, R. (2019). Prevalence of psychiatric disorders in adults with autism spectrum disorder: A systematic review and meta-analysis. *Research in Autism Spectrum Disorders, 59*, 22–33. https://doi.org/10.1016/j.rintellectual and developmental disability.2003.06.004

Lunsky, Y. (2004). Suicidality in a clinical and community sample of adults with mental retardation. *Research in Developmental Disabilities, 25*, 231–243. https://doi.org/

Lunsky, Y., Raina, P., & Burge, P. (2012). Suicidality among adults with intellectual disability. *Journal of Affective Disorders, 140*, 292–295. https://doi.org/10.1016/j.jad.2012.03.013

Lunsky, Y., Tint, A., Weiss, J. A., Palucka, A., & Bradley, E. (2018). A review of emergency department visits made by youth and adults with autism spectrum disorder from the parent perspective. *Advances in Autism, 4*, 10–18. https://doi.org/10.1108/AIA-08-2017-0019

Maddox, B. B., Trubanova, A., & White, S. W. (2017). Untended wounds: Non-suicidal self-injury in adults with autism spectrum disorder. *Autism, 21*, 412–422. https://doi.org/10.1177/1362361316644731

Maenner, M. J., Shaw, K. A., Baio, J., Washington, A., Patrick, M., DiRienzo, M., Christensen, D. L., Wiggins, L. D., Pettygrove, S., Andrews, J. G., Lopez, M., Hudson, A., Baroud, T., Schwenk, Y., White, T., Rosenberg, C. R., Lee, L-C., Harrington, R. A., Huston, M., … Dietz, P. M. (2020). Prevalence of autism spectrum disorder among children aged 8 Years: Autism and developmental disabilities monitoring network, 11 sites, United States, 2016. *MMWR Surveillance Summaries, 69*, 1–12. https://doi.org/10.15585/mmwr.ss6904a1

Maisel, M. E., Stephenson, K. G., South, M., Rodgers, J., Freeston, M. H., & Gaigg, S. B. (2016). Modeling the cognitive mechanisms linking autism symptoms and anxiety in adults. *Journal of Abnormal Psychology, 125*, 692–703. https://doi.org/10.1037/abn0000168

Malik, S., Kanwar, A., Sim, L.A., Prokop, L.J., Wang, Z., Benkhadra, K., & Murad, M.H. (2014). The association between sleep disturbances and suicidal behaviors in patients with psychiatric diagnoses: A systematic review and meta-analysis. *Systematic Reviews, 3*, Article e18. https://doi.org/10.1186/2046-4053-3-18

Maple, M., Wayland, S., Pearce, T., & Hua, P. (2018). *Services and programs for suicide prevention: An evidence check rapid review*. SAX Institute. https://www.saxinstitute.org.au/publications/programs-services-suicide-prevention

Marrus, N., Veenstra-Vanderweele, J., Hellings, J. A., Stigler, K. A., Szymanski, L., King, B. H., Carlisle, L. L., Cook, E. H., & Pruett, J. R., Jr. (2014). Training of child and adolescent psychiatry fellows in autism and intellectual disability. *Autism, 18*, 471–475. https://doi.org/10.1177/1362361313477247

Masi, A., DeMayo, M. M., Glozier, N., & Guastella, A. J. (2017). An overview of autism spectrum disorder, heterogeneity and treatment options. *Neuroscience Bulletin, 33*, 183–193. https://doi.org/10.1007/s12264-017-0100-y

Mazurek, M. O. (2014). Loneliness, friendship, and well-being in adults with autism spectrum disorders. *Autism, 18*, 223–232. https://doi.org/10.1177/1362361312474121

Mazzone, L., Postorino, V., Siracusano, M., Riccioni, A., & Curatolo, P. (2018). The relationship between sleep problems, neurobiological alterations, core symptoms of autism spectrum disorder, and psychiatric comorbidities. *Journal of Clinical Medicine, 7*, Article e102. https://doi.org/10.3390/jcm7050102

McCall, W. V., Benca, R. M., Rosenquist, P. B., Youssef, N. A., McCloud, L., Newman, J. C., Case, D., Rumble, M. E., Szabo, S. T., Phillips, M., & Krystal, A. D. (2019). Reducing suicidal ideation through insomnia treatment (REST-IT): A randomized clinical trial. *American Journal of Psychiatry, 176*, 957–965. https://doi.org/10.1176/appi.ajp.2019.19030267

Moseley, R.L., Gregory, N.J., Smith, P., Allison, C., & Baron-Cohen, S. (2020). Links between self-injury and suicidality in autism. *Molecular Autism, 11*, Article e14. https://doi.org/10.1186/s13229-020-0319-8

Nicolaidis, C., Raymaker, D. M., McDonald, K. E., Lund, E. M., Leotti, S., Kapp, S. K., Katz, M., Beers, L. M., Kripke, C., Maslak, J., Hunter, M., & Zhen, K. Y. (2020). Creating accessible survey instruments for use with autistic adults and people with intellectual disability: Lessons learned and recommendations. *Autism in Adulthood, 2*, 61–76. https://doi.org/10.1089/aut.2019.0074

O'Connor, R. C., & Portzky, G. (2018). Looking to the future: A synthesis of new developments and challenges in suicide research and prevention. *Frontiers in Psychology, 9*, 2139–2139. https://doi.org/10.3389/fpsyg.2018.02139

Office of the Surgeon General and the National Action Alliance for Suicide Prevention. (2012). *National strategy for suicide prevention: Goals and objectives for action.* US Department of Health and Human Services. https://www.ncbi.nlm.nih.gov/books/n/surgnssp/pdf
Olfson, M., Wall, M., Wang, S., Crystal, S., Gerhard, T., & Blanco, C. (2017). Suicide following deliberate self-harm. *American Journal of Psychiatry, 174*, 765–774. https://doi.org/10.1176/appi.ajp.2017.16111288
Oliveras-Rentas, R. E., Kenworthy, L., Roberson, R. B., 3rd., Martin, A., & Wallace, G. L. (2012). WISC-IV profile in high-functioning autism spectrum disorders: Impaired processing speed is associated with increased autism communication symptoms and decreased adaptive communication abilities. *Journal of Autism and Developmental Disorders, 42*, 655–664. https://doi.org/10.1007/s10803-011-1289-7
Osman, A., Bagge, C. L., Gutierrez, P. M., Konick, L. C., Kopper, B. A., & Barrios, F. X. (2001). The Suicidal Behaviors Questionnaire-Revised (SBQ-R): Validation with clinical and nonclinical samples. *Assessment, 8*, 443–454. https://doi.org/10.1177/107319110100800409
Patja, K., Iivanainen, M., Raitasuo, S., & Lönnqvist, J. (2001). Suicide mortality in mental retardation: A 35-year follow-up study. *Acta Psychiatrica Scandinavica, 103*, 307–311. https://doi.org/10.1034/j.1600-0447.2001.00019.x
Pelton, M. K., & Cassidy, S. A. (2017). Are autistic traits associated with suicidality? A test of the interpersonal-psychological theory of suicide in a non-clinical young adult sample. *Autism Research, 10*, 1891–1904. https://doi.org/10.1002/aur.1828
Pelton, M. K., Crawford, H., Robertson, A. E., Rodgers, J., Baron-Cohen, S., & Cassidy, S. (2020). A measurement invariance analysis of the Interpersonal Needs Questionnaire and Acquired Capability for Suicide Scale in autistic and non-autistic adults. *Autism in Adulthood, 2*, 193–203. https://doi.org/10.1089/aut.2019.0055
Pinder-Amaker, S. (2014). Identifying the unmet needs of college students on the autism spectrum. *Harvard Review of Psychiatry, 22*, 125–137. https://doi.org/10.1097/HRP.0000000000000032
Poorolajal, J., Haghtalab, T., Farhadi, M., & Darvishi, N. (2016). Substance use disorder and risk of suicidal ideation, suicide attempt and suicide death: A meta-analysis. *Journal of Public Health, 38*, 282–291. https://doi.org/10.1093/pubmed/fdv148

Productivity Commission. (2019). *Mental health, draft report*. Australian Government. https://www.pc.gov.au/inquiries/completed/mental-health/draft

Rabiee, A., Samadi, S. A., Vasaghi-Gharamaleki, B., Hosseini, S., Seyedin, S., Keyhani, M., Mahmoodizadeh, A., & Kermani, F.R. (2019). The cognitive profile of people with high-functioning autism spectrum disorders. *Behavioral Sciences, 9*, Article e20. https://doi.org/10.3390/bs9020020

Rai, D., Culpin, I., Heuvelman, H., Magnusson, C. M. K., Carpenter, P., Jones, H. J., Hannah, J., Emond, A. M., Zammit, S., Golding, J., & Pearson, R. M. (2018a). Association of autistic traits with depression from childhood to age 18 years. *JAMA Psychiatry, 75*, 835–843. https://doi.org/10.1001/jamapsychiatry.2018.1323

Rai, D., Heuvelman, H., Dalman, C., Culpin, I., Lundberg, M., Carpenter, P., & Magnusson, C. (2018b). Association between autism spectrum disorders with or without intellectual disability and depression in young adulthood. *JAMA Network Open, 1*, Article e181465. https://doi.org/10.1001/jamanetworkopen.2018.1465

Reichow, B., Steiner, A. M., & Volkmar, F. (2013). Cochrane review: Social skills groups for people aged 6 to 21 with autism spectrum disorders (ASD). *Evidence-Based Child Health: A Cochrane Review Journal, 8*, 266–315. https://doi.org/10.1002/ebch.1903

Reiss, S., Levitan, G. W., & Szyszko, J. (1982). Emotional disturbance and mental retardation: Diagnostic overshadowing. *American Journal of Mental Deficiency, 86*, 567–574.

Reynolds, A. M., Soke, G. N., Sabourin, K. R., Hepburn, S., Katz, T., Wiggins, L. D., Schieve, L. A., & Levy, S. E. (2019). Sleep problems in 2- to 5-year-olds with autism spectrum disorder and other developmental delays. *Pediatrics, 143*, Article e20180492. https://doi.org/10.1542/peds.2018-0492

Ribeiro, J. D., Witte, T. K., Van Orden, K. A., Selby, E. A., Gordon, K. H., Bender, T. W., & Joiner, T. E., Jr. (2014). Fearlessness about death: The psychometric properties and construct validity of the revision to the acquired capability for suicide scale. *Psychological Assessment, 26*, 115–126. https://doi.org/10.1037/a0034858

Rodriguez-Seijas, C., Gadow, K. D., Rosen, T. E., Kim, H., Lerner, M. D., & Eaton, N. R. (2020). A transdiagnostic model of psychiatric symptom co-occurrence and autism spectrum disorder. *Autism Research, 13*, 579–590. https://doi.org/10.1002/aur.2228

Roth, G. A., Abate, D., Abate, K. H., Abay, S. M., Abbafati, C., Abbasi, N., Abbastabar, H., Abd-Allah, F., Abdela, J., Abdelalim, A., Abdollahpour, I., Abdulkader, R. S., Abebe, H. T., Abebe, M., Abebe, Z., Abejie, A. N., Abera, S. F., Abil, O. Z., Abraha, H. N., & Murray, C. J. L. (2018). Global, regional, and national age-sex-specific mortality for 282 causes of death in 195 countries and territories, 1980: A systematic analysis for the Global Burden of Disease Study 2017. *The Lancet, 392*, 1736–1788. https://doi.org/10.1016/S0140-6736(18)32203-7

Ruzich, E., Allison, C., Smith, P., Watson, P., Auyeung, B., Ring, H., & Baron-Cohen, S. (2015). Measuring autistic traits in the general population: A systematic review of the Autism-Spectrum Quotient (AQ) in a nonclinical population sample of 6,900 typical adult males and females. *Molecular Autism, 6*, Article e2. https://doi.org/10.1186/2040-2392-6-2

Ruzzo, E. K., & Geschwind, D. H. (2016). Schizophrenia genetics complements its mechanistic understanding. *Nature Neuroscience, 19*, 523–525. https://doi.org/10.1038/nn.4277

Samson, A. C., Phillips, J. M., Parker, K. J., Shah, S., Gross, J. J., & Hardan, A. Y. (2014). Emotion dysregulation and the core features of autism spectrum disorder. *Journal of Autism and Developmental Disorders, 44*, 1766–1772. https://doi.org/10.1007/s10803-013-2022-5

Sanchez-Gistau, V., Baeza, I., Arango, C., González-Pinto, A., de la Serna, E., Parellada, M., Graell, M., Paya, B., Llorente, C., & Castro-Fornieles, J. (2013). Predictors of suicide attempt in early-onset, first-episode psychoses: A longitudinal 24-month follow-up study. *The Journal of Clinical Psychiatry, 74*, 59–66. https://doi.org/10.4088/JCP.12m07632

Sedgewick, F., Leppanen, J., & Tchanturia, K. (2021). Gender differences in mental health prevalence in autism. *Advances in Autism, 7*, 208–224. https://doi.org/10.1108/aia-01-2020-0007

Smith, J. M., Alloy, L. B., & Abramson, L. Y. (2006). Cognitive vulnerability to depression, rumination, hopelessness, and suicidal ideation: Multiple pathways to self-injurious thinking. *Suicide and Life-Threatening Behavior, 36*, 443–454. https://doi.org/10.1521/suli.2006.36.4.443

Stagg, S. D., & Belcher, H. (2019). Living with autism without knowing: Receiving a diagnosis in later life. *Health Psychology and Behavioral Medicine, 7*, 348–361. https://doi.org/10.1080/21642850.2019.1684920

Stanley, I. H., Day, T. N., Gallyer, A. J., Shelef, L., Kalla, C., Gutierrez, P. M., & Joiner, T. E. (2021). Autism-related traits and suicide risk among active duty U.S. military service members. *Psychological Services, 18*, 377–388. https://doi.org/10.1037/ser0000418

Steenfeldt-Kristensen, C., Jones, C. A., & Richards, C. (2020). The prevalence of self-injurious behaviour in autism: A meta-analytic study. *Journal of Autism and Developmental Disorders, 50*, 3857–3873. https://doi.org/10.1007/s10803-020-04443-1

Strauss, G. P., Nuñez, A., Ahmed, A. O., Barchard, K. A., Granholm, E., Kirkpatrick, B., Gold, J. M., & Allen, D. N. (2018). The latent structure of negative symptoms in schizophrenia. *JAMA Psychiatry, 75*, 1271–1279. https://doi.org/10.1001/jamapsychiatry.2018.2475

Uljarević, M., Hedley, D., Cai, R.-Y., Hardan, A. Y., & South, M. (2020a). Anxiety and depression from adolescence to old age in autism spectrum disorder. In F. R. Volkmar (Ed.), *Encyclopedia of Autism Spectrum Disorders*. Springer. https://doi.org/10.1007/978-1-4614-6435-8_102432-1

Uljarević, M., Hedley, D., Foley, K. R., Magiati, I., Cai, R. Y., Dissanayake, C., Richdale, A., & Trollor, J. (2020b). Anxiety and depression from adolescence to old age in autism spectrum disorder. *Journal of Autism and Developmental Disorders, 50*, 3155–3165. https://doi.org/10.1007/s10803-019-04084-z

Uljarević, M., Richdale, A. L., McConachie, H., Hedley, D., Cai, R. Y., Merrick, H., Parr, J. R., & Le Couteur, A. (2018). The hospital anxiety and depression scale: Factor structure and psychometric properties in older adolescents and young adults with autism spectrum disorder. *Autism Research, 11*, 258–269. https://doi.org/10.1002/aur.1872

Upthegrove, R., Abu-Akel, A., Chisholm, K., Lin, A., Zahid, S., Pelton, M., Apperly, I., Hansen, P. C., & Wood, S. J. (2018). Autism and psychosis: Clinical implications for depression and suicide. *Schizophrenia Research, 195*, 80–85. https://doi.org/10.1016/j.schres.2017.08.028

van Orden, K. A., Cukrowicz, K. C., Witte, T. K., & Joiner, T. E. (2012). Thwarted belongingness and perceived burdensomeness: Construct validity and psychometric properties of the Interpersonal Needs Questionnaire. *Psychological Assessment, 24*, 197–215. https://doi.org/10.1037/a0025358

van Spijker, B. A. J., Batterham, P. J., Calear, A. L., Farrer, L., Christensen, H., Reynolds, J., & Kerkhof, A. J. F. M. (2014). The Suicidal Ideation Attributes Scale (SIDAS): Community-based validation study of a new scale for the measurement of suicidal ideation. *Suicide and Life-Threatening Behavior, 44*, 408–419. https://doi.org/10.1111/sltb.12084

van Steensel, F. J., Bogels, S. M., & Perrin, S. (2011). Anxiety disorders in children and adolescents with autistic spectrum disorders: A meta-analysis. *Clinical Child and Family Psychology Review, 14*, 302–317. https://doi.org/10.1007/s10567-011-0097-0

Vanbergeijk, E., Klin, A., & Volkmar, F. (2008). Supporting more able students on the autism spectrum: College and beyond. *Journal of Autism and Developmental Disorders, 38*, 1359–1370. https://doi.org/10.1007/s10803-007-0524-8

Vogan, V., Lake, J. K., Tint, A., Weiss, J. A., & Lunsky, Y. (2017). Tracking health care service use and the experiences of adults with autism spectrum disorder without intellectual disability: A longitudinal study of service rates, barriers and satisfaction. *Disability and Health Journal, 10*, 264–270. https://doi.org/10.1016/j.dhjo.2016.11.002

Wark, S., McKay, K., Ryan, P., & Müller, A. (2018). Suicide amongst people with intellectual disability: An Australian online study of disability support staff experiences and perceptions. *Journal of Intellectual Disability Research, 62*, 1–9. https://doi.org/10.1111/jir.12442

Weinheimer, L. A. (2018). *A mixed method exploration of suicidality in clients with intellectual disability.* (Doctor of Psychology). University of Toronto, Canada. https://hdl.handle.net/1807/91979

Wigham, S., Rodgers, J., South, M., McConachie, H., & Freeston, M. (2015). The interplay between sensory processing abnormalities, intolerance of uncertainty, anxiety and restricted and repetitive behaviours in autism spectrum disorder. *Journal of Autism and Developmental Disorders, 45*, 943–952. https://doi.org/10.1007/s10803-014-2248-x

Wood, J. J., Kendall, P. C., Wood, K. S., Kerns, C. M., Seltzer, M., Small, B. J., Lewin, A. B., & Storch, E. A. (2020). Cognitive behavioral treatments for anxiety in children with autism spectrum disorder: A randomized clinical trial. *JAMA Psychiatry, 77*, 474–483. https://doi.org/10.1001/jamapsychiatry.2019.4160

4

Advance Care Planning with and for People Who Have Intellectual and Developmental Disabilities

Jacqueline McGinley and Deborah Waldrop

Madeline's* Story

Madeline, age 88 years old, had been hospitalised with pneumonia after her group home manager felt that "something just wasn't right" and persisted through multiple trips to the doctor until a diagnosis was rendered. Madeline's cousins visited frequently and her caregivers sat vigil as she lay in the intensive care unit, her oxygen saturation steadily worsening and requiring at first a facemask and then a ventilator. Madeline's

J. McGinley (✉)
Department of Social Work, Binghamton University, Binghamton, NY, USA
e-mail: jmcginle@binghamton.edu

D. Waldrop
School of Social Work, University at Buffalo, Buffalo, NY, USA
e-mail: dwaldrop@buffalo.edu

R. J. Stancliffe et al. (eds.), *End of Life and People with Intellectual and Developmental Disability*, https://doi.org/10.1007/978-3-030-98697-1_4

passing, just 12 days after being admitted to the hospital, marked the first death of an older adult with intellectual disability for her community agency, which was formed in the 1960s by a group of parent-advocates whose children experienced congenital rubella syndrome. Calls were made to notify family, friends, caregivers, and administrators that Madeline had died; unknown was the fact that others around the world were observing the emergence of an unprecedented population of people with intellectual and developmental disabilities who were reaching advanced age.

My encounters with Madeline as she died occurred less than two years after earning my Master of Social Work (MSW) degree. Several years later, while pursuing a doctoral degree, I found myself reflecting on her passing, along with the four other older adults I supported and who had died in the time since. I ultimately went back to the community agency where I began my career and partnered with them for my dissertation; together, we reconstructed eight individuals' last years of life with the goal of better understanding what happens when ageing adults with intellectual and developmental disabilities become seriously ill and reach life's end.

I interviewed one caregiver who had been with the agency for almost 20 years. About halfway through the interview, she said:

> *So, there was never really no plan, and we don't talk about that. We talk about all the good and all the bad, like in behaviors and these things. But we never talk about the fact that they are older, and they're getting older. Even when they're in their fifties, it's like they're in their seventies. They seem to age much quicker … they do and no one is ever prepared for that.*

As I listened to her, I recalled feeling similarly when Madeline died. There was an unwavering commitment to high-quality care through the duration of Madeline's life, but there seemed to be no real plan to help guide me, her other caregivers, or her family during the end of Madeline's life and on through bereavement. Others whom I interviewed would share similar reflections; each person's death felt like uncharted territory even as more and more of the adults with intellectual and developmental disabilities that we cared for aged and died.

JMM

*Assumed names are used in this story. It articulates the important elements of lived experiences.

More than ten years since Madeline's passing, as we write this chapter, we know now that plans do exist for people with intellectual and developmental disabilities. In fact, advance care planning (ACP) with people who have intellectual and developmental disabilities has been advocated for consistently (Friedman & Helm, 2010; Gaventa & Coulter, 2005; Kirkendall, et al., 2017; Perkins & van Heumen, 2018). Organisations, from The ARC (2008) to the Institute of Medicine (IOM, 2014), have endorsed and explicated strategies to ensure the inclusion of people with intellectual and developmental disabilities in future planning, particularly as they age, become seriously ill, and reach life's end. There has been a proliferation of free or low-cost tools to facilitate these conversations and document the decisions that emerge from them (see Resources at the end of this chapter). ACP with people who have intellectual and developmental disabilities is now regarded as both possible and essential.

Yet, ACP continues to exist squarely at the intersection of multidimensional policy debates and emotionally laden personal discourse, not just within the disability community but across society. This has given rise to important challenges that impact widespread implementation and impose limitations that demand attention from scholars and sensitivity among practitioners. We recognise the legitimate concerns expressed by some regarding the potential for ACP to be misused as a justification to withhold or withdraw treatment from people with intellectual and developmental disabilities, and we acknowledge that the COVID-19 pandemic has only heightened these concerns as the discriminatory rationing of care has been both widely feared and on some rare occasions observed (American Academy of Developmental Medicine and Dentistry (AADMD), 2020; McCallion, 2020) (see also Chapter 15). However, we hope to make clear that this chapter is *not* about the passive or active voluntary euthanasia of people with intellectual and developmental disabilities. While this chapter elucidates the broader socio-political context of the ACP process, we ultimately seek to offer a review of recent research related to critical issues and best practices for the purposes of informing those who seek the ethical use of ACP, with the goal of empowering people who have intellectual and developmental disabilities at life's end.

Review of the Literature

Overview

The United States' National Hospice and Palliative Care Organization (NHPCO, 2016) defines *advance care planning* as "…making decisions about the care you would want to receive if you become unable to speak for yourself" (para. 1). Similarly, the United Kingdom's National Palliative and End of Life Care Partnership (2015) suggests that to engage in ACP is to "express preferences for care and set personal goals for the time left" and states that "the process should include the possibility of recording preferences that might guide others if the person were to lose the mental capacity to make their own decisions" (p. 14). Across nations and definitions, ACP is consistently regarded as a person-centred, family-focused process that includes considering healthcare goals; getting information about diagnosis and prognosis; discussing wishes and values with relatives, caregivers, and healthcare providers; making decisions; sharing decisions; and completing advance directives (Carr & Luth, 2017; Watchman et al., 2018).

Advance directives and medical orders can be employed as instruments in the ACP process by instructing healthcare decisions when a person becomes seriously ill and/or unable to communicate for themselves. *Advance directives* are documents initiated by the patient and/or surrogate and include such things as a living will, healthcare proxy, or durable power of attorney. *Medical orders* are initiated and authorised by healthcare professionals for someone who is seriously ill and are legally binding documents that other healthcare providers must follow (IOM, 2014). The terminology, documentation, and legislative frameworks related to ACP may vary by nation and locality; it is thus imperative that those who facilitate the ACP process be familiar with regional policies and any special considerations in terms of how/if advance directives and medical orders can be used with people who have intellectual and developmental disabilities. Table 4.1 defines specific types of advance directives and medical orders.

ACP requires a reworking of processes, systems, tools, and approaches for people with intellectual and developmental disabilities to ensure access; however, it is, in many ways, not effectively different than for other subgroups of the population (McKenzie et al., 2017). The broad terms *future planning* and *person-centred planning* are often used interchangeably with ACP within the field of intellectual and developmental disabilities, although it is important to note that ACP is specific to the time period surrounding death (Heller et al., 2015). The use of these broader terms is useful; in that it encourages planning beyond strictly legal matters and treatment decisions (McKenzie et al., 2017; Watchman et al., 2018). ACP, in this broader sense, can encompass where people wish to die and what items they want at their bedside; who they want present as they die and what role they wish these individuals to assume; and how they wish to be remembered and what they wish to bequeath (Kirkendall et al., 2017; Stancliffe et al., 2016).

Over two decades ago, the use of advance directives by people with intellectual and developmental disabilities was found to be low, often attributed to the complexity of decision-making about end-of-life choices (Freedman, 1998). In a recent study, ACP discussions were found to occur more frequently especially in acute situations and after problems have emerged in the delivery of care (Voss et al., 2017). In their subsequent analysis of medical files for 30 people with intellectual and developmental disabilities in the palliative phase, Voss and colleagues (2019) found few were involved in planning and that the majority (93%) did have do-not-attempt-resuscitation (DNR) orders and a smaller subset (43%) to have documented agreements between professionals and relatives regarding future care, such as life-sustaining treatment. Emerging research is beginning to suggest a trend towards increased utilisation of advance directives particularly among seriously ill people with intellectual and developmental disabilities (McGinley et al., 2017), but further research is needed to ascertain if these trends hold across all people

Table 4.1 Types of advance directives and medical orders

Advance directive/medical order	Definitions
Healthcare Proxy Alternatives: Durable Medical Power of Attorney; Healthcare Agent; Healthcare Surrogate; Healthcare Representative; Healthcare Attorney-in-fact; Patient advocate	A document that identifies someone who is trusted to express a seriously ill persons' wishes if they became unable to do so for themselves
Living Will	A legal document that identifies the life-sustaining treatment (e.g., cardiopulmonary resuscitation [CPR], mechanical ventilation, intravenous hydration and nutrition, antibiotic and antiviral medication) a person does or does not want to utilise should they become seriously ill and unable to speak for themselves. These written documents can also articulate other care preferences (e.g., palliative or comfort care) and organ/tissue donations
Do Not Resuscitate (DNR) & *Do Not Intubate (DNI) Orders* Alternatives: Do-Not-Attempt-Resuscitation (DNAR)	Medical orders, often written but acceptable if delivered orally in some nations, that direct the healthcare team to not administer cardiovascular resuscitation (CPR) if a person's heart stops beating or not intubate (i.e., insert a tube, such as in the trachea to support ventilation) should a life-threatening event occur, respectively
Physician Orders for Life-sustaining Treatment (POLST) Alternatives: Medical Orders for Life-Sustaining Treatment (MOLST); Provider Orders for Life-Sustaining Treatment (POLST)	A form that provides doctor-ordered instructions to ensure, in the event of an emergency, someone who is seriously ill receives their preferred treatment. The POLST serves to mirror and enforce existing advance directives

Note The terminology is largely US-specific. Terms may be different in other countries with intellectual and developmental disabilities and among younger and healthier subsets of the population.

Many barriers to ACP exist, even as the rates of participation seem to be increasing. Healthcare providers do not typically communicate directly about serious illness with people who have intellectual and developmental disabilities due to the perceived lack of capacity, emotionality, and difficulty with the topic (McEnhill, 2008) (see also Chapter 5). Further, healthcare providers tend to be more reactive than proactive when it comes to ACP for these individuals, thus often delaying conversations until the final stages of an advance illness in response to worsening prognosis and/or a change in care needs (Voss et al., 2019). Caregivers, including both ageing parents and adult siblings, reported feeling similarly reluctant to engage in future planning because they lacked an understanding of the process, had fears related to their own mortality, and perceived a shortage within or held a distrust of their local disability system (Lee et al., 2019). Healthcare providers and caregivers also had challenges distinguishing their distinctive roles and associated tasks within the ACP process, which often led to delays or avoidance of the topic (Voss et al., 2017). It is thus not surprising that Stancliffe and colleagues (2016) found people with intellectual and developmental disabilities to be less self-determined in the domain of end-of-life planning when compared to people without disabilities.

In spite of these barriers, opportunities for meaningful ACP exist for people with intellectual and developmental disabilities and those who support them. Later in this chapter, best practices for improving the quality of the ACP process are described. Chief among them is fostering effective relationships between people with intellectual and developmental disabilities, their caregivers, and healthcare providers whereby the entire team regards the wishes of the person as central to the decision-making process (AAIDD, 2012; Voss et al., 2017). It is also important for the team to regard ACP as an ongoing process and not a singular act; research suggests that ACP should be revisited at minimum annually or upon a significant status change (IOM, 2014). Ultimately, the aim of ACP is to improve the quality of the dying experience for both the person who is dying and their caregivers by increasing the likelihood of adherence to the person's wishes when they become seriously ill and reach life's end (Voss et al., 2017). The outcomes for those who are dying and those who care for them seem to benefit significantly when the ACP process is carried out effectively (IOM, 2014; Wiese et al., 2015).

Decision-Making

An understanding of decision-making in the context of ACP with people who have intellectual and developmental disabilities is predicated on two distinct concepts, capacity and competency.

Capacity, in the context of ACP, asks the question: Does this person have the ability to understand their prognosis and provide informed consent supporting or refusing their medical care? Capacity can be exercised while receiving extra assistance or support (e.g., chunking key concepts, using assistive technology, engaging a support person). A person who is provided with a tablet that offers videos and pictures to support communication with their healthcare provider as they make complex medical decisions is exercising their capacity, just as is someone who makes these same decisions without the use of assistive technology (Kirkendall et al., 2017; Tuffrey-Wijne & Watchman, 2015).

Although there is widespread agreement that capacity is situational and person-specific, there is not consensus nor is there a standardised, universally available measure for assessing capacity in healthcare decision-making in serious illness with people who have intellectual and developmental disabilities (Freedman, 1998; Kirkendall et al., 2017). There is, however, widespread agreement that decision-making capacity should be assumed unless determined otherwise without prejudice based exclusively upon factors such as age or diagnosis. When determinations of capacity are made, they should not be considered static or applicable across all decision-making scenarios; these determinations must be considered situational and subject to change accordingly over the lifespan (Levy & van Stone, 2010). The IOM (2014) noted that people with intellectual and developmental disabilities may lack decision-making capacity for certain medical decisions, particularly those that are complex or high risk; yet, they retain capacity for other decisions, including the selection of a proxy or healthcare agent. In practice, however, it is not uncommon for people with intellectual and developmental disabilities, particularly those who have communication or sensory impairments, to be inaccurately presumed to lack capacity (Tuffrey-Wijne et al., 2010).

Competency, in the context of ACP, asks the question: Is this person 18 years of age or older and considered legally competent under the law to make their own medical decisions? As this question suggests, competency is presumed unless legally determined otherwise (Johnson, 2010; Turnbull, 2005). Competency is a legal concept that varies significantly from country to country. The variability includes how competence, and by way of this incompetence, is both defined and determined (Kirkendall et al., 2017). This variability also includes who determines if a person is competent and who is appointed to make decisions or has the authority to make decisions when a person lacks the ability to do so for themselves. Standards for surrogate decision-making can be unclear and often vary from setting-to-setting and case-to-case (Johnson, 2010). Familiarity with local policies and practices around competency determinations is essential for healthcare providers and surrogate decision-makers who engage people with intellectual and developmental disabilities in ACP and make decisions that ideally reflect their expressed and documented wishes.

Surrogate decision-makers can be relatives, friends, caregivers, or healthcare providers. A surrogate may be referred to as a power of attorney, guardian, proxy, or healthcare agent depending upon regional terminology. The problematic nature of surrogate decision-makers has been well-documented. Surrogates are not flawless translators of their loved ones' preferences; they are influenced by their own hopes and the social context. As a result, surrogates have been found to commonly assess patients' wishes incorrectly and apply them inconsistently (Vig et al., 2011).

Surrogates who seek to mitigate these challenges are advised by the ethical principles of: beneficence, non-maleficence, autonomy, and justice (Johnson, 2010; Turnbull, 2005).

- Upholding the principle of *beneficence* in the context of ACP requires surrogates to carefully consider the potential benefits and consequences of healthcare decisions and do both right and good by the person with intellectual and developmental disabilities.
- The principle of *non-maleficence* requires the same actions that do not cause harm and improve well-being.
- Individuals, regardless of disability, have the right to *autonomy* and, when it is exercised, it should be recognised and respected.
- Upholding *justice* assures that treatment options are provided equally to people with or without intellectual and developmental disabilities.

Individuals should be engaged by their surrogates and healthcare providers in the decision-making process, regardless of whether or not they have capacity to make all decisions or if they are legally competent to authorise such decisions. Barriers to this engagement should not be used as a justification for foregoing this essential task, but rather supports should be offered to ensure the person is given the right to exercise their autonomy and for their expressed wishes be a central part of the decision-making process (see also Chapter 13). Ethical principles help surrogates and healthcare providers to ensure equitable treatment of people with intellectual and developmental disabilities in the ACP process and in end-of-life decisions.

Facilitating Advance Care Planning

There are several important considerations when engaging people with intellectual and developmental disabilities in the ACP process, including when to have, who to engage, and how to go about having these conversations about goals and wishes. ACP is an individualised process; it thus requires professionals who engage in it to be flexible and responsive to the needs of the person and their caregivers (Voss et al., 2017). The resources provided at the end of this chapter offers a variety of tools for facilitating these conversations; those who facilitate ACP may thus want to consider these anew every time they engage in the process. There is not a "one-size-fits-all" approach for every individual and circumstance;

different levels of baseline knowledge must be accounted for and an individualised approach to intervention should be implemented (Stancliffe et al., 2016).

Current research suggests that ACP should be conducted early and often (Carr & Luth, 2017). Some healthcare providers and scholars believe that ACP should start as early as possible following a serious illness diagnosis (McKenzie et al., 2017), while others have suggested an upstream approach. Tuffrey-Wijne and colleagues (2015) suggested that ACP could occur before there is even a need for palliative care so long as it is appropriate to those involved. Researchers and practitioners from the International Summit on Intellectual Disability and Dementia warn against waiting until health needs change or cognition begins to decline, suggesting instead that ACP should be undertaken during early adulthood. Billings and Bernacki (2014) described the *Goldilocks phenomenon* regarding the best time to engage in the ACP process:

> Timing is important because the completion of an advance directive too far from or too close to the time of death can lead to end-of-life decisions that do not optimally reflect the patient's values, goals, and preferences: a poorly chosen target patient population that is unlikely to need an AD in the near future may lead to patients making unrealistic, hypothetical choices, while assessing preferences in the emergency department or hospital in the face of a calamity is notoriously inadequate. (p. 620)

In the case that ACP occurs too early, it is often because the person or their family found it challenging to understand the prognosis, recognise the imminence of the end stage of the disease trajectory, and become comfortable with the growing inevitability of death. This can be particularly true in the case of people with intellectual and developmental disabilities who have had multiple experiences of serious illness from which they have recovered; in these cases, the person and their family may find it hard to comprehend the current situation as distinctly different from times when the outcome had not been death. It can be important in these cases for ACP, when possible, to be delayed until involved individuals become familiar with the situation and the possible plans of care (Voss et al., 2019).

The provider who facilitates ACP (such as, a nurse manager from a healthcare organisation, a social worker from a disability organisation, or a bioethicist from an ethics committee) is most qualified when they are well-trained, knowledgeable of, and comfortable with conversations related to death and dying, and skilled at communicating about these subjects with people who have intellectual and developmental disabilities and their relatives. There is also tremendous variability in legal frameworks that inform ACP, so facilitators must also understand organisational policies and myriad policies from the other various systems of care the person with intellectual and developmental disabilities will encounter (Voss et al., 2017). They must consider what other individuals may be involved in decision-making and the delivery of care to ensure they are also engaged in the planning process. This includes bridging the divide that can exist between disability service providers and healthcare providers that span the care continuum, including those in primary and specialty care, hospitals, and palliative care settings (McKenzie et al., 2017; Tuffrey-Wijne & Watchman, 2015). The ACP facilitator must be sure to challenge healthcare providers' misconceptions around capacity and consent as they arise and support them in communicating difficult or complex information to people with intellectual and developmental disabilities and their families (Kirkendall et al., 2017). The facilitator must also thoughtfully and intentionally engage the person and their relatives so that they have the information needed to make informed decisions, understand their roles in the process, and ultimately have a positive experience (McKenzie et al., 2017; Voss et al., 2019).

People with intellectual and developmental disabilities can and should be engaged in the ACP process. Recent research has found these individuals to have the capacity and interest to participate in ACP, especially when person-centred accommodations are made and shared decision-making is facilitated by healthcare providers (IOM, 2014; Savage et al., 2015). Self-direction and person-centred planning inform many aspects of supports and services for people with intellectual and developmental disabilities (Kingsbury, 2010). Person-centred planning assumes that people are authorities in their own lives and, as such, deserve agency in deciding what services they receive and how they are provided (Heller et al., 2015). This imperative extends to ACP. To this end, the onus is on the person facilitating the ACP process, surrogate decision-makers, and healthcare providers to identify strategies for ensuring people with intellectual and developmental disabilities are engaged in the ACP process in such a way that meaningful participation is valued by the team and made accessible to the person.

There is no single form or policy that guides ACP with people who have intellectual and developmental disabilities. Policies differ significantly from nation-to-nation, region-to-region; and from organisation-to-organisation. The Georgetown University Center for Child and Human Development has proposed a three-step model (Think, Ask, Plan) for end-of-life decisions that seems broad enough to guide those involved in ACP while also allowing for the variability that exists within current practices. This model can be used by those leading ACP conversations to help guide the process but also encourage all individuals who may be involved. The first step, "Think", encourages those involved to learn about and reflect upon ethical frameworks that inform the decision-making process. The second step, "Ask", allows for the gathering of important information from multiple sources in preparation for ACP conversations with healthcare providers. The third and final step, "Plan", provides the opportunity for ACP tools to be used to assess various decisions, consider their impact on the quality of life, and ultimately document decisions (Georgetown University Center for Child & Human Development, 2020).

Additional best practices have begun to emerge for guiding ACP conversations. Multiple, ongoing conversations may be required and information may need to be "chunked" to facilitate building blocks of knowledge until sufficient understanding about prognosis and treatment is built such that decisions can be rendered (Tuffrey-Wijne & Watchman, 2015). The American Association on Intellectual and Developmental Disabilities (AAIDD) (2012) developed a Position Statement entitled *Caring at the End of Life* that advocates for advance care planning to include discussions of care preferences that are general yet specific enough to provide practice guidance. ACP should include discussions about what treatments the person would or would not want if they had a life-limiting illness (National Hospice and Palliative Care Organization, 2016). These conversations may also be extended to include discussions about the person's intentions for the future, where they wish to live, their finances and retirement, their support needs for both day-to-day and more complex decisions, and opportunities for social connectedness in later life (The ARC, 2020). It may also include equally important conversations about what music they want softly playing during their final days or whether or not they want a beloved pet near them as they die. Finally, ACP can be an emotional and frightening process for those who are involved, particularly for the person who is seriously ill and for their relatives. Follow-up is an important but often overlooked step in the process; those leading these conversations should check-in shortly afterwards and periodically thereafter to ensure that people feel sensitively supported and connected to the services necessary to help them cope with grief and bereavement.

We offer two stories that illustrate the implications of ACP that occurs early and often versus ACP that happens amidst a complex medical crisis. We invite the reader to reflect upon how these contexts might facilitate or impede person-centred, family-focused care to people with intellectual and developmental disabilities as they age, become seriously ill, and reach life's end.

Carlos J's* Story

Carlos J is a 59-year-old man who was diagnosed with Down syndrome shortly after birth. Carlos is generally healthy; he has Type 1 Diabetes which is managed with medication and routine primary care. He lives at home with his ageing mother who is increasingly worried about what will happen to Carlos should their bond be broken by her death. She has expressed these concerns and also Carlos's wish to retire with his social worker and nurse manager. Anticipating Carlos's annual care plan meeting next month, the social worker and nurse manager attend a webinar from The ARC's *Center for Future Planning* and begin reviewing relevant organisational and government policies related to advance care planning. The nurse manager begins visiting Carlos and his mother more frequently asking more questions related to his healthcare goals, their wishes and values for later life, and decision-making, including who might support Carlos and where he might live if his mother were to die. Carlos, his mother, day habilitation support staff, social worker, primary care physician, and nurse manager then convene for his annual care planning meeting. In addition to discussing Carlos's care needs for the coming year, the team discusses and documents Carlos's wishes to retire in the next 3–5 years and makes a plan to file the necessary paperwork to transfer guardianship to his cousin, with whom he would live, should his mother become seriously ill and reach the end of life. The social worker follows-up with Carlos and his mother a week after the meeting; they share that it was sometimes uncomfortable and sad to think about the future but that they felt less anxious now that their wishes were better understood by the team and documented in Carlos's support plan. They asked to schedule another meeting with the team after completing a living will and filing the paperwork for guardianship successorship.

*Assumed names are used in this story.

John Paul S' Story*

John Paul S, who goes by J.P., is a 43-year-old man who was diagnosed with Down syndrome within the first few weeks of life. J.P. understands most receptive communication and uses voice in short sentences and phrases to express himself, but he relies on his sister/guardian for support with some activities of daily living and most major decisions. J.P. is transported from his family home by ambulance to the hospital with severe symptoms of COVID-19; he arrives alone and with only limited personal information. J.P. appears distressed; he is having difficulty in breathing and appears confused by his surroundings and why he must constantly wear a breathing mask. The healthcare providers are able to locate the contact information for J.P.'s sister and utilise Zoom to contact her; she is able to explain to J.P. what is happening and coach the healthcare team about how to best support him. As the week progresses, the healthcare team relies on J.P.'s sister to comfort her brother remotely, explain to him what is happening, and to make healthcare decisions with him. The team confirms that J.P. has not engaged in advance care planning nor does he have advance directives, so they rely on his guardian to serve as a proxy for all major decisions. One day, his treating physician needs consent for catherisation. After multiple attempts to reach his sister, another relative contacts the hospital to inform them that his sister had also contracted COVID-19 and died within two days of the onset of symptoms. The team is devastated by J.P.'s losses, and they begin working with the remaining family to establish a new proxy and plan for his long-term care now that his beloved caregiver has died.

*Assumed names are used in this story.

Ethical Considerations

After publishing my first manuscript on ageing and dying with intellectual and developmental disabilities, a disability rights advocate emailed me and asked to speak via telephone. She told me that a colleague had shared the article with her, and it left her wondering what brought me to this work and what my intentions were for it.

As we spoke, she wove her own lived experiences with the disability rights movement. She shared with me the history of marginalisation, particularly in healthcare and research settings, that she and others had experienced throughout their lives. She then moved to more recent times, where she had grown increasingly worried might be used to justify withholding of essential care and the hastening of death.

As our conversation came to a close, she asked me where I stood on these issues. I assured her that my work would never endorse the passive or active voluntary euthanasia of people with intellectual and developmental disabilities, just as we did at the start of this chapter. But I also conceded that I had failed to consider the complexity of the issues I was writing about; it had not occurred to me that my paper might arouse strong feelings from people with disabilities and those who cared for them. I vowed to work diligently to address this gap in my own understanding and increase the sensitivity with which I approached this work.

> Our conversation made me acutely aware of the historical context that must always be at the front of mind when engaging people with intellectual and developmental disabilities and their relatives in the ACP process. I shared this with you so that you may learn from my experiences.
> JMM

Situating ACP for people with intellectual and developmental disabilities within a broader socio-political context facilitates understanding of the long-standing history of marginalisation that has led many in the disability community to be sceptical of advance directives for fear of rationing and prioritising care that could potentially lead to the premature termination of life-supportive care and hasten death. Concerns about marginalisation were articulated by the International Association for the Scientific Study of Intellectual Disabilities (IASSID) (2002) that observed:

> Achievement oriented societies tend to devalue people with intellectual disabilities as unproductive, dependent, even deviant. As they grow older, they encounter other negative perceptions which reinforce marginalisation. For example: ageism compounded with handicapism often is prevalent in many societies and encourages even more discriminatory attitudes against ageing adults with intellectual disabilities. To their credit, enlightened societies have attempted to overcome negative stereotypes associated with age and disability and have enabled their older members to live productive lives. Those that haven't have even further marginalised ageing people with intellectual disabilities often by relegating them to inferior old age services.

Marginalisation was most egregiously evident in the movement towards institutionalisation in the mid-1800s following centuries of people with intellectual and developmental disabilities living with family in communities. In these settings, there existed no form of ACP; death was considered an inevitable part of the institutionalised experiences and little if any concern was given to the circumstances in which it occurred. Individuals who died in this setting were met with the dehumanising practice of being placed in graves with only numbers; roommates and friends who witnessed these practices continue—even as we write this—to experience complicated grief as a result (Botsford & King, 2010). Even following the deinstutionalisation movement that came on the heels of the civil rights movement, people with intellectual and developmental disabilities have continued to experience disparities in the care and treatment of serious illness. Discriminatory practices continue in healthcare and other settings where quality of life in the context of disability is not well understood; emphasis is placed on cost-savings or cost-cutting strategies; and (often inaccurate) presumptions are made about the correlation between cognitive ability, communication, and capacity for informed consent.

The human rights perspective—although having origins in concepts dating back to the Babylonians—is derived from the United Nation's Universal Declarations of Human Rights, which was proclaimed in 1948 following the atrocities of the Holocaust and World War II (United Nations, 1948). A working group, inclusive of non-governmental organisations and people with disabilities, more recently drafted the Convention on the Rights of Persons with Disabilities (CRPD); it was adopted by consensus of the United Nations General Assembly in 2006 and subsequently opened for signatures (United Nations, 2006). In 2009, the United States signed the CRPD but the Senate failed in 2013 to reach the two-thirds majority necessary for ratification (United States International Council on Disabilities (USICD), 2016).

The CRPD, which is inclusive of people with intellectual and developmental disabilities, supports opportunities for autonomy, choice, and equality that are inherent to and necessary for widespread adoption of ACP. A human rights perspective provides a social-philosophical argument for the rights of all people—regardless of ability and capacity—to act autonomously, make choices free of undue influence, and receive appropriate and equitable care that minimises pain and suffering (United Nations, 1948). For conceptualising the barriers and challenges faced by people with intellectual and developmental disabilities as it relates to ACP, the human rights perspective helps to make clear how national and international policies converge over time to inform how systems perceive, act upon, and respond to people with intellectual and developmental disabilities even at the end of life.

ACP can bring to the surface deep-seated and well-founded worry among people with intellectual and developmental disabilities and those who care for them. While we continue to endorse ACP as an important process for empowering people to be involved in healthcare decision-making, we also acknowledge the risk for misuse and abuse particularly in eras, such as during the COVID-19 pandemic, where providers and policy makers ration care (see also Chapter 9). It is imperative that those involved in the ACP process educate themselves on the historical context in which these feelings are grounded and work diligently within themselves and their organisations to ensure all medical options are made available and care is rendered justly and in accordance with expressed wishes.

Conclusion

The field of intellectual and developmental disabilities has grasped the importance of ACP but knowledge of and facility with interventions is still nascent. The context of carrying out ACP across intersecting systems, such as disability services, healthcare, and palliative care, create a challenging level of complexity. Further, the multimorbidity and great propensity of multiple chronic conditions among people with intellectual and developmental disabilities can create additional complex care pathways. We are, however, at the threshold of opportunity now. The application of the principles of beneficence, non-maleficence, autonomy, and justice may seem challenging in a society that is fragmented, but we are in a time that calls for things to be considered and reconsidered. The time is rife with possibility for improving care for people with intellectual and developmental disabilities who are diagnosed with serious illnesses through ACP that supports education, ethical decision-making, communication, and compassion.

Reflection

Mary's* Story
Mary P. is a 68-year-old woman who was diagnosed with Fragile X syndrome during her early childhood. She has lived in a community residence for people with intellectual and developmental disabilities for the past 10 years, following the death of her father. Mary, who has heart disease and is oxygen dependent, develops pneumonia with laboured breathing and is transported to Memorial Hospital's Emergency Department. Mary's sister Patricia, who is her guardian, is called during the transport. Because Mary does not have a living will and her breathing deteriorates en-route to the hospital, the team decides that intubation is the only alternative. When Patricia arrives, she is told that her sister's condition is grave and that she is on a respirator. Patricia is very distressed by this news and tells the providers that this would not have been what either she or Mary would have chosen. Patricia is told that there is nothing that can be done immediately, and she will have to go through the local Department of Health to get permission to withdraw life support. Mary is transferred to the hospital's intensive care unit. While Patricia is trying to get paperwork completed, she is approached about artificial hydration, nutrition and resuscitation if Mary's heart stops. Patricia pleads, "This is not what she would have wanted. She wanted to be at home with her cat, not stuck in a bed with all these tubes. Is there anyone who can help us?".

*Assumed names are used in this story.

Mary P.'s situation is referred to the hospital Ethics Committee of which you are the Chair. Consider how you will guide the committee in upholding Mary's rights and the principles of autonomy, beneficence, non-maleficence, and justice.

1. What is the *right* course of action: legally, morally, ethically? Explain what has informed and guided your decision-making around this course of action.

2. What can the hospital do to support Patricia during this process? How might Patricia's bereavement be impacted by her experiences of Mary's end of life?
3. How might this situation have been different had there been a living will? What can systems of care, such as the community residence where Mary P. resides, do to engage individuals with intellectual and developmental disabilities and their surrogates in the advance care planning process ahead of crises?
4. What hospital policies could be developed to support the rights of people who have intellectual and developmental disabilities and become suddenly and critically ill? How might these policies anticipate and address concerns related to capacity, such that people with intellectual and developmental disabilities are engaged in decision-making processes?

Resources

Unless otherwise indicated by an asterisk (*), the resources listed below are designed specifically for people with disabilities:

1. *Advance Care Planning**. A range of guides to help people talk about their wishes for care. Though designed for the general community, these could be adapted for people with intellectual and developmental disabilities. https://www.nhdd.org/public-resources#where-can-i-get-an-advance-directive
2. *Five Wishes**. An easy-to-use legal document to help express wishes ahead of a serious illness. https://www.agingwithdignity.org/five-wishes
3. *Future Planning*. Learning about and building a plan for the future. Includes resources and short videos on why and how individuals have made plans. https://futureplanning.thearc.org/
4. *How to break bad news to people with intellectual disabilities. A guide for carers and professionals*. London: Jessica Kingsley. Guidelines for practitioners, families and carers to ease the process of explaining bad news. http://www.breakingbadnews.org

5. *Making My Own Healthcare Decisions for People with Disabilities*. A guide to writing a letter to the doctor about preferred healthcare decisions. https://thearc.org/wp-content/uploads/forchapters/Making-My-Own-Health-Care-Decisions--A-Letter-for-to-My-Doctors.pdf
6. *Palliative Care and People with Learning Disabilities (PCPLD) Network*. A network of people, including those with intellectual disability, professionals, and families with the aim of ensuring quality palliative care. On the Resources tab, there is a section on planning ahead/advance care planning. https://www.pcpld.org
7. *People Planning Ahead*. Guidance to ensure loved ones receive care that conforms to their personal, cultural, and religious beliefs. Available for purchase from third-party site (e.g., Amazon).
8. *Respecting Choices: Person-Centred Care**. Guidance for organisations and communities about evidence-based practices to ensure individual healthcare choices are known about and honoured. https://respectingchoices.org/
9. *The Conversation Project Starter Kits**. A tool to help have a conversation with a family member, friend, or loved one about wishes regarding end-of-life care. Available in several languages. https://theconversationproject.org/starter-kits/
10. *Thinking Ahead: My Way, My Choices, My Life at the End*. A workbook to make decisions about life support treatment and other end-of-life choices. https://mn.gov/mnddc/honoring-choices/Thinking-Ahead-English-web.pdf
11. *When I Die*. An example of a completed end-of-life plan by a person with intellectual and developmental disability. http://www.pcpld.org/wp-content/uploads/when_i_die_2_0.pdf.

References

American Academy of Developmental Medicine and Dentistry (AADMD). (2020). *Advocacy*. www.aadmd.org/advocacy

American Association on Intellectual and Developmental Disabilities (AAIDD). (2012). *Caring at the end of life*. https://www.aaidd.org/news-policy/policy/position-statements/caring-at-the-end-of-life

Billings, J. A., & Bernacki, R. (2014). Strategic targeting of advance care planning interventions: The goldilocks phenomenon. *JAMA Internal Medicine, 174*(4), 620–624. https://doi.org/10.1001/jamainternmed.2013.14384

Botsford, A. L., & King, A. (2010). End-of-life policies and practices. In S. L. Friedman & D. T. Helm (Eds.), *End-of-life care for children and adults with intellectual and developmental disabilities* (pp. 13–30). AAIDD.

Carr, D., & Luth, E. A. (2017). Advance care planning: Contemporary issues and future directions. *Innovation in Aging, 1*(1), 1–10. https://doi.org/10.1093/geroni/igx012

Freedman, R. I. (1998). Use of advance directives: Facilitating health care decisions by adults with mental retardation and their families. *Mental Retardation, 36*(6), 444–456. https://doi.org/10.1352/0047-6765(1998)036%3c0444:UOADFH%3e2.0.CO;2

Friedman, S. L., & Helm, D. T. (Eds.). (2010). *End-of-life care for children and adults with intellectual and developmental disabilities*. AAIDD.

Gaventa, W. C., & Coulter, D. L. (2005). *End-of-life care: Bridging disability and aging with person-centered care*. Haworth Pastoral Press.

Georgetown University Center for Child and Human Development. (2020). *Think-Plan-Act*. https://gucchd.georgetown.edu

Heller, T., Gibbons, H. M., & Fisher, D. (2015). Caregiving and family support interventions: Crossing networks of aging and developmental disabilities. *Intellectual and Developmental Disabilities, 53*(5), 329–345. https://doi.org/10.1352/1934-9556-53.5.329

Institute of Medicine (IOM). (2014). *Dying in America: Improving quality and honoring individual preferences near the end of life*. The National Academies Press.

International Association for the Scientific Study of Intellectual Disabilities. (IASSID). *Fact sheet: Ageing and intellectual disabilities*. https://www.iassidd.org/wp-content/uploads/2019/04/aging-factsheet.pdf

Johnson, B. B. (2010). Practical guide to health care decision making. In S. L. Friedman & D. T. Helm (Eds.), *End-of-life care for children and adults with intellectual and developmental disabilities* (pp. 133–146). AAIDD.

Kingsbury, L. A. C. (2010). Use of person-centred planning for end-of-life decision making. In S. L. Friedman & D. T. Helm (Eds.), *End-of-life care for children and adults with intellectual and developmental disabilities* (pp. 275–289). American Association on Intellectual and Developmental Disabilities.

Kirkendall, A., Linton, K., & Farris, S. (2017). Intellectual disabilities and decision making at end of life: A literature review. *Journal of Applied Research in Intellectual Disabilities, 30*(6), 982–994. https://doi.org/10.1111/jar.12270

Lee, C.E., Burke, M.M., & Stelter, C.R. (2019). Exploring the perspectives of parents and siblings toward future planning for individuals with intellectual and developmental disabilities. *Intellectual and Developmental Disabilities, 57*(3), 198–211. https://doi.org/ https://doi.org/10.1352/1934-9556-57.3.198

Levy, J., & van Stone, M. (2010). Ethical foundations and legal issues. In S. L. Friedman & D. T. Helm (Eds.), *End-of-life care for children and adults with intellectual and developmental disabilities* (pp. 31–49). AAIDD.

McCallion, P. (2020). Social work response needed to the challenge of COVID-19 for aging people with intellectual and developmental disabilities. *Journal of Gerontological Social Work, 63*(6–7), 595–597. https://doi.org/10.1080/01634372.2020.1769791

McEnhill, L. S. (2008). Breaking bad news of cancer to people with learning disabilities. *British Journal of Learning Disabilities, 36*(3), 157–164. https://doi.org/10.1111/j.1468-3156.2008.00517.x

McGinley, J., Waldrop, D. P., & Clemency, B. (2017). Emergency medical services providers perspective of end-of-life decision making for people with intellectual disabilities. *Journal of Applied Research in Intellectual Disabilities, 30*(6), 1057–1064. https://doi.org/10.1111/jar.12363

McKenzie, N., Mirfin-Veitch, B., Conder, J., & Brandford, S. (2017). "I'm still here": Exploring what matters to individuals with intellectual disability during advance care planning. *Journal of Applied Research in Intellectual Disabilities, 30*(6), 1089–1098. https://doi.org/10.1111/jar.12355

National Hospice and Palliative Care Organization (NHPCO). (2016). *Advance care planning.* http://www.nhpco.org/advance-care-planning

National Palliative and End of Life Care Partnership. (2015). *Ambitions for palliative and end of life care.* www.endoflifecareambitions.org.uk

Perkins, E., & van Heumen, L. (2018). When the journey ends: Parental caregiving and later life planning. In B. S. Nelson Goff & N. Piland Springer

(Eds.), *Intellectual and developmental disabilities: A roadmap for families and professionals,* (pp. 216–228), Routledge, Taylor & Francis Group.

Savage, T. A., Moro, T. T., Boyden, J. Y., Brown, A. A., & Kavanaugh, K. L. (2015). Implementation challenges in end-of-life research with adults with intellectual and developmental disabilities. *Applied Nursing Research, 28*(2), 202–205. https://doi.org/ https://doi.org/10.1016/j.apnr.2014.10.002

Stancliffe, R. J., Wiese, M. Y., Read, S., Jeltes, G., & Clayton, J. M. (2016). Knowing, planning for, and fearing death: Do adults with intellectual disability and disability staff differ? *Research in Developmental Disabilities, 49–50*, 47–59. https://doi.org/10.1016/j.ridd.2015.11.016

The ARC. (2008). *Position statement: Aging*. http://www.thearc.org/document.doc?id=3652

The ARC. (2020). *The center for future planning*. https://futureplanning.thearc.org/

Tuffrey-Wijne, I., Bernal, J., & Hollins, S. (2010). Disclosure and understanding of cancer diagnosis and prognosis for people with intellectual disabilities: Findings from an ethnographic study. *European Journal of Oncology Nursing, 14*(3), 224–230. https://doi.org/10.1016/j.ejon.2010.01.021

Tuffrey-Wijne, I. & Watchman, K. (2015). Breaking bad news to people with learning disabilities and dementia. *Learning Disability Practice, 18*(7), 16–23. https://doi.org/ https://doi.org/10.7748/ldp.18.7.16.e1672

Turnbull, H. R. (2005). What should we do for Jay? In W. C. Gaventa & D. L. Coulter (Eds.), *End-of-life care: Bridging disability and aging with person-centered care.* (pp. 1–26). Haworth Pastoral Press.

United Nations. (1948). *Universal Declaration of Human Rights, United Nations General Assembly, 10 December 1948, resolution 217 A.* http://www.un.org/en/universal-declaration-human-rights/index.html

United Nations. (2006). *Convention on the Rights of Persons with Disabilities (CRPD), United Nations General Assembly, 13 December 2006, resolution A/RES/61/106*. http://www.un.org/disabilities/convention/conventionfull.shtml

United States International Council on Disabilities (USICD). (2016). *Convention on the rights of persons with disabilities.* http://usicd.org/index.cfm/crpd

Vig, E. K., Sudore, R. L., Berg, K. M., Fromme, E. K., & Arnold, R. M. (2011). Responding to surrogate requests that seem inconsistent with a patient's living will. *Journal of Pain and Symptom Management, 42*(5), 777–782. https://doi.org/10.1016/j.jpainsymman.2011.08.003

Voss, H., Vogel, A., Wagemans, A. M., Francke, A. L., Metsemakers, J., Courtens, A. M., & de Veer, A. J. (2017). Advance care planning in palliative care for individuals with intellectual disabilities: A systematic review. *Journal of Pain and Symptom Management, 54*(6), 938–960. https://doi.org/10.1016/j.jpainsymman.2017.04.016

Voss, H., Vogel, A., Wagemans, A. M., Francke, A. L., Metsemakers, J. F., Courtens, A. M., & de Veer, A. J. (2019). Advance care planning in the palliative phase of people with intellectual disabilities: Analysis of medical files and interviews. *Journal of Intellectual Disability Research, 63*(10), 1262–1272. https://doi.org/10.1111/jir.12664

Watchman, K., Janicki, M. P., Udell, L., Hogan, M., Quinn, S., & Berankova, A. (2018). Consensus statement of the International Summit on Intellectual Disability and Dementia on valuing the perspective of persons with intellectual disability. *Journal of Intellectual Disabilities, 23*(2), 266–280. https://doi.org/10.1177/1744629517751817

Wiese, M., Stancliffe, R.J., Read, S., Jeltes, G., & Clayton, J.M. (2015). Learning about dying, death, and end-of-life planning: Current issues informing future actions. *Journal of Intellectual and Developmental Disability, 40*(2), p. 230–235. https://doi.org/10.3109/13668250.2014.998183

5

Decision-Making at the End of Life: Challenges and Opportunities for People with Intellectual and Multiple Disabilities in Residential Homes in Germany

Sabine Schäper

In both the fields of health care and disability care, there is a long tradition of patients or residents being treated as an object of professional care. This chapter is based on a broad understanding of decision-making and highlights the relevance of attributions and professional self-conceptions as well as aspects of organisational culture to enable a person to be respected with full legal capacity, including at the end of life.

S. Schäper (✉)
Department of Social Services, Catholic University of Applied Sciences North Rhine-Westphalia, Münster, Germany
e-mail: s.schaeper@katho-nrw.de

R. J. Stancliffe et al. (eds.), *End of Life and People with Intellectual and Developmental Disability*, https://doi.org/10.1007/978-3-030-98697-1_5

From Paternalistic to Participative Decision-Making in Health Care

In the health care sector, the paternalistic model of the physician–patient dyad was first questioned during the 1950s. Among others, Balint (1957) advocated for shifting from a subject-object-relationship to mutual participation. This advocacy questioned the widespread idea of an inactive and incompetent patient on the one hand, against the focus on evidence-based medicine on the other, with the former devaluing any kind of subjective appraisal. The paternalistic model was supported by dynamics of (self-) disempowerment, where patients, dependent on the decision-making of health professionals, experienced personal subjective anxiety arising from their lack of autonomy. With the rise of medical ethics since the 1950s, the idea of the autonomous patient became increasingly popular. Consequently, current decisions on treatment and care are ethically and legally permitted only on the base of the informed consent of the patient. The idea of informed consent was formalised as a principle of medical ethics in the 1960s for all medical research projects involving human beings (World Medical Association General Assembly, October 2013). In the following decades, a broad consensus arose that the relationship between the patient and the physician should be optimised with *shared* or *participative decision-making*. Patient and physician as partners assume though, that physicians have the requisite communication skills and expertise in psychosocial issues so that decision-making occurs as a truly interactional procedure.

Loh et al. (2007) explored the complexity of participative decision-making and described the nuanced steps, starting with the (shared) notification of an upcoming decision and the agreement on equal rights for the partners. The nuance of even the early steps demonstrates that participative decision-making depends on an equal balance of power between patient and physician. The risk for sham participation exists at every step of the decision-making procedure, as the evaluation of decision-making models by Wirtz et al. (2006) showed. Who decides on the set of options declared as available in the current situation, and who can evaluate pros and cons of the options from a completely neutral perspective? Moreover, some important steps are missing in the procedure. What about the right to choose *not* to know about details of one's own health status and perspective?

Limitations in Implementing Participative Decision-Making

The preference for participatory decision-making has reached broad consensus, the more so as studies have shown that participatory decision-making offers positive outcomes including reduction of conflicts, higher satisfaction with the treatment and better treatment compliance (Loh et al., 2007). The actual implementation of participatory decision-making though remains largely unexplored. In Germany, shared decision-making is far from routine, even in the general health sector (Danner et al., 2020). Participation is seen as time consuming in an increasingly cost-contained health care system. A wide range of limiting factors exist at an organisational, economic and contextual level, and personal patient factors are viewed as of secondary importance.

The Idea of the Autonomous Patient

The shift from a paternalistic to autonomous patient model mirrors the shift from focussing on the *welfare* of the patient as chief priority to primarily considering the *will* of the patient themselves (Geisler, 2002). This approach gathered momentum as the health sector came under cost containment measures on the one hand, and the rising tendency to avoid liability risks on the other. With specific reference to the end of life, the idea of decision-making by the patient as autonomous implies an autonomy-focussed concept of medical ethics, in which every person might be expected to take decisions around dying into their own hands. This though ignores inter-subjective values as important elements of end-of-life care. The ethics of care reminds us that the relationship between physician and patient should not primarily be a contractual one, but instead is shaped by trust and care (Tronto, 1993). Human relations, more than normative reflections, are the source of knowing what is good or bad and the best locus for decision-making should be conducted in dialogue and relationality. This illustrates how the implementation of models of decision-making is influenced by different conceptions of autonomy and dependency, or rather conceptions of humankind.

On closer reflection, the underlying concepts of humankind of each of these models become apparent: The paternalistic model is based on a belief of differentiation, which Foucault (1982) identified as an aspect of power relations. Starting from a reconstruction of the history of psychiatry, hospitals and prisons, Foucault analysed typical ways power is exercised in these institutions, based on the assumption of quasi-ontological differences, "a system of differentiations which permits one to act upon the actions of others" and "which are at the same time its conditions and its results" (Foucault, 1982, p. 792).

Foucault argued that the modern Western state took up a specific kind of power technique from the Christian cultural history, which he defined as *pastoral power* (1982). One of the most important characteristics of pastoral power was that "…it cannot be exercised without knowing the inside of people's minds, without exploring their souls, without making them reveal their innermost secrets. It implies a knowledge of the conscience and an ability to direct it" (p. 783). Foucault proposed that pastoral power, akin to that exercised by a shepherd upon his sheep, at the same time protects each individual while also controlling it in a very subtle manner, so that the individual person feels safe and free. Despite the decreasing influence of the Christian church as a social and political power over time, pastoral power spread across society, a useful approach to govern people effectively by a sort of self-government.

Using Foucault's analysis, one could be sceptical that in the health care shared decision-making model a special sort of pastoral power is exercised, *masked* as participation. In reality though, pastoral power is maintained because the quasi-ontological difference between physician and patient has not been addressed.

Decision-Making, Self-Determination and Participation

In the field of disability care, the terms that reflect Foucault's analysis of power include *othering*, *ableness* (as a positive ontology) and *disabled* (as a negative ontology), each juxtaposed in a way that potentiates the difference—and the social distance between people (Campbell, 2005, 2008). Originating from the critical cultural sciences, the concept of *othering* includes a critique against subtle forms of power, confirming a hierarchy between *us* and *them*, positioning *the others* or *them* as lesser. As both Foucault and Goffman analysed in their research on institutions, the history of disability services can be understood as a construction of otherness with a legitimisation of exclusion and punishment.

From this point of view, even the notion of the autonomous patient is open to suspicion. It allows distinction between persons, some who are deemed to have the ability to decide autonomously using their full cognitive capacity, and others with limited capacity to do so. The foundation of this distinction again is *ableism,* a conception of humankind depending on full ability, "a network of beliefs, processes, and practices that produce a particular kind of self and body...that is projected as the perfect, as the species-typical, and, therefore, as essential and fully human" (Campbell, 2005, p. 127). *Ableism,* therefore, justifies non-participation in decision-making procedures by those with lesser assumed ability. As a counter model to any kind of obvious or subtle paternalism, the rights-based approach of the United Nations Convention on the Rights of Persons with Disabilities (UNCRPD; United Nations, 2006) advocates respecting the full legal capacity of every human being despite any kind of individual impairment (see also Chapter 4).

As decision-making models depend on the concept of personhood, the practice of decision-making by people with intellectual disabilities is embedded within a given socio-historical context. The following overview on the aetiology of participation and self-determination as principles in German disability policy and practice shows the relevance of disability history and practice for people with disabilities to autonomously make end-of-life decisions.

Legal Capacity of Persons with Intellectual Disability in Germany as a Framework for End-of-Life Decision-Making

Disability services began to question paternalistic cultures of care in Germany in the 1970s, precipitated by a report of a commission of experts and self-advocates to the German Parliament on the living conditions of psychiatric patients (Finzen, 2015). Up to the end of the 1970s, many people with intellectual disabilities lived as long-term patients in psychiatric hospitals or large institutions and the report offered an important milestone, providing a dramatic insight into daily life hidden behind walls. Within Europe, the Scandinavian countries served as catalysts for the shift to self-determination, participation and *normal* living conditions, shaping the development of disability services around Europe. However, in Germany the tradition of paternalistic care remained persistent. Today, large institutions still exist, with 42% of adults with intellectual disabilities living in residential homes (Thimm et al., 2018). A quarter of these are residents of institutions with more than 80 inhabitants and a differentiated internal service portfolio, indicated by institution-centred instead of person-centred approaches and attitudes (Thimm et al., 2018). The specific scripts and professional self-conceptions of staff members in such institutions reflect persistent paternalistic traditions of organisational cultures.

Moreover, until the 1990s the legal status of persons with intellectual disabilities in Germany was weak. Legal guardianship was designed to reflect total lack of capacity to make decisions so that the scope for individuals to make any decisions was limited. In 1992, a legal reform opened up the restrictive legal framework, defining specific duties and legal competencies of the personal guardian. However, the guardian's right to decide *instead* of the person themselves remained possible by the legal instrument of "reservation of consent" to the legal guardian (Federal Ministry of Justice & Consumer Protection, 2013, Sect. 1903). Close to twenty years later, Germany's ratification of the UNCRPD in 2009 (United Nations Human Rights Treaty Bodies, n.d.) became an important political driver for awareness raising of the right of people with disabilities to decide for themselves.

The United Nations Human Rights Office of the High Commissioner Committee on the Rights of Persons with Disabilities (2014) in its *List of issues* responding to the initial report of Germany called for "changes to the existing custodianship system, including details of measures taken to provide supported decision-making where substituted decision-making currently takes place" (para 6). Especially, Germany had to "clarify how the term 'reservation of consent' complied with Article 12 of the Convention" (para 7). In its comment to the first periodic state report, the German Institute for Human Rights, leading the monitoring procedures of the UNCRPD in Germany, pointed out that some main principles of the legal framework contradict the UNCRPD. The Institute reported that guardianship may be installed against the *natural will* of the person (Federal Ministry of Justice & Consumer Protection, 2013, Sect. 1903), and the declaration of consent may be reserved to the legal guardian in defined areas of personal issues (Federal Ministry of Justice & Consumer Protection, 2013, Sect. 1903). Moreover, the idea of *welfare* is still used as a guiding principle for decisions instead of the will and preference of the person (§1901 Abs. 2 German Civil Code) (Deutsches Institut für Menschenrechte, 2015, p. 19). However, in the next periodic report the German state rejected the need to change the legal framework related to legal guardianship or fundamental structural changes regarding the cooperation of all relevant actors in this field in the first instance (Bundesministerium für Arbeit und Soziales [BMAS], 2019). It took another two years to meet the necessity of a reform of legal guardianship and assistance. In 2020, the discussion lead to a draft law, based on a research project on behalf of the Federal Ministry of Justice and Consumer Protection in 2017 about the quality of legal guardianship (Bundesministerium für Justiz und Verbraucherschutz [BMJV], 2018). The German Institute for Human Rights highlighted that supported decision-making has to be open for everyone who is in need for support, regardless of impairment. The support has to be accessible, adequate and available for all (see also Chapter 16) (Deutsches Institut für Menschenrechte, 2020). Again, the reform does not go far enough, as it does not proceed to a clear rejection of the reservation of consent to the legal guardian, thereby maintaining limits to the full legal capacity of persons under custodianship. As the German Institute for Human Rights emphasised, other regulations could have been conceivable (Deutsches Institut für Menschenrechte, 2020, p. 7). At least, the draft of the new law claims that substituted decision-making by others is to be flanked by an acceptable justification. Arguably though, absolute clarity is needed; that any decision needs to be legitimised by the will of the person in question, with no exception.

Respecting Personal Choices and Legal Capacity

From the legal point of view, the claim of Article 12 of the UNCRPD seems to be unambiguous: legal capacity has no exemption. Even the International Classification of Functioning, Disability and Health (ICF; World Health Organization, 2001) points out that making choices is key to realising full and equal participation. These positions though continue to be contested. For example, Freeman et al. (2015) argued that the UN Committee went too far in advocating that every individual person, at all times, has the capacity to make decisions. Many practical indicators reflect how the notion of legal capacity is not unanimously agreed upon or understood, and instead is open to interpretation risk:

1. The legal framework in many countries continues to deny the individual right to decide in all areas of life. The actual framework in Germany focuses on the right to autonomy, but in detail the legal guardian has the right to substituted decision-making even in the new legislation.
2. The main challenge seems to be not only to choose, but also to *realise effective* shared decision-making. Attitudes of legal guardians may jeopardise the claim of Article 12 of the UNCRPD. Research in Germany suggests that nearly a quarter of legal guardians believe that only very few (or none) of their clients could be strengthened in their autonomy in any way (Bundesministerium für Justiz und Verbraucherschutz (BMJV), 2018). Legal guardians often do not trust

the capacity of the persons they represent. Therefore, legal reform has to be accompanied by awareness raising about the guardian's role in capacity building of the person.

3. Meadows (2017) identified a lack of consequent change in organisational culture contributing to lack of movement towards shared decision-making in the field of mental health. Even more, shared decision-making can be misused to conceal injustice in contexts, where the options for individual choices are limited.
4. In Germany, there is a debate about how to determine legal capacity of people with intellectual disabilities and other cognitive impairments like dementia. The default position is often *incapable of giving consent* (in German referred to as *einwilligungsunfähig).* This position again reflects the attitude of *ableism* and *othering,* or as Campbell (2008) argued, perpetuates the distinction "between perfected naturalized humanity and the aberrant, the unthinkable, quasi-human hybrid and therefore non-human" (para. 9). Degener (2016) also proposed that the "incapacity approach to disability" reflects the continued presence of the medical model (p. 3), which though of decreased influence, persists in legal jurisdictions. Deneger (2016) argued instead that "the human rights model of disability defies the presumption that impairment may hinder human rights capacity" (p. 4). The perception that "impairment can foreclose legal capacity" (p. 3) is used to legitimise substitute decision-making, contrary to the position of the UNCRPD, which reaffirms the universality of individual human rights for all persons irrespective of the presence of disability.
5. The frequently proposed distinction between *legal* capacity and *mental* capacity perpetuates the idea that people with intellectual disability are incapable of giving consent. Some authors highlight legal capacity as an inherent and inalienable right and add the construct of mental capacity to focus the individual skills for decision-making. However, even mental capacity as an individual ability of a person to give consent is a social construct, for it is not possible to assess the individual capacity of making a decision in its social context at that specific time. During the past two decades, there has been a shift from a categorical to a more functional approach to mental

capacity, by developing assessment instruments. However, some still point out that the empirical evidence and theoretical concept of mental capacity are weak. Scientific effort to define standards to evaluate mental capacity leads to a level of rigour that even persons without disabilities will fail (Klie et al., 2014). Further, attempts to categorise persons as either competent or incompetent in and of itself maintains continued discrimination.

(Dis)respecting Personal Choices: Examples from Lived Experience

Research in Germany has suggested that respecting individual choices decreases with age and increases with care dependency. Where the person will receive end-of-life care and die are important decisions with far-reaching consequences for quality of life and quality of care. Moves from residential homes for people with intellectual disabilities to nursing homes in the last phase of life occur quite often (Thimm et al., 2018). A recent mixed-methods research project exploring the conditions of end-of-life care for people with intellectual and multiple disabilities in Germany was completed with staff members and frontline managers in three German regions (Schäper et al., 2019, 2020). Of 79 reported deaths in disability services, in 62.5% of these the person themself was reportedly involved in decisions in some way, but in more than a third of the cases there was no participation at all. Subsequent interviews with staff members revealed an ambivalent picture. Professional staff repeatedly expressed the wish to support residents until the end of their life, but structural reasons forced decisions for transition to nursing home or hospital care. Further, some staff were reluctant to entertain any real possibility of individual decision-making, again using competence as a justification for lack of participation. To illustrate, one interviewee stated: "*Well, it's hard to say. So, yes, I think so, it would be nice if they could determine their place of death on their own. However, that will never be possible for many people, let's say, because of their cognitive limitations, that they can decide by themselves. Yes, it's a really difficult subject, I must say*" (Schäper et al., 2019, 2020). Commentary such as this yet again demonstrates the tendency of *othering*, constructing an insurmountable difference between *us* and *them*, and perpetuating the position that people with disability are not competent to decide by themselves.

Also evident in Schäper et al.'s (2019, 2020) research was a lack of initiative to approach questions about the end of life due to fear of negative emotional consequences. One staff, in response to interview questions about the possibility of person-centred planning for the end of life answered evasively: "*No, this is not yet a part of our routine, it is only a project on the fringe of our daily life, but is never realised. Because of the lack of time – you need a little bit of patience and a calm atmosphere – it is not consequently implemented in this way. You would have to take a lot of time – and people do not have that much concentration. You should really take an hour for one resident or so to talk about these issues …but this is not realistic – unless they come with some questions on their own. That's just the way it is – you have to use these moments*". Responses such as these suggest the attribution of responsibility to the dying individual on the one hand, and the limitations of the residential environment on the other. Staff themselves did not perceive that they have agency in actively preparing the individual for decision-making at the end of life, this reflecting a typical mechanism in institutions. Staff members construct themselves as being powerless in a powerful institution, as West-Leuer (2002) found in her qualitative study on psychodynamic mechanisms in institutions. This "construction of powerlessness" (West-Leuer, 2002, p. 444) helps to make both one's own competencies *and* limitations invisible. By deflecting individual responsibility, the staff member also renounces responsibility to actively shape, develop and reflect their own professional position and practice. This seems to be the other side of ableism in institutions for the professional staff members; they themselves remain in a passive condition. This mechanism can be understood as Foucault's previously described *pastoral power*: a sort of self-governance, including the active participation of the individual subject in its own "government" according to the aims of the state or of a given organisation. Thus, the circle of power and powerlessness continues and no one professional changes their practice, while organisations continue their institutionalised persistence, and staff and residents behave in a complementary manner. Perhaps this practice is particularly pronounced where new challenges to organisations and professionals present themselves, as in end-of-life care. Dying itself is largely an experience of powerlessness. The challenge ahead is for care organisations to carefully manage power, to unmask disempowerment and instead include strategies of decision-making which meaningfully respect individual preferences and choices.

Schäper et al.'s (2020) research on conditions of end-of-life care for people with intellectual disabilities in Germany revealed a wide range of professional practice attitudes and strategies in residential homes. The relationship between organisational culture and professional self-conceptions of staff allows a paradigm understanding of end-of-life care in disability services, summarised in Table 5.1.

Arising from the paradigm range presented in Table 5.1, it is possible to derive some recommendations focussing on organisational and professional practice in end-of-life care:

- Disability residential services should evaluate and revise conditions of care to better prepare for the increasing numbers of elderly people with disabilities requiring end-of-life care so those individuals can die at their preferred place.
- Staff members need supervision and capacity building to promote a rights-based view about individual decision-making on dying and death.
- Staff members need time to realise person-centeredness as a professional routine integral to quality end-of-life care.
- Professional practice needs to be reflected as a matter of course and as part of staff professional attitude to resist the temptation towards own powerlessness and thereby powerlessness of persons with disabilities.

To conclude this section, the impact of organisational structures and professional attitudes should be highlighted to ensure a participatory and person-centred culture of end-of-life care in residential services for people with intellectual and multiple disabilities.

Table 5.1 Strategies and characteristics of disability services in end-of-life care: Paradigm range (Schäper et al., 2020)

Paradigm	Delegating paradigm	Dichotomising paradigm	Functional paradigm	Integrating paradigm
Typical strategies to deal with end-of-life care	Outsourcing (to nursing homes, hospitals)	Contrasting individual attitudes versus organisational culture	Professional routines dominate individual perspectives	High level of reflection of attitudes and conditions of care
Characteristics	End-of-life care does not occur in organisational concepts and mission statements Management does not regard disability services as responsible for end-of-life care Negative emotional experiences in end-of-life care including fear, helplessness and excessive demands	Staff members focus on their own role within the institution Organisation is perceived as rigid and controlling Own professional practice is regarded as a contrast insofar as staff members 'do their best' for the residents	Functional quality of care is guaranteed by a high level of professional standards and rules Emotional participation of staff members is limited Professional routines dominate relations to residents Collective strategies predominate compared to person-centred care	Management and staff try to balance the dimensions of person-centred care, the reality of death and dying and high quality of care They reflect professional practice and organisational strategies regularly and are open-minded, acknowledging their own limitations instead of declaring residents as incapable

Support for Decision-Making as a Multicomponent Concept of Professional and Organisational Development

Seeking to describe what is meant by supported decision-making, one can refer to The United Nations Human Rights Office of the High Commissioner Committee on the Rights of Persons with Disabilities (2014) General Comment on Article 12. The committee recommended that supported decision-making must be available to *all*, must be based on the individual will and preferences (instead of a presumed best interest) and must be accessible, matching the individual's preferred communication needs. Importantly too, the person must have the right to refuse support or the support person at any time. Moreover, the committee emphasised that "perceived or actual deficits in mental capacity must not be used as justification for denying legal capacity" (para. 13).

This approach to decision-making support overcomes the notion of *othering* with explicit acknowledgement that we as human beings *all* sometimes rely on others to help make decisions, particularly in critical stages of life. We all are at times dependent on care, as the concept of care ethics points out. Instead of pointing out otherness, we all share the need for support with every human being. Decision-making then can be seen as a *relational* process.

A Hermeneutic Approach: Relational Decision-Making

Instead of functional assessment of competencies and standardisation of procedures, a hermeneutic approach translates to a relational concept of decision-making.

There is little empirical evidence on the real effects of support in decision-making procedures. Kohn et al. (2013) indicated that supported decision-making may not always be empowering but can result in disempowerment when the process is subject to manipulation or abuse. Organisational culture and procedures and professional attitudes influence decision-making support offered to a person. Alternatively, experience in person-centred planning can change organisational attitudes about supported decision-making (Adler et al., 2017). Quality of communication and encounter within a reciprocal relationship is essential. Supported decision-making is not primarily a technique or functional routine on an individual level, but more an experience of resonance between persons, encouraged by a culture of openness for shared participation and understanding. The concept of support *for* decision-making (Bigby & Douglas, 2020) invites carers not to wait for legal reforms but instead to start to enable clients, organisations and oneself to better respect individual choices within existing practices. Bigby and Douglas (2020) offer a toolbox that guides supporters to reflect upon their role and find ways of supporting decision-making, even with those persons with more severe disabilities. This approach reflects the hermeneutic, relational intention of support for decision-making, "characterised by trust, genuine positive regard and honest interpersonal interactions" (p. 15). The story of Mr A below provides an example of support for decision-making

Mr A's story*: Relational Decision-Making

Mr A was a man of 85 years with intellectual disability who took a very active part in life. In the past, his social worker responsible for planning end-of-life support had frequently initiated conversations about his living will and possible treatment decisions. Mr A had refused these talks, always clearly stating: "I want to live, not die". Even during a severe kidney disease 8 years prior, his attitude did not change. Thus, no written document existed about his end-of-life decisions. There was also no legal guardian, because he placed great value on his autonomy.

When he fell ill with severe COVID-19, he initially emphasised very strongly that he absolutely wanted to live. At the same time, he insisted on being able to stay in his residential group home and not go to hospital. For some time it looked as if he would recover. One staff member, who was the most consistent caregiver due to limited care numbers to reduce infection spread, became his most valued support. The staff member sensitively talked with Mr A about all the issues bothering him while ill. Even when Mr A became increasingly weak and could hardly express himself clearly, the staff member gently tried to explain to him what was going on and to verbalise his suspected state of mind. As the disease progressed and Mr A became increasingly plagued by pain and restlessness, he repeatedly and clearly expressed, "Please, please, sleep". Based on the close conversations and intimate knowledge of Mr A, the staff member was able to interpret this statement as a request for help to alleviate pain and restlessness. The staff member made sure he had correctly interpreted Mr A's communication and informed the palliative doctor, who prescribed sedation. Mr A became more relaxed. The staff member felt reconciled that Mr A's request was respected and that he was able to then die peacefully.

*Assumed names are used in this story. From research conducted by Schäper et al. (2021).

Ethical Deliberation as Training in Ethical Reasoning

In the health care context, the concept of ethical deliberation has been established in recent years especially for situations characterised by ethical dilemmas (Steinkamp & Gordijn, 2003). This concept is not as widespread in the field of disability services but could be helpful. Establishing ethical deliberation as a routine in disability services ensures space to reflect on concrete cases. At the same time, ethical deliberation as an organisational practice may enhance ethical competence of staff in general, well evidenced in the field of social work (Großmaß, 2013). The complexity of care at the end of life demands a reflexive competence where staff needs to all at once question their individual moral orientation, the ethical implications of professional concepts, organisational mission statement and professional standards. Knowledge of basic ethical positions in moral philosophy is needed to understand ethical principles, including justice, responsibility and care, and human rights. This ethical reasoning (Großmaß, 2013) enables professional carers to realise support for decision-making in asymmetric relationships. It also raises awareness about typical dilemmas and protects against potential abuse of power.

The story below about Mrs B shows that ethical case conferencing has a capacity-building effect and works to ensure "the cornerstone of...participative [clinical] ethics, in which all those involved in health care delivery are being taken seriously as subjects of ethical reflection" (Steinkamp & Gordijn, 2003, p. 245), including the dying person themself.

Mrs B's story*: Ethical Case Conference as a Means of Empowerment
During a routine medical examination, the personal doctor of Mrs B, a 68-year-old woman with intellectual disability, diagnosed a malignant tumour in her abdomen. The doctor then recommended a prompt operation to remove the tumour. At the same time, it was clear that metastases already existed, so a complete cure seemed very unlikely. Mrs B protested strongly against going to a hospital because of bad experiences in previous hospital stays. The staff of the residential home tried to persuade her, and Mrs B's sister was torn between concern and the will to respect her wishes. The sister suggested an ethical case conference, but stressed that it was important that Mrs B also participated in the conference. The panel of experts in the conference, which included the doctor, her sister and her main support worker from the residential home, initially frightened Mrs B. However, with careful moderation, the counsellor present reduced the complexity of the situation for all involved and focussed on the main ethical dilemma about the risk to die very soon and the Mrs B's right to informed self-determination. It was possible to motivate all those involved to listen to Mrs B very carefully and encourage her to express her own will. This enabled Mrs B to understand what was at stake for her and to express her preference very clearly. In response to the sister's statement, "But then there is nothing more we can do for you", Mrs B said, "Yes I know that. I have to die". It was hard for all to realise this as Mrs B's conscious and clear living will. Nevertheless, this was exactly

what helped everyone to gain acceptance for her decision. Mrs B's participation in the case conference had an enabling and empowering effect, for all.

*Assumed names are used in this story. From research conducted by Schäper et al. (2021).

The research of Schäper et al. (2021) showed that the regular use of this kind of ethical case discussion supported by a counsellor could be routinely used as part of end-of-life person-centred planning in residential care homes. All residents may choose to use counselling, but it is also respected if they do not accept the offer, as the decision to not engage with the topic of death "in itself is a choice" (Wiese & Tuffrey-Wijne, 2020, p. 317). Unfortunately, compared to the general population in Germany, only a small proportion of persons with disabilities have a directive for end-of-life decisions, pointing to a clear need for timely *preventive* counselling as a basis for effective decision-making about the end of life (Schäper et al., 2019).

Decision-Making During the COVID-19 Pandemic

There is very little knowledge about the way people with disabilities are affected by the COVID-19 pandemic and the social consequences of political and institutional risk strategies. The available research focus has been on infection rates and mortality, with limited research about individual experiences and approaches to decision-making. Heslop et al. (2020), reporting on available UK data of people with learning (intellectual) disabilities who had died from COVID-19, showed that essential elements of participation in decision-making processes including mental capacity assessment, discussion, family involvement and documentation were insufficiently implemented.

Across the world, prevailing political strategies still reflect the attitude of *othering*, maintaining the substantial difference between *us* as citizens, who are given full rights to own our decisions, and people with intellectual disabilities as *them*, perceived as not capable to do so. The recent COVID-19 Disability Rights Monitor survey about the impact of the pandemic on persons with disabilities worldwide "received alarming testimonies that persons with disabilities were denied or deprived of life-saving treatment for COVID-19 on the basis of disability, resulting in many preventable deaths. These are discriminatory decisions which give rise to grave human rights violations" (Brennan, 2020, p. 41). In contravention to the UNCRPD, responses during the COVID-19 pandemic suggest that for people with intellectual and developmental disability, legal capacity, individual choice and preference are disregarded, and participation in end-of-life planning and decision-making is limited on both an individual and political level.

Conclusion: Future Needs

The results from research conducted in Germany together with experiences during the COVID-19 pandemic show that shifts towards respecting the will and preferences of people with intellectual disabilities at the end of life are still far from routine, or even consistently considered by either health or disability care services.

To conclude, the following are the needs for the future:

- Legal reforms: The full legal capacity of persons regardless of disability or individual impairment should be strengthened by developing less restrictive alternatives to current legal guardianship.
- Change of attitudes: Professional carers need to reflect on the prevailing attitudes of *othering*. There is a critical need for professional capacity building about legal capacity and support for decision-making as a method to ensure the respect of choices.
- Programmes for supported decision-making: Disability services should develop multicomponent decision-making programmes, across all

organisational levels with a commitment to full and equal participation by all.

- Empowerment: People with intellectual and multiple disabilities need education about health care and their rights to self-determined decision-making. Peer support could be a good addition to share experience and encourage each other to decide for themselves even in complex issues.
- Legal guardians need to develop awareness about their responsibility to respect the full legal capacity of their clients and to adopt the principles of supported decision-making. They also need sufficient time to exercise their role. So too, people with intellectual and multiple disabilities need education to understand the role of guardianship, and their own role and rights within that relationship.
- Ethical deliberation should be implemented in disability services to cultivate ethical reasoning as a professional competence.

This broad range of options to ensure end-of-life decision-making, adding in organisational requirements, shows that the capacity to decide does not primarily depend on the intellectual capacity of persons with disabilities. Instead, it depends on the availability of support for decision-making, offered in a relational way.

Reflection

Supporting participation for self-determined decision-making should take several dimensions of action and reflection into account: The person, the level of support, the organisational culture and networking with others. Some recommendations for supporting self-determined decision-making are offered below, each rounded off with reflection questions for consideration.[1]

[1] Taken from selected results of the research project "Palliative Care for People with Intellectual and Multiple Disabilities" (Schäper et al., 2021). See Resources at end of this Chapter.

Self-determined communication is promoted. People with intellectual and multiple disabilities use their own means of expression about their own wishes and needs. They are given the opportunity to exchange their own experiences with dying, death and mourning. Staff have knowledge of communication development and are trained in the field of augmentative and alternative communication.

1. Do we recognise and respond to basic and unusual forms of expression as opportunities for communicative exchange and self-determination?
2. What methods and materials do we need to find better and more creative access to individual forms of expression? Where do we experience limits?

Health literacy enables people with intellectual and multiple disabilities to recognise, communicate and cope with their own physical changes and illnesses. Methods of biography work and personal future planning enable people to develop their own self-concept and reconstruct their own life story. Relatives are taken seriously as bearers of biographical knowledge. They can be helpful in recognising and supporting possibilities for self-determination in the last phase of life.

1. How do we perceive people with intellectual and multiple disabilities? How do we value situations of dependency? Do we legitimate limitations of self-determination in certain situations?
2. What forms of support in decision-making processes do we already use? What do we need to better support self-determined decisions?
3. What services do we provide so that people can learn more about health and dying? How can we design information to enable people with intellectual and multiple disabilities to give informed consent?

People with intellectual and multiple disabilities are given the opportunity to express their own wishes and ideas (even before their own last phase of life). Decisions at the end of life are well prepared.

1. Do we actively offer conversations about health and end-of-life planning? Do we make sure that conversation about the end of life is voluntary?
2. Do we have good means of documenting end-of-life planning and decisions?
3. Do we actively encourage the person to reconsider and reformulate end-of-life decisions and plans?

Organisations develop a person-centred and end-of-life sensitive organisational culture. Structures and procedures on the organisational level are consistently oriented towards the needs of the individual. Organisations develop a culture of trust and handing over responsibility. There are many possibilities for choice and decision-making in the organisation.

1. How well do we succeed in understanding the wishes and interests of the person?
2. What barriers exist in the organisation that hinders us from consistently orienting ourselves to the wishes of the person? How can these barriers be changed? What can I concretely contribute to respect choices?

The person at the end of life is at the centre of the network of care. In the networking structures of all those involved, the person concerned is at the centre of what is happening. Organisations develop methods and tools to shape multi-perspective decision-making processes when critical issues arise.

1. Who do we need to involve when it comes to complex end-of-life decisions and the person cannot decide alone? In decision-making, how do we prevent the person themself being moved from the centre and others interests becoming dominant?
2. What can I contribute to better understanding and considering the voice of the person? How can I motivate others in the support network to do so?
3. How can the person benefit from the cooperation with palliative services to better realise self-determined decision-making?

Resources

1. *Bonn Lighthouse (Future planning at the end of life: What I want).* Living will in plain language.

   Original of website in German: https://bonn-lighthouse.de/pv/
   Translation to English: https://bonn--lighthouse-de.translate.goog/pv/?_x_tr_sl=de&_x_tr_tl=en&_x_tr_hl=en&_x_tr_pto=nui,sc

2. *Netzwerk Persönliche Zukunftsplanung (Network Personal Future Planning).* A network of people and organisations from Germany, Luxembourg and Austria to promote future planning as a method to realize self-determination for people with intellectual disabilities.

   Original of website in German: https://www.persoenliche-zukunftsplanung.eu/neuigkeiten.html
   Translation to English: https://www.persoenliche-zukunftsplanung.eu/neuigkeiten.html

3. *PiCarDi (Palliative Care for People with Intellectual and Multiple Disabilities).* A research project conducted in Germany about providing the best possible palliative care and hospice support for people with intellectual and multiple disabilities. Includes easy-read information and resources about advance care planning:

   Original of website in German: https://www.picardi-projekt.de
   Translation to English: https://www-picardi--projekt-de.translate.goog/?_x_tr_sl=de&_x_tr_tl=en&_x_tr_hl=en&_x_tr_pto=nui,sc

4. *Support for decision making: A practice framework: La Trobe University Living with Disability Research Centre.* A guide to ensuring people with cognitive disability have support to make decisions. https://www.firah.org/upload/notices3/2015/support-for-decision-making.pdf

References

Adler, J., & Georgi-Tscherry, P. (2017). *Persönliche Zukunftsplanung mit Menschen mit körperlicher und intellektueller Beeinträchtigung und ihr Beitrag zu Veränderungen [Future planning with people with physical and intellectual disabilities and its impact on changing everyday life]* . Interkantonale Hochschule für Heilpädagogik. https://www.vereinigung-cerebral.ch/fileadmin/media/Dachverband/Dokumente/docs_Tat/docs_PZP/Persoenliche_Zukunftsplanung_Studie_def.pdf

Balint, M. (1957). *The doctor, his patient, and the illness*. Pitman.

Bigby, C., & Douglas, J. (2020). Supported decision making. In R. J. Stancliffe, M. L. Wehmeyer, K. A. Shogren & B. H. Abery (Eds.), *Choice, preference, and disability. Promoting self-determination across the lifespan* (pp. 45–66). Springer.

Brennan, C. S. (2020). *Disability rights during the pandemic. A global report on findings of the COVID-19 Disability Rights Monitor*. https://www.internationaldisabilityalliance.org/sites/default/files/disability_rights_during_the_pandemic_report_web_pdf_1.pdf

Bundesministerium für Arbeit und Soziales [BMAS]. (2019). *Combined second and third periodic report of the Federal Republic of Germany on the United Nations Convention on the Rights of Persons with Disabilities.* BMAS. https://www.gemeinsam-einfach-machen.de/GEM/DE/AS/UN_BRK/Staatenpruefung/Zweite_Staatenpruefung/Bericht_englisch.pdf

Bundesministerium für Justiz und Verbraucherschutz [BMJV]. (2018). *Qualität in der rechtlichen Betreuung [The quality of legal guardianship].* BMJV. https://www.bmjv.de/SharedDocs/Downloads/DE/Service/Fachpublikationen/Forschungsbericht_Qualitaet_rechtliche_Betreuung.pdf

Campbell, F. K. (2005). Legislating disability. Negative ontologies and the government of legal identities. In S. Tremain (Ed.), *Foucault and the government of disability* (pp. 108–130). University of Michigan Press.

Campbell, F. K. (2008). Refusing able(ness): A preliminary conversation about ableism. *M/C Journal, 11*(3). https://doi.org/10.5204/mcj.46

Danner, M., Geiger, F., Wehkamp, K., Rueffer, J. U., Kuch, C., Sundmacher, L., Skjelbakken, T., Rummer, A., Novelli, A., Debrouwere, M., & Scheibler, F. (2020). Making Shared Decision-Making (SDM) a reality: Protocol of a large-scale long-term SDM implementation programme at a Northern German University Hospital. *British Medical Journal Open, 10*(10), e037575https://doi.org/10.1136/bmjopen-2020-037575

Degener, T. (2016). Disability in a human rights context. *Laws, 5*(3), 35. https://www.mdpi.com/2075-471X/5/3/35

Deutsches Institut für Menschenrechte. (2015). *Parallelbericht an den UN-Fachausschuss für die Rechte von Menschen mit Behinderungen anlässlich der Prüfung des ersten Staatenberichts Deutschlands gemäß Artikel 35 der UN-Behindertenrechtskonvention [Parallel report to the UN-Committee on the Rights of Persons with Disabilities on the occasion of the initial report by the German state according to Article 35 UN-CRPD].* https://www.institut-fuer-menschenrechte.de/fileadmin/Redaktion/Publikationen/Parallelbericht/Parallelbericht_an_den_UN-Fachausschuss_fuer_Rechte_Menschen_mit_Behinderungen_Pruefung_1_Staatenbericht.pdf

Deutsches Institut für Menschenrechte. (2020) *Stellungnahme zum Referentenentwurf des Bundesministeriums der Justiz und für Verbraucherschutz „Entwurf eines Gesetzes zur Reform des Vormundschafts—und Betreuungsrechts" [Statement by the German Institute for Human Rights on the "Draft law on the reform of guardianship" of the Federal Ministry of Justice and Consumer Protection]* https://www.institut-fuer-menschenrechte.de/fileadmin/Redaktion/Publikationen/Stellungnahmen/Stellungnahme_Referentenentwurf_BMJV_Entwurf_Gesetzes_Reform_Vormundschafts_Betreuungsrecht.pdf

Finzen, A. (2015). Auf dem Wege zur Reform—die Psychiatrie-Enquete wird 40 [The "Psychiatrie-Enquete"—The German report on the state of psychiatry in 1975]. *Psychiatrische Praxis, 42*(7), 392–396. https://doi.org/10.1055/s-0034-1387589

Foucault, M. (1982). The subject and power. *Critical Inquiry, 8*(4), 777–795.

Freeman, M.C., Kolappa, K., Calas de Almeida, M., Kleinman, A., Makhasvili, N., Phakiahi, S., Saraceno, B., & Thornicroft, G. (2015). Reversing hard won victories in the name of human rights: A critique of the General Comment on Article 12 of the U.N. Convention on the Rights of Persons with Disabilities. *Lancet Psychiatry, 2*(9), 844–50. https://doi.org/10.1016/S2215-0366(15)00218-7

Geisler, L. (2002). *Arzt-Patient-Beziehung im Wandel—Stärkung des dialogischen Prinzips [Doctor-patient relationship in transition - strengthening the dialogue principle].* http://www.linusgeisler.de/art2002/0514enquete-dialogisches.html

Federal Ministry of Justice and Consumer Protection. (2013). *German civil code.* Available as translation to English: https://www.gesetze-im-internet.de/englisch_bgb/englisch_bgb.pdf

Großmaß, R. (2013). Ethical reasoning—Ethik in der beruflichen Praxis [Ethical reasoning—Ethics in professional practice]. In R. Großmaß & R. Anhorn (Eds.), *Kritik der Moralisierung [Critique of moralisation]* (pp. 209–226). Springer.

Heslop, P., Byrne, V., Calcin, R., Huxor, A., Sadoo, A., & Sullivan, B. (2020): *Deaths of people with learning disabilities from COVID-19: A comparator analysis of 206 deaths.* University of Bristol. https://www.bristol.ac.uk/media-library/sites/sps/leder/Deaths%20of%20people%20with%20learning%20disabilities%20from%20COVID-19.pdf

Klie, T., Vollmann, J., & Pantel, J. (2014). Autonomie und Einwilligungsfähigkeit bei Demenz als interdisziplinäre Herausforderung für Forschung, Politik und klinische Praxis [Autonomy and capacity to consent in dementia as an interdisciplinary challenge for research, policy and clinical practice]. *Informationsdienst Altersfragen, 41*(4), 5–15.

Kohn, N. A., Blumenthal, J. A., & Campbell, A. T. (2013). Supported decision-making: A viable alternative to guardianship? *Penn State Law Review, 117*(4), 2013. https://doi.org/10.2139/ssrn.2161115

Loh, A., Simon, D., Kriston, L., & Härter, M. (2007). Patientenbeteiligung bei medizinischen Entscheidungen: Effekte der partizipativen Entscheidungsfindung aus systematischen Reviews [Patient participation in medical decision-making: Effects of participatory decision making from systematic reviews]. *Deutsches Ärzteblatt, 104*(21). https://www.aerzteblatt.de/archiv/55786/Patientenbeteiligung-bei-medizinischen-Entscheidungen

Meadows, G. (2017). Shared decision making: A consideration of historical and political contexts. *World Psychiatry, 16*(2), 154–155. https://doi.org/10.1002/wps.20413

Schäper, S., Fellbaum, K., Fricke, C., Heusner, J., Jennessen, S., Schlichting, H., Schroer, B., Thimm, A. & Weithardt, M. (2021). *"Am Ende geht es um Da-Sein ...". Palliative Versorgung und hospizliche Begleitung von Menschen mit geistiger und schwerer Behinderung (PiCarDi)* ["In the end, it's about being there ...". Palliative care for people with intellectual and multiple disabilities]. https://www.picardi-projekt.de/

Schäper, S., Schroer, B. & Levin, C. (2020, October 7–9). *Conditions of end-of-life care for people with intellectual disabilities.* The impact of professional self-conceptions and organisational culture in disability services [Poster presentation]. 11th European Association for Palliative Care World Research Congress Online.

Schäper, S., Thimm, A., & Schroer, B. (2019). End of life care for persons with intellectual disabilities: Concepts and challenges in disability services in three German regions. *Journal of Intellectual Disability Research, 63*(7), 646.

Steinkamp, N., & Gordijn, B. (2003). Ethical case deliberation on the ward: A comparison of four methods. *Medicine, Health Care and Philosophy, 6*(3), 235–246.

Thimm, A., Rodekohr, B., Dieckmann, F., & Hassler, T. (2018). *Wohnsituation Erwachsener mit geistiger Behinderung in Westfalen-Lippe und Umzüge im Alter [Residential settings of adults with intellectual disabilities in Westphalia-Lippe and moving in old age]*. Catholic University of Applied Sciences North-Rhine Westphalia. https://www.katho-nrw.de/fileadmin/primaryMnt/KatHO/Teilhabeforschung/Forschungsprojekte/2018_Erster_Zwischenbericht_Forschungsprojekt_MUTIG.pdf

Tronto, J. (1993). *Moral boundaries: A political argument for an ethic of care.* Routledge.

United Nations. (2006). *United Nations Convention on the rights of persons with disabilities.* https://www.un.org/development/desa/disabilities/convention-on-the-rights-of-persons-with-disabilities.html

United Nations Human Rights Office of the High Commissioner Committee on the Rights of Persons with Disabilities. (2014). *General Comment No. 1 on Article 12: Equal recognition before the law.* CRPD/C/GC/1. https://www.ohchr.org/en/hrbodies/crpd/pages/gc.aspx

United Nations Human Rights Treaty Bodies. (n.d.) *Ratification status for Germany.* https://tbinternet.ohchr.org/_layouts/15/TreatyBodyExternal/Treaty.aspx?CountryID=66&Lang=EN

West-Leuer, B. (2002). *Interaktive Beratung und Schulentwicklung. Eine Untersuchung zum Ist-Zustand* [Interactive counselling and school development. An investigation into the current state]. http://sundoc.bibliothek.uni-halle.de/diss-online/02/02H115/prom.pdf

Wiese, M., & Tuffrey-Wijne, I. (2020). End-of-life choices. In R. J. Stancliffe, M. L. Wehmeyer, K. A. Shogren, & B. H. Abery (Eds.), *Choice, preference, and disability: Promoting self-determination across the Lifespan* (pp. 317–335). Springer.

Wirtz, V., Cribb, A., & Barber, N. (2006). Patient-doctor decision-making about treatment within the consultation: A critical analysis of models. *Social Science & Medicine, 62*(1), 116–124.

World Health Organization. (2001). *International classification of functioning, disability and health (ICF).* https://apps.who.int/iris/handle/10665/42407

World Medical Association General Assembly. (October, 2013). Declaration of Helsinki. Ethical principles for medical research involving human subjects. Adopted June 1964 and amended by the 64th WMA General Assembly, Fortalez, Brazil, October, 2013. *JAMA, 310*(20). https://www.wma.net/wp-content/uploads/2016/11/DoH-Oct2013-JAMA.pdf

6

How People with Intellectual Disability Are Dying and Implications for Quality Care

Jane Bernal, Michele Y. Wiese, and Stuart Todd

There is growing interest about how people with intellectual disability can be supported well at the end of their lives. Little research attention has been given to the patterns of their dying, the extent to which it was anticipated, the availability of and timely access to planning and quality care, and the places where death and dying occurred. Understanding patterns of death and mortality is critical, offering important implications for end-of-life care. In this chapter, the international literature is used to provide a picture of dying and death among people with

J. Bernal · S. Todd
University of South Wales, Cardiff, Wales, UK
e-mail: stuart.todd@southwales.ac.uk

M. Y. Wiese (✉)
School of Psychology, Western Sydney University,
Penrith, NSW, Australia
e-mail: m.wiese@westernsydney.edu.au

R. J. Stancliffe et al. (eds.), *End of Life and People with Intellectual and Developmental Disability*, https://doi.org/10.1007/978-3-030-98697-1_6

intellectual disability, with wherever possible, comparison to the general population. The implications for end-of-life care are highlighted.

Cause of Death

The cause of death of people with intellectual disability offers much information about disease risk and health, particularly if comparisons can be drawn with the general population (Heslop & Hoghton, 2019). Data from the UK, USA and Australia broadly agree that the leading causes of death experienced by people with intellectual disability are predominantly respiratory and circulatory diseases, and cancers (Bernal et al., 2021; Glover et al., 2017; Landes, Stevens et al., 2021; Salomon & Trollor, 2019; Todd et al., 2020, 2021; Trollor et al., 2017). Comparisons to the general population reveal two important differences. Firstly, that leading causes of death vary between people with intellectual disability and the general population, and secondly that there is higher relative mortality for people with intellectual disability of deaths with the same cause (O'Leary et al., 2018). Regarding cause-of-death variation, Australian data showed that in a Sydney population of people with intellectual disability the leading cause of death was respiratory illness, accounting for 35% of deaths, while for the comparable general population the leading death cause was cancer, representing almost 39% of deaths. Concerning higher relative mortality of deaths from the same cause, the same research showed that for deaths caused by respiratory illness, the mortality rate was 35% for people with intellectual disability but just 3% in the general population (Durvasula & Beange, 2002). UK research reflects similar variation when comparing populations (Hosking et al., 2016).

Cause-of-death data highlight some particularities about the deaths of people with intellectual disability, pointing to the need for better attention to potentially avoidable deaths and the quality of care (see *Potentially Avoidable Deaths* below). Just a few of these particularities are noted here to illustrate. People with Down syndrome have been reported to have higher relative mortality. The available research suggests that having Down syndrome is a special case, with risk of death up to four times more likely compared to adults with other intellectual disability types (Esbensen et al., 2007). This is likely explained at least in part by the syndrome's characteristic health problems including thyroid disorder, congenital circulatory defects and increased risk of young onset dementia of Alzheimer type (Strydom et al., 2019).

Another particularity is the number of deaths due to epilepsy, evidenced by data from the UK, Canada and Australia (Durvasula & Beange, 2002; Glover et al., 2017; Ouellette-Kuntz et al., 2015). People with intellectual disability experience a higher lifetime prevalence of epilepsy, with a more complex presentation than do the general population (Alvarez, 2014; Robertson et al., 2015). Epilepsy is also treatable and therefore is a potentially avoidable cause of death (see *Potentially Avoidable Deaths* below). Epilepsy can also present as part of a terminal illness, particularly dementia, common in middle-aged adults with Down syndrome (Holland et al., 1998).

There are also particularities around cause of death due to external factors, wherein external causes have a different pattern compared to the general population. The evidence suggests that whereas accidental poisoning, choking, falls and pedestrian accident are more common causes of death for people with intellectual disability, for the general population land transport accidents, suicide and homicide predominate (Australian Institute of Health & Welfare, 2021; Hosking et al., 2016; O'Leary et al., 2018). These differing patterns suggest two conclusions. First, that people with intellectual disability still lead rather different lives, and while they are protected from some external factors, they are at higher risk of others. Second, that all the external causes of death for both populations are potentially avoidable, preventable by public health measures. In the case of intellectual disability, these measures could include falls prevention and swallowing risk assessments (Perez et al., 2015; Smulders et al., 2013). It is important to note too that cause-of-death patterns vary both within and across both populations due to individual factors like age at death, presence of underlying disability (see also Chapters 3, 7 and 13), as well as lifestyle factors including smoking and alcohol intake, body weight and physical exercise (O'Leary et al., 2018; Paulsen et al., 2017; Salomon & Trollor, 2019).

The variability in cause of death between people with intellectual disability and the wider population will be discussed further later in this chapter. For now, the evidence clearly shows different dying trajectories, calling for detailed attention to the implications for end-of-life care.

Cause-of-death researchers have repeatedly highlighted problems about accuracy of the available international data. Inaccurate recording on death certificates results in unreliable data about underlying cause of death, hampering public health and preventative healthcare efforts (Landes et al., 2019; see *Potentially Avoidable Deaths* below). The available research repeatedly reports how terms like *intellectual disability* or *Down syndrome* have been erroneously listed as cause of death, neither of which are legitimate primary causes of death (Esbensen et al., 2007). Inaccurate recording also means that diagnostic overshadowing may obscure the real cause of death. Death certificates also do not necessarily or reliably give information about the deceased person's intellectual disability status (Dunwoodie Stirton & Heslop, 2018).

Researchers have therefore used a range of strategies to identify the population of people with intellectual disability in which to examine mortality and cause of death (Heslop & Glover, 2015). These include national and local intellectual disability registers (McCarron et al., 2015; McGuigan et al., 1995; Patja et al., 2001; Tyrer & McGrother, 2009), recipients of funded care (Lauer & McCallion, 2015; Stankiewicz et al., 2018), primary healthcare records (Glover et al., 2017; Hosking et al., 2016), hospital data (Ouellette-Kuntz et al., 2015; Watkins, 2021) and requesting notification of deaths from multiple sources, for example, the Confidential Inquiry into the Deaths of People with Learning Disabilities (CIPOLD) (Heslop et al., 2014) and the Learning Disabilities Mortality Review (LeDeR) (Heslop et al., 2019). All these sources have problems with the completeness of the living population within which the deaths occurred and the reliability of the diagnosis of intellectual disability. The available data tend to be skewed towards those with more severe or more obvious disabilities, as do studies based on well-defined living populations using specific services or groups of services (Chaney & Eyman, 2000; Florio & Trollor, 2015; Landes, Stevens et al., 2021; Todd et al., 2020; Trollor et al., 2017; Wiese et al., 2018). Those who live independently or with family and not receiving services are likely excluded, and therefore not represented in most studies. Although beset by accuracy problems, these remain our only data sources (Heslop & Hoghton, 2019).

Anticipating Death

A critical step in understanding how to offer good end-of-life care is to determine *when* it may be needed. Can anticipation be improved to offer earlier access to supportive services? May there be some deaths that are harder or impossible to anticipate, calling for different service responses?

The available evidence indicates that up to half of deaths of adults with intellectual disability may be unexpected. For example, in a study by Todd and colleagues in UK intellectual disability accommodation settings, staff reported they had not anticipated 54% of 146 deaths occurring across an 18-month period (Todd et al., 2020). Hunt et al. (2020), further analysing the Todd et al. (2020)'s data set, showed that unexpected deaths were significantly higher compared to the general population (17% against 8%), and fewer were reported to have been ill in the year leading up to death (20% against 44%). Longer lasting illness was associated with increased levels of staff anticipation, but even when the final illness had lasted for 3 months or more a quarter of the deaths were unanticipated. The delivery of end-of-life care depends initially on some recognition that a person with intellectual disability may be dying. The data above suggest that since many people with intellectual disability may die less anticipated deaths, they may then die with no end-of-life care at all, or with care that has been hurriedly put together. Such findings also raise questions about potentially preventable deaths (see *Potentially Avoidable Deaths* below).

Examining death expectancy as a single outcome variable can oversimplify patterns of death. Lack of anticipation may be related to people with learning disabilities dying from causes that are harder to anticipate; from a failure to recognise relevant signs and symptoms; from failures in effective communication between the person with intellectual disability, carers and healthcare professionals; or from a combination of these factors. Bernal et al.'s (2021) and Todd et al.'s (2020) research showed that the extent to which death was anticipated influenced qualities of end-of-life care. The following relationships are notable:

- Anticipation varied according to the cause of death. Deaths from cancer or associated with dementia were most anticipated. Deaths from cardiovascular or respiratory illness were least anticipated.
- End-of-life care plans were in place for 83% of deaths that were more anticipated compared to 47% of less anticipated deaths.
- End-of-life care planning was more likely to include contributions from the dying person and from external professionals when the death was anticipated.
- Death in the accommodation setting was more likely for anticipated deaths. Of the less anticipated deaths, sudden deaths were at home or in the most acute parts of the hospital. For other less anticipated deaths, death in hospital was more likely.
- For well-anticipated deaths, disability staff reported they felt sufficiently supported from external professional services.

While these findings offer critical information that could predict when end-of-life care could and should be initiated, the complexity of identifying *when* death is coming cannot be overestimated. Literature from both the intellectual disability and general communities suggests that caregivers are not always able to identify an oncoming death and a range of variables add additional complexity. Dying trajectories for illnesses other than cancer are less predictable, especially in people who have multiple health problems or are otherwise frail, even when the conditions for that death might otherwise be expected (Breen et al., 2018; Teggi, 2018; Todd et al., 2021). Prognosis is known to be less precise where there are repeat episodes of acute illness followed by apparent recovery, gradual organ failure (renal, cardiac or respiratory) or accumulation of deficits, dwindling or frailty (Teggi, 2018). This difficulty applies to the whole population, not just those with intellectual disability. Even in general community aged-care residential settings, where death is more commonplace and caregivers more practised, deaths are not always accurately predicted (Persson et al., 2018).

As people with intellectual disability age and outlive their family caregivers, their usual home is likely to be a disability accommodation setting (Brennan et al., 2018). Intellectual disability support staff have reported feeling under-equipped to identify indicators of approaching death (Wiese et al., 2012). Given the compromised communication typically experienced by people with intellectual disability, their ability to communicate symptoms and their reliance on familiar third parties, the sophistication of the skill set required of this third party to anticipate death is becoming obvious. Disability support professionals are largely untrained in the provision of end-of-life care (Wiese et al., 2012), let alone the skills to identify indicators of oncoming death, though tools that may facilitate recognition of deteriorating health and onset of dying are developing (Scott et al., 2019; Vrijmoeth et al., 2018). Further, death is likely to be a rare event in small disability accommodation settings, limiting opportunities to learn from experience (Todd et al., 2020). The dying trajectories for adults with intellectual disability are only now beginning to be described and understood, together with the complexities of anticipating when death is likely (Bernal et al., 2021; Todd et al., 2021; Vrijmoeth et al., 2018). These trajectories may also have altered with the impact of COVID-19 (see *Lessons from Deaths During a Pandemic* below). Though the variables affecting the ability to anticipate death are complex, one implication is clear. If caregivers are not anticipating deaths, many people with intellectual disability are disadvantaged in their dying because access to end-of-life care is denied or delayed.

It is also essential to consider the extent to which the dying person with intellectual disability realises they are dying, and is supported to understand and anticipate their own death (see also Chapter 2). Hunt et al. (2020) reported that in UK intellectual disability accommodation settings for adults, staff said about 22% of dying people with intellectual disability in their care were either certainly or probably aware of their dying, substantially less than 61% awareness reported for the general population during the same period (Office for National Statistics, 2015). Perhaps not surprisingly, those who were judged to have capacity to understand were significantly more likely to be aware of their approaching death. In a more detailed analysis using Hunt et al.'s (2020) data, Bernal et al. (2021) showed that more dying people with intellectual disability were reported by staff to be *not aware* they were dying than *were aware*, though there were also high levels of reported uncertainty about whether the person had known. Other research echoes these findings (Ryan et al., 2011; Tuffrey-Wijne et al., 2010, 2020).

The evidence is clear that people with intellectual disability do have some understanding about what death is as a concept, with better understanding for those with milder compared to more significant intellectual disability (Stancliffe et al., 2016). A UK-wide survey of intellectual disability staff supporting people with intellectual disability who had died or were known to be terminally ill found that 53% said the person had been told their diagnosis, but only 18% were informed they were dying (Tuffrey-Wijne et al., 2020). The reasons the staff gave for not telling the person included that they themselves had not realised the person would die or were unsure of prognosis, concerns the dying person would not understand, how they might react, or because family had asked for them not to be told. These findings raise the question of whether understanding ones' diagnosis is necessarily cognitively less complex than understanding one is dying. Plans about disclosure must consider who is best placed to initiate and continue the process. Research shows that specialist palliative care professionals report difficulties communicating with people with intellectual disability about their deaths (Foo et al., 2021) while disability staff report fear and distress about discussing death, and variable organisational support (Tuffrey-Wijne et al., 2017). It appears that few feel adequately equipped, therefore, to approach disclosure with people with intellectual disability. While resources are available to assist with this task (see Resources at the end of this chapter), more research is needed about how to develop open conversations with people with intellectual disability about their own death. Further, the professional development of those charged with the skilled and sensitive task of disclosure requires urgent attention (see also Chapter 2).

Potentially Avoidable Deaths

Information about the extent to which individual deaths are avoidable is an important indicator of the health system's effectiveness. Potentially avoidable deaths are those *amenable* to appropriate healthcare, *preventable* via public health measures or *both* (Hosking et al., 2016; O'Leary et al., 2018). They are deaths which likely would not have occurred had the health system worked effectively (Australian Institute of Health & Welfare, 2020). The seminal findings of the UK Mencap reports (Mencap, 2007, 2012) followed by the CIPOLD provided a platform for better understanding and addressing potentially avoidable deaths (Glover et al., 2017; Heslop & Hoghton, 2019; Heslop et al., 2014), and led to the establishment of LeDeR in England (Heslop et al., 2019).

People with intellectual disability are more likely to die prematurely and from causes that may have been amenable to better healthcare, than the comparable general population. The overall rate of avoidable deaths is higher, and the pattern also differs. Avoidable deaths in the general population are largely preventable, while for people with intellectual disabilities deaths which would have been amenable to better care predominate (Glover et al., 2017; Hosking et al., 2016). To illustrate, Glover et al. (2017) showed that for the intellectual disability population in the UK, potentially avoidable deaths were typified by respiratory illness (about half from pneumonia and about 21% from aspiration), ischaemic heart disease, epilepsy and cerebrovascular disease, potentially amenable to healthcare treatment. In contrast, avoidable deaths in the comparable general population were characterised by suicide, accidents and alcohol-related events, potentially preventable by public health measures. Hosking et al.'s (2016) research in England showed that the rate of deaths amenable to care of people with intellectual disability was almost six times higher than a comparable general population. Australian data echo this pattern, showing that about 38% of reported deaths were potentially avoidable compared to 17% in a comparable general population cohort (Trollor et al., 2017).

The summary of evidence about potentially avoidable deaths links to other health data for this group (Liao et al., 2021). We know that older people with intellectual disability experience multiple complex health morbidities including arthritis, diabetes, cardiovascular disease, asthma, carcinomas and mental health disorders (Hussain et al., 2020). These morbidities occur against a backdrop of inequities in both socioeconomic position and healthcare access (Brameld et al., 2018; Trollor et al., 2017). Clearly, the problems associated with multiple health comorbidities continue as death approaches and do influence patterns of death. Importantly though, mortality risk research reminds us that simply equating comorbidity with risk of death offers an incomplete picture. Instead, the higher risk of death by people with intellectual disability, especially the higher risk from conditions amenable to care, highlights the need for healthcare (a) tailored to presenting disease/s, and (b) offered early in disease presentation, to slow mortality rate for these avoidable deaths (Brameld et al., 2018; Gustavson et al., 2005; Hussain et al., 2020; Landes, Stevens, et al., 2021; Segerlantz et al., 2020; Trollor et al., 2017).

As for other research discussed so far in this chapter, the extent to which the available data are representative is unclear. Inconsistencies in definition of the term *potentially avoidable death* mean compromised ability to compare research. A corollary of this problem is compromised reliable data about deaths which would have been either amenable to care or preventable (O'Leary et al., 2018). A more reliable picture about the pattern of avoidable deaths of people with intellectual disability would assist future identification of individuals who may benefit from care, either curative or palliative (Bernal et al., 2021).

Place of Death

Where people die informs policy about place of care, particularly in a landscape which increasingly encourages ageing in place. It should reflect the balance of individual preferences, caregiver capacity, available supports and presenting care needs. With respect to the deaths of people with intellectual disability, the interplay of these factors and the place where death occurs are not yet well understood.

Comparing the place of death of people with intellectual disability to the general population demonstrates some variation. While acknowledging that people die across a range of settings (including nursing homes and hospices) (Bigby et al., 2008; McCarron et al., 2017), when comparing both populations in this chapter only data from *usual home* and *hospital* are summarised, with the rationale that they are the two most commonly reported places of death in the available literature. McCarron et al.'s (2017) comparison of the deaths of 37 adults with intellectual disability and 57 deaths from the general population in Ireland showed that most deaths of those with intellectual disability occurred in their usual home (54%), with 35% in hospital. Hunt et al.'s. (2020) findings from the UK are broadly similar. For the intellectual disability population, place of death was approximately equal (49.7%) in the usual home and in hospital. Most deaths for the general population had occurred in hospital (49%), with 22% in their usual home. Todd et al (2020) reported that when a death had been expected, people were more likely to die in the setting in which they usually lived. Less expected deaths more often occurred in hospital. Segerlantz et al.'s (2020) Swedish study of deaths of older cancer patients showed that most people with intellectual disability died in their usual home, compared to those from the general population who more typically died in hospital. The evidence suggests that disability care settings can be places for dying as well as living and if well equipped, are able to support people to die at home, perhaps especially when the death is anticipated.

The finding that people with intellectual disability more commonly die in their usual home, than do the general population, is not fully understood. For some deaths, it may reflect the capacity of disability accommodation settings to adapt to new challenges; for others, it may reflect problems with access to healthcare. As people with intellectual disability are more likely to die prematurely for reasons amenable to healthcare, it is possible that some deaths at home may reflect inequitable access to appropriate hospital care. It is also plausible that some less anticipated deaths happened in hospital because the person was appropriately receiving potentially curative treatment.

Place of death as an indicator of appropriate care cannot be addressed in isolation from the range of variables that affect where a person dies.

As has already been discussed, these include the extent to which the death was expected, individual choice, pre-existing health comorbidities and healthcare access. Other variables also influence where the person dies. Wiese et al.'s (2018) Australian study of deaths in out-of-home care showed that those who were taking five or more medications (polypharmacy) at the time of death were more likely to die in hospital, as were those who had one or more hospital admissions in the year leading up to death. Another Australian study (this though of *disability deaths*, wherein intellectual disability was subsumed) showed too that time of day and season had a role, with more deaths occurring at the usual home at night and more in winter compared to summer. Further, those who died in their usual home tended to be younger, while those who were older tended to die in hospital (Salomon & Trollor, 2019).

Asking questions about place of death leads to questions about the preferences of the dying person. Research about where people with intellectual disability might prefer to die is new. McCarron et al. (2017) via caregiver reports in Ireland showed that of six adults with intellectual disability who had expressed their preference about place of death, all stated a preference for the familiarity of their own home. Hunt et al.'s (2020) UK study of 152 deaths showed firstly that only 12% of the intellectual disability sample articulated a preference about place of death compared to 35% of the general population. Secondly, like McCarron et al. (2017), they found those with intellectual disability who expressed a preference unanimously said they would prefer to die at home, while 81.5% of a general population sample did so. The emerging research suggests that while people with intellectual disability are often excluded from decision-making, they seem to share similar views to the rest of the community; most people would prefer to die at home (Ali et al., 2019).

Segerlantz et al. (2020) remind us that while honouring preferred place of death is increasingly viewed as an important indicator of end-of-life care, place of death should not be prioritised at the expense of other quality indicators. Preferences may change for the dying person as their condition changes (Gomes et al., 2013; Tuffrey-Wijne, 2009), suggesting the need for an individualised and nimble approach to the deployment of healthcare resources. Place of death is itself affected by a range of interacting variables and therefore cannot be taken as a proxy measure for the quality of end-of-life care.

Age at Death

Together with cause of death, understanding patterns about age at death offers useful information about potentially premature deaths (Heslop & Hoghton, 2019). Mortality studies show an increased rate of death of people with intellectual disability relative to the general population, most marked in infancy and childhood and decreasing as both populations age. Life expectancy for people with intellectual disability is shorter than the general population, and the mean and median ages at death are younger (Florio & Trollor, 2015; Hosking et al., 2016; McCallion & McCarron, 2014; McCarron et al., 2015; Patja et al., 2001; Stankiewicz et al., 2018; Trollor et al., 2017). Though the available research reflects variation in sampling and measurement methods, these findings remain broadly consistent across studies.

Level of intellectual disability also influences mortality and life expectancy. While those with mild intellectual disability have a life expectancy that is almost similar to the general population, more significant levels of intellectual disability are associated with reductions in life expectancy (Patja et al., 2000). So too is the divide influenced by other conditions previously discussed, including pre-existing comorbidities, phenotype and healthcare access (Liao et al., 2021; McCarron et al., 2017).

Evidence from across high income countries consistently shows higher rates of premature death among people with intellectual disability, whether this is expressed in terms of life expectancy, mean or median age at death. In the USA, the average age at death has been reported at approximately 59 to 62 years for adults with intellectual disability, compared to 78.5 years (at birth) and 84.2 years (after 20 years) in a comparable general population (Lauer et al., 2015). Data across two German regions showed that women with intellectual disability have a life expectancy of between approximately 70 and 73 years, while for men it is between 65 and approximately 71 years. This compared to the general population of 82 years for women and 77 years for men (Dieckmann et al., 2015). UK and Australian data demonstrate similar patterns, with age at death substantially younger than the general population (Glover et al., 2017; Heslop et al., 2014; Salomon & Trollor, 2019; Todd et al., 2021) and younger than other disability types (Australian Institute of Health & Welfare, 2020).

Similar to data reliability problems reported in previous sections of this chapter, gaining reliable information about age at death is limited by the quality of available data (Landes et al., 2019). Notwithstanding these limitations, the pattern remains overwhelmingly clear—people with intellectual disability continue to die at a younger age than their non-disabled counterparts.

End-of-Life Care

So far in this chapter the findings demonstrate important differences between the deaths of people with intellectual disability and the general population. The distribution of death causes differs, the deaths of people with intellectual disability are more likely to be premature, amenable to healthcare and less likely to be anticipated; in short, their patterns of dying are disadvantaged. In this section, we comment on three specific issues: current access to end-of-life care, palliative care needs and meaningful engagement in care by the dying person.

To the first issue about access to care. There is apparent inequality in access to specialist palliative care services, who consistently report that few adults with learning disabilities are referred (Adam et al., 2020; Tuffrey-Wijne et al., 2016). The reasons for this could be many, including the higher relative proportion of unanticipated deaths meaning no or little time for referral (Bernal et al., 2021; Hunt et al., 2020; Todd et al., 2020; see also *Anticipating Death* above), organisational barriers to healthcare (Adam et al., 2020; Wark et al., 2017) and differences in cause of death and dying trajectories (Bernal et al., 2021; see also *Cause of Death* above). Inequitable access to end-of-life care for adults with intellectual disability could also be a legacy of the origins of palliative care. Historically, palliative care services stem from cancer care; therefore, those with other life-limiting conditions such as respiratory or circulatory illnesses may less likely be referred, irrespective of the presence of intellectual disability (Rosenwax et al., 2016).

To the second issue of palliative care needs, a recent systematic review of palliative care needs is instructive (Adam et al., 2020). Of the 52 included studies comprising a total of 2970 participants in this review, just 31 adults with intellectual disability acted as participants (1% of the total sample), despite the key research question being about the palliative care needs of that group. Instead, the data reflected a strong bias towards the experience of healthcare professionals, from either a specialist intellectual disability or specialist palliative care background. While their perspective is critical, healthcare professionals report is proxy to and does not necessarily reflect the voice of those with lived experience of intellectual disability. Recent research has shown that people with intellectual disability themselves do have expertise about their care needs at the end of life, with findings emphasising the importance of attention to physical care and monitoring, involvement in one's own care, maintenance of social connections and activities with family and friends, and spiritual reconciliation (including prayer, solace seeking and resolving unsettled relationships) (Cithambarm et al., 2021; Tuffrey-Wijne et al., 2007).

On the issue of meaningful engagement in the dying person's own care, the evidence presented earlier in this chapter (see *Anticipating Death* above), that people with intellectual disability are not routinely told about their upcoming death, shows lack of meaningful engagement from the outset of the dying journey. Further, the demonstrable paucity of research about preferred place of death reflects similar lack of engagement (see *Place of Death* above). While the literature consistently advocates that people with intellectual disability should be meaningfully engaged in their own end-of-life care, in practice a pattern of paternalism and exclusion prevails (see also Chapters 2 and 16).

Lessons from Deaths During a Pandemic

Thus far, the chapter's focus has been to offer some typical patterns of death experienced by people with intellectual disability. The world's experience of COVID-19 from the end of 2019 offers a unique opportunity to examine whether the conditions of death of people with intellectual disability reflect a pre-existing pattern or are somehow distinctive in this historical moment. As currently available literature is necessarily limited, a cautious approach to conclusions is needed.

Research is available from the UK and USA on the first wave of COVID-19 infections in the spring and summer of 2020 (Baksh et al., 2021; Gleason et al., 2021; Glover, 2020; Heslop et al., 2021; Landes, Turk et al., 2021; Watkins, 2021). Although these studies varied in methodological approach, taken together they show that people with intellectual disability were substantially over-represented among COVID-19 deaths and that the ages at which those deaths occurred were younger than for the general population. The pandemic may have widened the pre-existing gap in mortality between people with intellectual disability and their contemporaries in the general population. Although differential risk of death from COVID-19 for people with intellectual disability appeared elevated at all ages, this was more marked in younger people.

Examining the detail of COVID-19 deaths of people with intellectual disability, analysis in England conducted by Heslop et al. (2021) showed that of the 1347 deaths of adults with intellectual disability referred to the LeDeR for the period March to June 2020, 615 (46%) were from suspected or confirmed COVID-19. Among the conclusions arising from the analysis, the following are noteworthy:

- Increased risk of COVID-19 death by people with intellectual disability was not limited to older age, emphasising the importance of vaccination access irrespective of age for this group.
- Pre-existing health comorbidities for people with intellectual disability translated to a greater risk of death from COVID-19, highlighting the need for preventative measures to reduce health inequalities and potentially avoidable deaths.

- Recognising deteriorating illness was problematic, likely exacerbated by communication difficulties and compromised self-reporting ability.

Inequitable treatment of hospitalised patients with intellectual disability and COVID-19 has also been described during the first wave of the pandemic in the UK (Baksh et al., 2021). Those with intellectual disability were more seriously ill at hospital admission but were less likely to be offered non-invasive respiratory support or be transferred to intensive care. They also died more rapidly than COVID-19 patients without disability. COVID-19 has shone a light on existing societal disparities that were a common experience for people with intellectual disability well prior to the pandemic. Advocates argue that the COVID-19 experience presents an opportunity to address long-term disparities in the care of people with intellectual disability (Fudge Schormans et al., 2021; Mladenov & Brennan, 2021).

The Death Patterns in Summary

Taken together, the evidence presented in this chapter clearly indicates that patterns of death for adults with intellectual disability are characterised by disadvantage and inequity. Compared to the general population, people with intellectual disability experience:

- Higher premature mortality rates.
- Increased proportion of deaths seen as amenable to care.
- Higher mortality of deaths from the same cause.
- Higher prevalence of multiple health comorbidities.
- Substantially higher rates of unexpected death.
- Poorer access to timely healthcare, specialist palliative care and end-of-life care planning.
- Decreased chances of being told about their approaching death.
- Deaths more commonly in the usual home compared to hospital, the reasons for which are unclear.
- Death at a younger age, both compared to the general population and other disability types.

- Overrepresentation in COVID-19 deaths.
- Inequitable healthcare access during COVID-19 illness.

There is also clear evidence of problems with the quality of available data. This represents a significant shortfall in the reliable knowledge about patterns of death and requires urgent attention. Pertaining to the most common data source, death certificates, the problems include:

- Unreliable, inaccurate or incomplete recording of the cause of death.
- No reliable way of knowing which deaths were of people with intellectual disability, translating to lack of population-specific data.

Pertaining specifically to anticipating death, data problems are reflected by:

- A paucity of reliable, objective indicators that flag death may be coming.
- Capacity and skill of caregivers to reliably anticipate death, particularly if they have little experience with oncoming death.
- Capacity and willingness of healthcare professionals and caregivers to communicate effectively with people with intellectual disability about their dying.

People with intellectual disability face disadvantaged dying as do other marginalised populations such as the very old, those with dementia and the institutionalised (Teggi, 2018). Bindley et al. (2019) remind us that vulnerability at the end of life arises "from unequal social status as well as exposure to social inequity" (p. 14). The patterns of death experienced by people with intellectual disability reflect a recurring theme of inequity and disadvantage across the lifespan. While people with disability continue to be marginalised throughout life, so too will be the conditions of their death. The evidence indicates a crucial message—the need to do death better, and to reduce and remove inequity and disadvantage at all stages of life.

Reflection

Jed's* story

Jed was 60 when he died. He had lived in a group home with four other people with intellectual disability for fifteen years. He paid regular visits to his mother who lived with dementia in a generic elder care home. His father, to whom he had been close, had died in his early 70 s of a heart attack. Jed visited the grave weekly. He had a sister who lived in another part of the country.

Jed enjoyed going to local rugby matches and car racing where the other spectators greeted him as a friend. He was a shy, retiring man who seldom initiated a conversation. When he did speak, it tended to be about immediate practical matters. Staff at his home believed he understood more than he let on.

He had an annual health check. He also had a general healthcare plan but no end-of-life care plan. He was overweight and had type 2 diabetes but seemed to be in generally good health until the morning when he awoke unable to move his left side.

He was taken to hospital where a CT scan showed a large cerebral thrombosis. He was very distressed, both by his symptoms and being in hospital. Staff from his home were able to be with him, though the manager commented after that this would have become difficult if he had lived longer. Over the next two days, he had a series of further strokes, the last of which proved fatal. The cause of death noted on the death certificate was cerebral infarction. There was no mention of underlying causes or of his intellectual disability.

*Assumed names are used in this story.

1. In relation to the key topics covered in this chapter, what are the issues that stand out for you about Jed's death?
2. Setting aside Jed's intellectual disability, are there actions you believe could or should have been taken?
3. Could his death have been preventable or amenable to better healthcare?

4. At what point might Jed have been involved in planning his own end-of-life care?

Resources

1. *Books Beyond Words*. Wordless picture stories for people who have trouble reading, to provide information and open up conversations about death. https://booksbeyondwords.co.uk/
2. *Palliative Care and People with Learning Disabilities (PCPLD) Network*. A network of people, including those with intellectual disability, professionals and families with the aim of ensuring quality palliative care. This website includes webinars, podcasts and resources on a range of palliative care topics. https://www.pcpld.org
3. Tuffrey-Wijne, I. (2013). *How to break bad news to people with intellectual disabilities: A guide for carers and professionals*. London: Jessica Kingsley. Guidelines for practitioners, families and carers to ease the process of explaining bad news. http://www.breakingbadnews.org

References

Adam, E., Sleeman, K. E., Brearley, S., Hunt, K., & Tuffrey-Wijne, I. (2020). The palliative care needs of adults with intellectual disabilities and their access to palliative care services: A systematic review. *Palliative Medicine, 34*(8), 1006–1018. https://doi.org/10.1177/0269216320932774

Ali, M., Capel, M., Jones, G., & Gazi, T. (2019). The importance of identifying preferred place of death. *BMJ Supportive & Palliative CaRe, 9*, 84–91. https://doi.org/10.1136/bmjspcare-2015-000878

Alvarez, N. (2014). Intellectual disability and epilepsy. In J. Merrick, D. E. Greydanus, & D. R. Patel (Eds.), *Intellectual disability: Some current issues* (pp. 149–176). Nova Science Publishers Inc.

Australian Institute of Health and Welfare. (2020). *Mortality patterns among people using disability services: 1 July 2013 to 30 June 2018: Technical report prepared by the AIHW for the National Disability Insurance Scheme Quality and Safeguards Commission (September 2020)*. https://www.aihw.gov.au/reports/disability-services/mortality-patterns-of-people-using-disability-serv/technical-report/technical-report

Australian Institute of Health and Welfare. (2021). *Deaths in Australia*. https://www.aihw.gov.au/reports/life-expectancy-death/deaths-in-australia

Baksh, R. A., Pape, S. E., Smith, J., & Strydom, A. (2021). Understanding inequalities in COVID-19 outcomes following hospital admission for people with intellectual disability compared to the general population: A matched cohort study in the UK. *British Medical Journal Open, 11*(10), e052482 https://doi.org/10.1136/bmjopen-2021-052482

Bernal, J., Hunt, K., Worth, R., Shear, J., Jones, E., Lowe, K., & Todd, S. (2021). Expecting the unexpected: Measures, outcomes and dying trajectories for expected and unexpected death in adults with intellectual disabilities in social care settings in the UK. *Journal of Applied Research in Intellectual Disabilities, 34*(2), 594–605. https://doi.org/10.1111/jar.12827

Bigby, C., Webber, R., Bowers, B., & McKenzie-Green, B. (2008). A survey of people with intellectual disabilities living in residential aged care facilities in Victoria. *Journal of Intellectual Disability Research, 52*(5), 404–414. https://doi.org/10.1111/j.1365-2788.2007.01040.x

Bindley, K., Lewis, J., Travaglia, J., & DiGiacomo, M. (2019). Disadvantaged and disenfranchised in bereavement: A scoping review of social and structural inequity following expected death. *Social Science & Medicine, 242*,112599. https://doi.org/10.1016/j.socscimed.2019.112599

Brameld, K., Spilsbury, K., Rosenwax, L., Leonard, H., & Semmens, J. (2018). Use of health services in the last year of life and cause of death in people with intellectual disability: A retrospective matched cohort study. *BMJ Open, 8*, e020268. https://doi.org/10.1136/bmjopen-2017-020268

Breen, L., Aoun, S. M., O'Connor, M., Howting, D., & Halkett, G. K. B. (2018). Family caregivers' preparations for death: A qualitative analysis. *Journal of Pain and Symptom Management, 55*(6), 1473–1479. https://doi.org/10.1016/j.jpainsymman.2018.02.018

Brennan, D., Murphy, R., Mccallion, P., & McCarron, M. (2018). "What's going to happen when we're gone?" Family caregiving capacity for older people with an intellectual disability in Ireland. *Journal of Applied Research in Intellectual Disabilities, 31*(2), 226–235. https://doi.org/10.1111/jar.12379

Chaney, R. H., & Eyman, R. K. (2000). Patterns in mortality over 60 years among persons with mental retardation in a residential facility. *Mental Retardation, 38*(3), 289–293. https://doi.org/10.1352/0047-6765(2000)038%3c0289:PIMOYA%3e2.0.CO;2

Cithambarm, K., Duffy, M., & Courtney, E. (2021). What constitutes good quality end-of-life care? Perspectives of people with intellectual disabilities and their families. *Journal of Policy and Practice in Intellectual Disabilities, 18*(3), 207–216. https://doi.org/10.1111/jppi.12376

Dieckmann, F., Giovis, C., & Offergeld, J. (2015). The life expectancy of people with intellectual disabilities in Germany. *Journal of Applied Research in Intellectual Disabilities, 28*(5), 373–382. https://doi.org/10.1111/jar.12193

Dunwoodie Stirton, F., & Heslop, P. (2018). Medical certificates of cause of death for people with intellectual disabilities: A systematic literature review. *Journal of Applied Research in Intellectual Disabilities, 31*(5), 659–668. https://doi.org/10.1111/jar.12448

Durvasula, S., & Beange, H. (2002). Mortality of people with intellectual disability in Northern Sydney. *Journal of Intellectual & Developmental Disability, 27*(4), 255–264. https://doi.org/10.1080/1366825021000029311

Esbensen, A. J., Seltzer, M. M., & Greenberg, J. S. (2007). Factors predicting mortality in midlife adults with and without Down syndrome living with family. *Journal of Intellectual Disability Research, 51*(12), 1039–1050. https://doi.org/10.1111/j.1365-2788.2007.01006.x

Florio, T., & Trollor, J. (2015). Mortality among a cohort of persons with an intellectual disability in New South Wales, Australia. *Journal of Applied Research in Intellectual Disabilities, 28*(5), 383–393. https://doi.org/10.1111/jar.12190

Foo, B., Wiese, M. Y., Curryer, B., Stancliffe, R. J., Wilson, N. J., & Clayton, J. (2021). Australian specialist palliative care staff's experiences of talking with people with intellectual disability about their dying and death: A thematic analysis of in-depth interviews. *Palliative Medicine, 35*(4), 738–749. https://doi.org/10.1177/0269216321998207

Fudge Schormans, A., Hutton, S., Blake, M., Kory, E., & Head, J. K. (2021). Social isolation continued: Covid-19 shines a light on what self-advocates know too well. *Qualitative Social Work, 20*(1–2), 83–89. https://doi.org/10.1177/1473325020981755

Gleason, J., Ross, W., Fossi, A., Blonsky, H., Tobias, J., & Stephens, M. (2021). The devastating impact of Covid-19 on individuals with Intellectual Disabilities in the United States. *NEJM Catalyst*. https://catalyst.nejm.org/doi/full/, https://doi.org/10.1056/CAT.21.0051

Glover, G. (2020). *Deaths of people identified as having learning disabilities with COVID-19 in England in the spring of 2020*. https://assets.publishing.ser

vice.gov.uk/government/uploads/system/uploads/attachment_data/file/933612/COVID-19__learning_disabilities_mortality_report.pdf
Glover, G., Williams, R., Heslop, P., Oyinlola, J., & Grey, J. (2017). Mortality in people with intellectual disabilities in England. *Journal of Intellectual Disability Research, 61*(1), 62–74. https://doi.org/10.1111/jir.12314
Gomes, B., Calanzani, N., Gysels, M., Hall, S., & Higginson, I. J. (2013). Heterogeneity and changes in preferences for dying at home: A systematic review. *BMC Palliative Care, 12*(7). Advance online publication. https://doi.org/10.1186/1472-684X-12-7
Gustavson, K. H., Umb-Carlsson, Õ., & Sonnander, K. (2005). A follow-up study of mortality, health conditions and associated disabilities of people with intellectual disabilities in a Swedish county. *Journal of Intellectual Disability Research, 49*(12), 905–914. https://doi.org/10.1111/j.1365-2788.2005.00728.x
Heslop, P., Blair, P. S., Fleming, P., Hoghton, M., Marriott, A., & Russ, L. (2014). The Confidential Inquiry into premature deaths of people with intellectual disabilities in the UK: A population-based study. *The Lancet, 383*(9920), 889–895. https://doi.org/10.1016/S0140-6736(13)62026-7
Heslop, P., Calkin, R., Huxor, A., Allen, C., Allen, L., Avis, M., Burnett, A., Cook, N., Farr, A., Gielnik, K., Gray, A., Hanford, D., Mepsted, K., Richards, J., Vergara, E., Williams, R., & Wistow, A. (2019). *The Learning Disability Mortality Review (LeDeR) Programme annual report 2018*. University of Bristol Norah Fry Centre for Disability Studies. https://www.hqip.org.uk/wp-content/uploads/2019/05/LeDeR-Annual-Report-Final-21-May-2019.pdf
Heslop, P., Byrne, V., & Calkin., Huxor, A., Sadoo, A., & Sullivan, B. (2021). Deaths of people with intellectual disabilities: Analysis of deaths in England from COVID-19 and other causes. *Journal of Intellectual Disability Research, 34*(6), 1630–1640. https://doi.org/10.1111/jar.12914
Heslop, P., & Glover, G. (2015). Mortality of people with intellectual disabilities in England: A comparison of data from existing sources. *Journal of Applied Research in Intellectual Disabilities, 28*(5), 414–422. https://doi.org/10.1111/jar.12192
Heslop, P., & Hoghton, M. (2019). Mortality in people with intellectual disability. In M. Scheepers & M. Kerr (Eds.), *Seminars in the psychiatry of intellectual disability: College seminar series* (pp. 102–114). Cambridge University Press.

Holland, A., Hon, J., Huppert, F. A., Stevens, F., & Watson, P. (1998). Population-based study of the prevalence and presentation of dementia in adults with Down's syndrome. *British Journal of Psychiatry, 172*, 493–498.

Hosking, F. J., Carey, I. M., Shah, S. M., Harris, M., DeWilde, S., Beighton, C., & Cook, D. (2016). Mortality among adults with intellectual disability in England: Comparisons with the general population. *AJPH Research, 106*(8), 1483–1490. https://doi.org/10.2105/AJPH.2016.303240

Hunt, K., Bernal, J., Worth, R., Shearn, J., Jarvis, P., Jones, E., Lowe, K., Madden, P., Barr, O., Forrester-Jones, R., Kroll, T., McCarron, M., Read, S., & Todd, S. (2020). End-of-life care in intellectual disability: A retrospective cross-sectional study. *BMJ Supportive & Palliative Care, 10*(4), 469–477. https://doi.org/10.1136/bmjspcare-2019-001985

Hussain, R., Wark, S., Janicki, M. P., Parmenter, T. R., & Knox, M. (2020). Multimorbidity in older people with intellectual disability. *Journal of Applied Research in Intellectual Disabilities, 33*(6), 1234–1244. https://doi.org/10.1111/jar.12743

Landes, S. D., Stevens, J. D., & Turk, M. A. (2019). Obscuring effect of coding developmental disability as the underlying cause of death on mortality trends for adults with developmental disability: A cross-sectional study using US mortality data from 2012 to 2016. *BMJ Open, 9*(2). https://doi.org/10.1136/bmjopen-2018-026614

Landes, S. D., Stevens, J. D., & Turk, M. A . (2021). Cause of death in adults with intellectual disability in the United States. *Journal of Intellectual Disability Research, 65*(1), Advance online publication. https://doi.org/10.1111/jir.12790

Landes, S. D., Turk, M. A., Damiani, M. R., Proctor, P., & Baier, S. (2021). Risk factors associated with COVID-19 outcomes among people with intellectual and developmental disabilities receiving residential services. *JAMA Network Open, 4*(6), e2112862. https://doi.org/10.1001/jamanetworkopen.2021.12862

Lauer, E., Heslop, P., & Hoghton, M. (2015). Identifying and addressing disparities in mortality: US and UK perspectives. In C. Hatton & E. Emerson (Eds.), *International review of research in developmental disabilities: Health disparities and intellectual disabilities* (pp. 196–245). Elsevier Science & Technology.

Lauer, E., & McCallion, P. (2015). Mortality of people with intellectual and developmental disabilities from select US state disability service systems and medical claims data. *Journal of Applied Research in Intellectual Disabilities, 28*(5), 394–405. https://doi.org/10.1111/jar.12191

Liao, P., Vajdic, C., Trollor, J., & Reppermund, S. (2021). Prevalence and incidence of physical health conditions in people with intellectual disability—A systematic review. *PLoS ONE, 16*(8), e0256294https://doi.org/10.1371/journal.pone.0256294

McCallion, P., & McCarron, M. (2014). Deaths of people with intellectual disabilities in the UK. *Lancet, 383*(9920), 853–855. https://doi.org/10.1016/s0140-6736(13)62190-x

McCarron, M., Burke, E. M., White, P., O'Dwyer, C., Ryan, K., O Farrell, J., & McCallion, P. (2017). *"He'd mind you, you mind him": Experiences of end of life care for people with an intellectual disability as perceived by staff carers.* https://www.tcd.ie/tcaid/research/endoflifepublications/ReportExperiencesofendoflife.pdf

McCarron, M., Carroll, R., Kelly, C., & McCallion, P. (2015). Mortality rates in the general Irish population compared to those with an intellectual disability from 2003 to 2012. *Journal of Applied Research in Intellectual Disabilities, 28*(5), 406–413. https://doi.org/10.1111/jar.12194

McGuigan, S. M., Hollins, S., & Attard, M. (1995). Age-specific standardized mortality rates in people with learning disability. *Journal of Intellectual Disability Research, 39*(6), 527–531. https://doi.org/10.1111/j.1365-2788.1995.tb00573.x

Mencap. (2007). *Death by indifference: Following up the Treat me right! report*. https://www.mencap.org.uk/sites/default/files/2016-06/DBIreport.pdf

Mencap. (2012). *Death by indifference: 74 deaths and counting*. https://www.mencap.org.uk/sites/default/files/2016-08/Death%20by%20Indifference%20-%2074%20deaths%20and%20counting.pdf

Mladenov, T., & Brennan, C. S. (2021). The global COVID-19 Disability Rights Monitor: Implementation, findings, disability studies response. *Disability & Society, Advance Online Publication.* https://doi.org/10.1080/09687599.2021.1920371

O'Leary, L., Cooper, S.-A., & Hughes-McCormack, L. (2018). Early death and causes of death of people with intellectual disabilities: A systematic review. *Journal of Applied Research in Intellectual Disabilities, 31*(3), 325–342. https://doi.org/10.1111/jar.12417

Office for National Statistics. (2015). *Statistical Bulletin: National Survey of Bereaved people (VOICES): 2104*. https://www.ons.gov.uk/peoplepopulationandcommunity/healthandsocialcare/healthcaresystem/bulletins/nationalsurveyofbereavedpeoplevoices/2015-07-09

Ouellette-Kuntz, H., Shooshtari, S., Balogh, R., & Martens, P. (2015). Understanding information about mortality among people with intellectual and developmental disabilities in Canada. *Journal of Applied Research in Intellectual Disabilities, 28*(5), 423–435. https://doi.org/10.1111/jar.12195

Patja, K., Iivanainen, M., Vesala, H., Oksanen, H., & Ruoppila, I. (2000). Life expectancy of people with intellectual disability: A 35-year follow-up study. *Journal of Intellectual Disability Research, 44*(5), 591–599. https://doi.org/10.1046/j.1365-2788.2000.00280.x

Patja, K., Mölsä, P., & Iivanainen, M. (2001). Cause-specific mortality of people with intellectual disability in a population-based, 35-year follow-up study. *Journal of Intellectual Disability Research, 45*(1), 30–40. https://doi.org/10.1111/j.1365-2788.2001.00290.x

Paulsen, J., Åskim, A., Mohus, R. M., Mehl, A., Dewan, A., Solligård, E., Damås, J. K., & Åsvold, B. O. (2017). Associations of obesity and lifestyle with the risk and mortality of bloodstream infection in a general population: A 15-year follow-up of 64 027 individuals in the HUNT Study. *International Journal of Epidemiology, 46*(5), 1573–1581. https://doi.org/10.1093/ije/dyx091

Perez, C. M., Ball, S. L., Wagner, A. P., Clare, I. C. H., Holland, A. J., & Redley, M. (2015). The incidence of healthcare use, ill health and mortality in adults with intellectual disabilities and mealtime support needs. *Journal of Intellectual Disability Research, 59*(7), 638–652. https://doi.org/10.1111/jir.12167

Persson, Å., Sandgren, A., Fürst, C.J., Ahlström, G., & Behm, L. (2018). Early and late signs that precede dying among older persons in nursing homes: The multidisciplinary team's perspective. *BMC Geriatrics, 18*(134). https://doi.org/10.1186/s12877-018-0825-0

Robertson, J., Hatton, C., Emerson, E., & Baines, S. (2015). Prevalence of epilepsy among people with intellectual disabilities: A systematic review. *Seizure, 29*, 46–62. https://doi.org/10.1016/j.seizure.2015.03.016

Rosenwax, L., Spilsbury, K., McNamara, B. A., & Semmens, J. B. (2016). A retrospective population based cohort study of access to specialist palliative care in the last year of life: Who is still missing out a decade on? *BMC Palliative Care, 15*(46). https://doi.org/10.1186/s12904-016-0119-2

Ryan, K., Guerin, S., Dodd, P., & McEvoy, J. (2011). Communication contexts about illness, death and dying for people with intellectual disabilities and life-limiting illness. *Palliative and Supportive Care, 9*, 201–208. https://doi.org/10.1017/s1478951511000137

Salomon, C., & Trollor, J. (2019). *A scoping review of causes and contributors to deaths of people with disability in Australia.* https://www.ndiscommission.gov.au/document/1881

Scott, S., Denton, L., Conway, F., & Kinley, J. (2019). Managing health changes for people with a learning disability in a residential care home setting. *International Journal of Palliative Nursing, 25*(11), 531–540. https://doi.org/10.12968/ijpn.2019.25.11.531

Segerlantz, M., Axmon, A., & Ahlström, G. (2020). End-of-life care among older cancer patients with intellectual disability in comparison with the general population: A national register study. *Journal of Intellectual Disability Research, 64*(5), 317–330. https://doi.org/10.1111/jir.12721

Smulders, E., Enkelaar, L., Schoon, Y., Geurts, A. C., & van Schrojenstein Lantman-de Valk, H. M. J., & Weerdesteyn, V. (2013). Falls prevention in persons with intellectual disabilities: Development, implementation, and process evaluation of a tailored multifactorial fall risk assessment and intervention strategy. *Research in Developmental Disabilities, 34*(9), 2788–2798. https://doi.org/10.1016/j.ridd.2013.05.041

Stancliffe, R. J., Wiese, M. Y., Read, S., Jeltes, G., & Clayton, J. M. (2016). Knowing, planning for and fearing death: Do adults with intellectual disability and disability staff differ? *Research in Developmental Disabilities, 49–50*, 47–59. https://doi.org/10.1016/j.ridd.2015.11.016

Stankiewicz, E., Ouellette-Kuntz, H., McIsaac, M., Shooshtari, S., & Balogh, R. (2018). Patterns of mortality among adults with intellectual and developmental disabilities in Ontario. *Canadian Journal of Public Health, 109*(5–6), 866–872. https://doi.org/10.17269/s41997-018-0124-8

Strydom, A., Curmi, A., & McQuillan, A. (2019). Physical health and clinical phenotypes. In V. P. Preasher & M. Janicki (Eds.), *Physical health of adults with intellectual and developmental disabilities.* (2nd ed., pp. 71–86). Springer.

Teggi, D. (2018). Unexpected death in ill old age: An analysis of disadvantaged dying in the English old population. *Social Science & Medicine, 217*, 112–120. https://doi.org/10.1016/j.socscimed.2018.09.048

Todd, S., Bernal, J., Shearn, J., Worth, R., Jones, E., Lowe, K., Madden, P., Barr, O., Forrester-Jones, R., Jarvis, P., Kroll, T., McCarron, M., Read, S., & Hunt, K. (2020). Last months of life of people with intellectual disabilities: A UK population-based study of death and dying in intellectual disability community services. *Journal of Applied Research in Intellectual Disabilities, 33*(6), 1245–1258. https://doi.org/10.1111/jar.12744

Todd, S., Bernal, J., Worth, R., Shearn, J., Brearley, S., McCarron, M., & Hunt, K. (2021). Hidden lives and deaths: The last months of life of people with intellectual disabilities living in long-term, generic care settings in the UK. *Journal of Applied Research in Intellectual Disabilities, 34*(6), 1489–1498. https://doi.org/10.1111/jar.12891

Trollor, J., Srasuebkul, P., Xu, H., & Howlett, S. (2017). Cause of death and potentially avoidable deaths in Australian adults with intellectual disability using retrospective linked data. *British Medical Journal Open, 7*, e013489. https://doi.org/10.1136/bmjopen-2016-013489

Tuffrey-Wijne, I. (2009). The preferred place of care for people who are dying. *Learning Disability Practice, 12*(6), 16–21. https://doi.org/10.7748/ldp2009.07.12.6.16.c7125

Tuffrey-Wijne, I., Bernal, J., Butler, G., Hollins, S., & Curfs, L. (2007). Using nominal group technique to investigate the views of people with intellectual disabilities on end-of-life care provision. *Journal of Advanced Nursing, 58*(1), 80–89. https://doi.org/10.1111/j.1365-2648.2007.04227.x

Tuffrey-Wijne, I., Bernal, J., & Hollins, S. (2010). Disclosure and understanding of cancer diagnosis and prognosis for people with intellectual disabilities: Findings from an ethnographic study. *European Journal of Oncology Nursing, 14*, 224–230. https://doi.org/10.1016/j.ejon.2010.01.021

Tuffrey-Wijne, I., Finlayson, J., Bernal, J., Taggart, L., Lam, C. K. K., & Todd, S. (2020). Communicating about death and dying with adults with intellectual disabilities who are terminally ill or bereaved: A UK-wide survey of intellectual disability support staff. *Journal of Applied Research in Intellectual Disabilities, 33*(5), 927–938. https://doi.org/10.1111/jar.12714

Tuffrey-Wijne, I., McLaughlin, D. M., Curfs, L., Dusart, A., Hoenger, C., McEnhill, L., Read, S., Ryan, K., Satgé, D., Straßer, B., Westergård, B.-E., & Oliver, D. (2016). Defining consensus norms for palliative care of people with intellectual disabilities in Europe, using Delphi methods: A White Paper from the European Association of Palliative Care. *Palliative Medicine, 30*(5), 446–455. https://doi.org/10.1177/0269216315600993

Tuffrey-Wijne, I., Rose, T., Grant, R., & Wijne, A. (2017). Communicating about death and dying: Developing training for staff working in services for people with intellectual disabilities. *Journal of Applied Research in Intellectual Disabilities, 30*(6), 1099–1110. https://doi.org/ https://doi.org/10.1111/jar.12382

Tyrer, F., & McGrother, C. (2009). Cause-specific mortality and death certificate reporting in adults with moderate to profound intellectual disability.

Journal of Intellectual Disability Research, 53, 898–904. https://doi.org/10.1111/j.1365-2788.2009.01201.x

Vrijmoeth, C., Echteld, M. A., Assendelft, P., Christians, M., Festen, D., & van Schrojenstein Lantman-de Valk, H., Vissers, K., & Groot, M. (2018). Development and applicability of a tool for identification of people with intellectual disabilities in need of palliative care (PALLI). *Journal of Applied Research in Intellectual Disabilities, 31*(6), 1122–1132. https://doi.org/10.1111/jar.12472

Wark, S., Hussain, R., Müller, A., Ryan, P., & Parmenter, T. (2017). Challenges in providing end-of-life care for people with intellectual disability: Health services access. *Journal of Applied Research in Intellectual Disabilities, 30*(6), 1151–1159. https://doi.org/10.1111/jar.12408

Watkins, A. (2021). *COVID-19-related deaths in Wales amongst people with learning disabilities from 1st March to 19th November 2020*. https://phw.nhs.wales/publications/publications1/covid-19-related-deaths-in-wales-amongst-people-with-learning-disabilities-from-1st-march-to-19th-november-2020/

Wiese, M., Stancliffe, R. J., Balandin, S., Howarth, G., & Dew, A. (2012). End-of-life care and dying: Issues raised by staff supporting older people with intellectual disability in community living services. *Journal of Applied Research in Intellectual Disabilities, 25*(6), 571–583. https://doi.org/10.1111/jar.12000

Wiese, M. Y., Stancliffe, R. J., Piepers, D. W., Durvasula, S., & Wilson, N. J. (2018). *Analysis of the Client Death Notification (CDN) database: Report to NSW family & community services*. [Unpublished manuscript]. Western Sydney University and University

7

Living and Dying Well with Dementia

Mary McCarron, Andrew P. Allen, Niamh Mulryan, Martina Leigh, Louise O'Reilly, Cathy McCarthy, Pamela Dunne, Evelyn Reilly, and Philip McCallion

Introduction

It is now widely acknowledged that people with intellectual disability are living longer. However, a gap of about 20 years in life expectancy remains between the general population and people with intellectual disability (Doyle et al., 2021; McCarron et al., 2015, 2019; O'Leary et al., 2018). The life expectancy of people with Down syndrome has

M. McCarron (✉) · A. P. Allen · N. Mulryan · C. McCarthy
Trinity Centre for Ageing and Intellectual Disability, Trinity College Dublin, Dublin, Ireland
e-mail: mccarrm@tcd.ie

A. P. Allen
e-mail: apallen@tcd.ie

M. Leigh
The Muiríosa Foundation Service, Monasterevin, Ireland

N. Mulryan · L. O'Reilly · C. McCarthy
Avista, Dublin, Ireland

R. J. Stancliffe et al. (eds.), *End of Life and People with Intellectual and Developmental Disability*, https://doi.org/10.1007/978-3-030-98697-1_7

though, increased from a mean age of 12 years in 1940 to many adults with Down syndrome now living to their 50s, 60s, and 70s. While this is to be celebrated, growing old for people with Down syndrome can bring unprecedented challenges for themselves, family, and staff caregivers, reflected by limited understanding and preparedness to deal with the challenges of living and dying with dementia.

With longer life comes the consequence of advanced age, the greatest risk factor for dementia (McCallion et al., 2019). Longitudinal studies of people with Down syndrome have estimated cumulative risk to be 90% by age 65 (McCarron, McCallion, Reilly, et al., 2017) and mean age of diagnosis to be 55 years (McCarron, McCallion, Carroll, et al., 2017; Sinai et al., 2018). There is though, large variation in the clinical presentation and diagnosis of Alzheimer's disease, with rate of disease progression and related decline often unclear.

Challenges with Diagnosis and Predicting Survival with Implications for End-of-Life Planning

Differential diagnosis is important but challenging. Several other clinical conditions common in people Down syndrome may mimic and or/coexist with dementia. It is important that a full comprehensive diagnostic work up is carried out to understand the nature and cause of presenting memory and functional concerns, for example hypothyroidism, vitamin B12 and folate deficiencies, depression, sensory difficulties, or adjustment to life events (Sheehan et al., 2014). Indeed, a study comparing three well-respected diagnostic criteria methods of DSM-IV, DC-LD, and ICD-10 found that prevalence rate estimates for dementia

P. Dunne · E. Reilly
Tallaght University Hospital, Dublin, Ireland

P. McCallion
School of Social Work, Temple University, Philadelphia, PA, USA
e-mail: philip.mccallion@temple.edu

in persons with Down syndrome differed depending upon the diagnostic criteria used (Strydom et al., 2007, 2013).

Understanding survival and related predictors and factors is also important. Sinai et al. (2018) reported a median survival time post-diagnosis of 3.78 years and median age at death of 60 years, with survival associated with severity of intellectual disability, living status, anti-dementia medications, and epilepsy. McCarron, McCallion, Reilly, et al. (2017) reported a median survival of 7 years for those with co-occurring late-onset epilepsy. Monitoring decline on an annual basis and recognising that both level of intellectual disability and the emergence of new onset seizures are predictors of survival in people with Down syndrome is important. Unfortunately, advanced dementia and end of life are instead often recognised late in people with Down syndrome. In addition, existing lifelong cognitive impairment further compromised due to dementia frequently means that the views of the person with Down syndrome are difficult to ascertain, although involvement in advanced planning for end of life before symptom onset is possible. Discussions around end-of-life care are not a simple undertaking and require staff and family caregivers to acknowledge and consider the person's will and preferences, understanding and communication style, and their involvement in life decisions prior to dementia.

Clinical Features of Late-Stage Dementia

In the general population, the clinical features of advanced dementia are characterised by significant losses in function with profound memory deficits (e.g. inability to recognise family), loss of verbal communication and ambulatory ability, inability to perform activities of daily living, and urinary and faecal incontinence. The common clinical complications are eating problems and infections, and these require management decisions (Mitchell, 2015). The clinical features of advanced dementia in people with Down syndrome are similar to the general population and are outlined in Table 7.1. One important exception is that among adults with Down syndrome rates of late-onset seizures (occurring in mid

Table 7.1 Clinical features of advanced dementia for persons with Down syndrome

Domain	Clinical features
Neurocognitive	Progressive worsening of memory Inability to verbally communicate Apathy—depression Confusion and disorientation (place, time and person) Delirium Unresponsiveness
Functional	Immobility with hypertonia Need for total assistance for activities of daily living Incontinence Frailty Weakness, fatigue
Nutritional	Loss of appetite Reduced ability to self-feed Swallowing difficulties Propensity to aspirate
Co-morbid intellectual disability health conditions	Seizures Constipation and complications of immobility Respiratory difficulties and repeat pneumonia

Sources Coppus et al. (2008), Cosgrave et al. (2000), McCarron et al. (2005, 2014), Prasher (2005), Strydom et al. (2010), and Visser et al. (1997)

to older adulthood) may range up to 70–80% (McCarron, McCallion, Carroll, et al., 2017).

Living with Advanced Dementia

Living with dementia, particularly approaching late and end-stages, requires a plan of care that anticipates the needed resources to maintain the person where he or she has always lived and/ or to plan for transition to a setting better able to cope with deteriorating health and death. If the person is to die at home, staff, family, and household peers need significant support to deal with the anticipation of, and caregiving in the days leading to death. If palliative care services and hospice care is to be utilised, transitions should be seamless, with continuing involvement of those important to the person and the provision of required supports. Maintaining linkages with people, places, possessions, and rituals that the person has valued is critical, as is continued contact with peers, friends, family, and staff who value and care about the person (Jokinen et al., 2013; McCarron et al., 2011). On an interpersonal level, communication ability will over time break down, including the ability to form words. Many individuals can understand more than they often express, so efforts at communication by others are still important. Strategies include talking even if there is no response, using touch, responding positively to sounds the person makes, and saying goodbye when leaving the room (McCallion, 1999).

Living with advanced dementia is also about the journey to a good death. Involving preferred supports and caregivers, honouring preferences and previously made plans for last days, and post-death rituals and practices are all important. As dementia advances the support of day-to-day living is often about identifying and responding to signs of distress, pain management, and prevention of deterioration of body functions. Strategies should include supporting dignity, reducing discomfort, protecting the integrity of the body, supporting nutrition and hydration, as well as preventing complications associated with inactivity (e.g. bedsores, pneumonia, etc.) (McCallion et al., 2011, 2012).

Relationship-Based Care, Advance Care Planning, and Palliative Care

Although there are many similarities in the end-of-life care needs of people with dementia, with and without Down syndrome, there is increasing appreciation about unique needs when intellectual disability is present. People with Down syndrome develop dementia at much earlier age, with an average age of onset at approximately 51–55 years (Margallo-Lana et al., 2007; McCarron, McCallion, Reilly, et al., 2017). Staff and family caregivers watch people experience multiple losses, including rapid loss of skills that they worked hard to achieve. Having conversations about end of life and death at this age is not without its challenges for staff and family caregivers, as well as for the person with Down syndrome themself. Additionally, many people with Down syndrome reside in residential settings where the primary caregivers are not family members. Some older people with intellectual disability may have lost parents through death and may not live in the same neighbourhood as extended family. McCausland et al. (2018), with data from the Intellectual Disability Supplement to The Irish Longitudinal Study on Ageing, reported that just one in five participants lived close to a family member (11% living with family, 10% living in the same neighbourhood). Younger participants with milder intellectual disability were more likely to live close to family, while those with severe-profound intellectual disability living in residential care were the least likely to live close to family. Consequently, staff caregivers become like family, some having supported the person for most of their adult life (see also Chapter 16). Wilkinson et al. (2005) reported that staff may under-report problems experienced by people with intellectual disability and dementia, often out of concern that a move to a different setting might result. There have been long-standing concerns that the same under-reporting occurs among families (McCallion & Ferretti, 2021). Given that people with Down syndrome living in group homes are increasingly using community-based health services, there is a growing need for general population health providers and memory clinics to be ready to meet the needs of people with intellectual disability presenting with memory concerns (McCarron et al., 2018). Challenges due to variations in age of disease onset, symptom presentation, and lack of appropriate assessment instruments translate to a need for resource training, dedicated liaison positions, and better preparation for both intellectual disability and general population services (McCarron et al., 2018; McCarron, McCallion, Reilly, et al., 2017).

People are often stigmatised when they have dementia (Benbow & Jolley, 2012; Swaffer, 2014) or intellectual disability (Ali et al., 2012; Scior et al., 2016). The form stigma takes may be quite particular when a person with Down syndrome develops dementia. A person-centred approach to care that empowers the individual to improve their quality of life can counteract perceptions that their individuality is lessened by their condition. Kitwood (1997) emphasised the importance of facilitating choice and the expression of feelings or preferences with the person. Kitwood also suggested that a such an approach should focus on comfort, attachment, inclusion, occupation, and identity. There is a need for person-centred principles and approaches that consider the unique relationships in the lives of people with Down syndrome and dementia.

Relationship-Based Care

The relationship-based approach to care emphasises the impact of relationships on quality of life, and how health care is imbued with meaning by those relationships. In relationship-based care, the care triad (Adams & Grieder, 2005) involves the person with Down syndrome, family caregivers, and paid caregivers, nested within a dementia care network and broader health system that impacts upon the individual and caregivers.

Relationship-based care has always had particular relevance for intellectual disability care settings where relationships, particularly with staff caregivers, have developed often over years (McCarron, McCallion, Carroll, et al., 2017). For people living with dementia, continued emotional connection to people they know and love as the disease progresses is often what makes living with dementia bearable. McCarron, McCallion, Reilly, et al. (2017) described care that embraces years and decades of knowing and working with someone, and the authenticity of those relationships, as the cornerstone of good person-centred dementia care and which ultimately contributes to a good death. For people with Down syndrome and dementia, these valued relationships may encompass staff as well as family.

Advance Care Planning

Given our understanding of the increased risk of dementia in people with Down syndrome as well as the shorter survival time from diagnosis to death, issues such as guardianship and advance care planning should be addressed as early as possible (see also Chapters 4 and 6). Unfortunately, important end-of-life discussions, like addressing nutrition and hydration concerns, pain management, seizure control, resuscitation, and place of death, often do not take place until there is a health-care crisis, where decisions are often charged by high emotions resulting in great distress for everyone involved (McCarron, McCallion, Carroll, et al., 2017). When supporting the person with Down syndrome and advanced dementia, such end-of-life discussions are likely to involve decisions about not doing excessively burdensome things; for example, not treating certain conditions like repeated pneumonia, but rather instituting comfort measures and treating conservatively. Care may be about starting treatments like syringe drivers to administer medications to reduce distress in the final days (McCarron et al., 2011).

Individuals with dementia and Down syndrome will present with different needs and differing levels of ability and willingness to engage in decision-making regarding end-of-life care. As noted in the report from the International Summit on Intellectual Disability and Dementia, advance care planning is recommended prior to extensive progression of dementia, and specialist training is required for staff who work in intellectual disability, palliative, hospice or dementia services (McCallion et al., 2017).

Of great concern is that dementia is often under, or late diagnosed. This means poor preparation of the person themselves, their peers, family, and staff caregivers, as well as lack of identification of additional resources to support changing care needs. More importantly, the opportunity to establish the desires of the person for the care they would like is lost (McCallion et al., 2017, 2018). The need for a proactive approach to decision-making regarding end-of-life care is critical. Research indicates that care staff in particular often lack training in talking about death with people with intellectual disability and may even fear discussing death with clients (Tuffrey-Wijne et al., 2017). This is likely to be also true for family caregivers. Yet, the evidence is clear that care is best supported when underpinned by person-centred, relationship-based, and family-focused strategies (Kirkendall et al., 2012). Educational resources about advanced dementia in people with an intellectual disability and end-of-life preparations are available for carers (see Resources at the end of this chapter).

Until relatively recently, there was a lack of research on advance care planning for people with intellectual disability that focused on the perspective of people with intellectual disability themselves (Voss et al., 2017; see also Chapter 13). However, people with mild to moderate intellectual disability are often able to articulate their wishes regarding end-of-life decision-making (Stancliffe et al., 2017). People with intellectual disability, including those with Down syndrome, have reported that advance care planning was positively influenced by adapting the pace and process of planning to suit them, supporting them in making their own choices, open and honest communication from support teams, and continued support in their current lifestyle and associated plans (McKenzie et al., 2017). Burke et al. (2017) developed a suite of accessible tools to support advance care planning and end-of-life conversations for person with intellectual disability (see Resources at the end of this chapter).

Palliative Care

Palliative care has been primarily designed to cater for people with diseases such as cancer (Fahey-McCarthy et al., 2008), where the typical trajectory of dying is understood well enough to prepare for death and model care accordingly (Clark, 2007). The model has not always been well-translated for dementia and for people with Down syndrome and dementia in end-stage disease. Interviews with care staff indicate ambiguity in understanding the difference between palliative care and high-quality, person-centred care in a general sense (McCarron et al., 2011). Equitable access to palliative care services for people with Down syndrome requires changes to the services themselves, recognition of differing communication needs, vigilance for signs of serious illness and end of life, and support for people who matter to the dying person. The people who matter must also be involved in symptom management, education, and training, in collaboration with service providers (European Association for Palliative Care, 2015; Tuffrey-Wijne et al., 2016). Palliative care should also take account of those caregivers who may be grieving the person's dying and death, including family, peers, and paid carers (Fahey-McCarthy et al., 2008).

Anticipation and Place of Death for People with Dementia

Anticipation and place of death add further complexity (see also Chapter 6). A population-based study in the UK found that just over a third of deaths were anticipated by care staff (Todd et al., 2020). If a death was anticipated, planning time was more likely afforded, and the dying person with dementia was more likely to die in their familiar home. Unanticipated deaths, however, added complexity, including complex grief of those left behind, likely due to the lack of a preparation period. Place of usual care had a major impact on outcomes. To illustrate, Todd et al. (2020) noted that supported living settings may deal with death less regularly, and so be less prepared for it, compared to registered care homes. At the same time, they noted the possibility that people may move out of intellectual disability specific care settings as they age, particularly when there is increased frailty or poorer health, increasing the likelihood of deaths being unanticipated in intellectual disability settings. People with Down syndrome are not a homogeneous group, with often wide variation in the age of clinical presentation of advanced dementia and subsequent rates of decline and patterns of dying. Yet there are knowable signs of impending death. Table 7.2 outlines these signs, any three of which likely indicates that the person's final days of life are approaching.

Practical Care Considerations

If no reversible causes of decline are identified and the person is felt to be approaching final days of life, the following six steps are a useful guide to assist with decision-making and care management:

Table 7.2 Signs that the end of life is approaching

Any three of the following symptoms may indicate the final days: – Experiencing rapid day-to-day and irreversible deterioration – Completely bedbound, requiring frequent interventions – Becoming less conscious, with lapses into unconsciousness, diminished or absent response to voices – Unable to swallow – Diminished or no urine secretion – An acute event that requires revision of care goals – Profound weakness – Changing breathing pattern – Gurgling or rattly breathing – Changing body temperature (hot and clammy, or cold) – Pale or mottled skin

1. Clarify the clinical situation and clearly communicate this with family and staff.
2. Establish primary goals of care which are now focused on comfort.
3. Evaluate current and anticipated physical symptoms, psycho-social needs, and spiritual care.
4. Present the treatment options and their risks and benefits.
5. Weigh the options against the person's values, will, and preferences.
6. Provide additional and ongoing support to family, peers, and care team (McCarron, 2009).

One of the most difficult challenges for the care team is often the issue of nutrition and hydration. The symbolic significance and meaning attached to eating and drinking, or lack thereof, should be carefully individualised, taking into account both the person, staff, and family's perspective. For caregivers, watching someone who is now unable to eat or drink is difficult, creating feelings of guilt and remorse. Some may equate this with the perception of starving to death. In these stressful circumstances starting or continuing artificial nutrition and hydration, including feeding tubes, may seem attractive, even life-saving. Others may feel that a drip or tube medicalises death and detracts from feeling human and the normality of the dying process.

Care literature from the general population indicates little evidence to support the use of tube feeding in persons with advanced dementia (Finucane et al., 1999; Mitchell et al., 2015). The evidence does not support arguments that tube feeding prevents malnutrition or pneumonia, reduces the risk of pressure sores, or promotes comfort and improved functional status (Biernacki & Barrett, 2001; McCarron, McCallion, Carroll, et al., 2017), all of which are often cited as reasons for such intervention. Good et al. (2014), in an updated Cochrane review of the role of artificial hydration in palliative care patients at end of life found no difference in patient survival when comparing those who did to those who did not receive artificial hydration. They also showed no benefit on thirst, with adverse effects including increased peripheral oedema, breathlessness, and chest secretions. The authors did query some potentially beneficial effect in reducing delirium, with less sedation and less myoclonus. A systematic review conducted by Kingdon et al. (2021) echoed these findings. Given the sensitivities around the symbolic meaning of eating and drinking at the end of life, combined with the lack of empirical research about artificial hydration if the dying person has Down syndrome and dementia, a strong argument is made for discussing this issue well before the person reaches their final days. Such discussions should include whether assisted hydration is clinically viable dependent on the place of care, and all decisions should be documented as part of the advance care plan.

Staff and family need to be supported in understanding that there is a lot they can do to support and maintain the comfort of the person in their final days. Being present with the person, and paying careful attention to communication, mouth care, hygiene and skin care, positioning, eating and drinking supports, pain management, medication management, and spiritual supports are all important for a good death. These issues are each detailed in Table 7.3.

Communication technology, such as telehealth, may be helpful to ensure continuity of care between healthcare professionals, family, friends, and all care staff. When distance requires, telehealth has become an important tool to reduce emotional distress and maintain connectedness and presence, so the person with advanced dementia does not die alone. While innovations like telehealth may be helpful for some, access and the ability to use technology may not be feasible for everyone. Technology use increased during the COVID-19 pandemic and associated lockdowns experienced by many countries (McCarron et al., 2020),

Table 7.3 Care principles and tasks to focus on during final days of life

Care principles	Tasks
Communication	When approaching the end of life, the person is likely to sleep a lot and drift in and out of consciousness. Remember the person may still be able to hear, so include them in the conversation. Keep a sense of normality, familiar voices, smells, music and textures (like loved clothing or blankets). Keep the environment calm and relaxed
Mouth care	The person will lose the ability to swallow and will rarely complain of thirst. However, dry mouth and chapped lips can be uncomfortable. Use lip balm and swab the mouth regularly, checking for white patches on the tongue or gums, which may indicate infection
Hygiene and skin care	Understand that the person will be too weak for a shower or bath, however, a bed bath will be refreshing and less distressing. Ensure the dignity of the person is upheld by using towels to drape the body and only expose the body parts that are being washed. Dry the skin gently but thoroughly
Positioning	Re-position regularly for comfort and to minimise pressure area compromise. Be alert for any reddening of the skin and particularly observe for pressure point areas such as the shoulders and elbows. The bed should have a pressure-relieving mattress
Eating and drinking	The person's need for food and fluid diminishes and they rarely feel hungry or thirsty. Food and fluids can prolong dying and cause additional distress. Careful weighing up of the benefits versus the burden for the dying person is important. Support sips of fluid if the person wishes and can manage. Offer chilled drinks, ice chips, flavours for pleasure. Avoid nil by mouth
Pain	Monitor for signs of pain and distress including nonverbal signs such as grimaces, groaning and restlessness. Change of positioning can sometimes help and get advice from nursing and medical team on medication
Medication	Discontinue any medication not in keeping with goals of care (e.g. statins, anti-hypertensives, anti-dementia drugs) Consider anticipatory medication for the management of common symptoms at end of life such as pain, nausea, agitation and restlessness, breathlessness, and secretions Consider availability and storage of medication, and the administration route (oral, sub-cutaneous, and/or rectal). If a sub-cutaneous line/syringe driver is required to facilitate administration and dose replacement of oral medication (e.g. background anti-epileptic medication) skilled nursing support and close monitoring will be required

(continued)

Table 7.3 (continued)

Care principles	Tasks
Spiritual care	Ideally well before the last days, natural routines offer opportunity for meaningful care conversations about dying and death. Purposeful attention will help establish personal values, desired ritual or spiritual practices surrounding the end of life, including funeral or memorial plans. Advance care planning meetings will help build on this information as will engagement with chaplains and spiritual caregivers to support the person, their family and staff caregivers during the dying process

but for some may add further stress if people are struggling to manage complex end-of-life care. Regardless of technology-assisted or in-person, it is important for those providing care to work with the individual, family, and usual staff to understand not only the desired care but also the possible limitations of care. It is helpful to think about the following when determining the possible limits to honouring desired care at the end of life. It is critical to be realistic and honest about what can and cannot be offered:

- Prepare: understand the wishes of the person, their will and preferences, their values, beliefs.
- Know your population and understand the best pathways to care.
- Have an advanced care plan and be proactive. Support the involvement of loved ones if wished by person and engage with family, staff, and important others.
- Make sure everything is clearly documented: who was involved, what was discussed, decisions made, signed, and dated.
- Engage with general practitioner and palliative care service and establish relationships.
- If the dying person is to receive care in a non-medical setting (e.g. family or group home), engage with the local nursing team for symptom management.
- Organise equipment to ensure safety (e.g. hoists, accessible bathroom).
- Ensure a care setting that is peaceful for the dying person, welcoming of loved ones and friends, but also private when it needs to be.
- Be aware that environmental familiarity is important for the dying person.
- Address the need for careful risk assessments to balance the desired care, limits to care and maintenance of safety for all.

The story offered below brings alive the range of issues thus far summarised in this chapter.

Joan's* story

Joan was a sixty-nine-year-old woman who had Down syndrome and a moderate intellectual disability. She had grown up at home with her parents and seven siblings. She moved into residential care at a young age, but family continued to visit and be involved. Over time, Joan developed relationships with peers and had long-standing key workers who interacted with her every day as they worked successfully to build a family-like home atmosphere. Joan had a personal plan which clearly documented how she wanted to live her life.

Joan was diagnosed with Alzheimer's type dementia in 2005, and from 2011 she resided in a bungalow within a residential setting with five other women all of whom required dementia care and full nursing support. As her dementia progressed, she developed new onset seizures in 2013. There was increased confusion and disorientation, difficulty in performing spatial tasks and noticeable aphasia. The clinical team helped family, key workers, and peers understand the changes that were happening, explained what was likely to happen as symptoms progressed, and supported those who had always cared for Joan to continue to support her quality of life. In 2015, the symptoms associated with Alzheimer's disease were becoming significantly more apparent, with her body becoming increasingly fragile, difficulty with weight bearing, increase in falls, and changes in appetite and dietary intake. The risk of aspiration and aspiration pneumonia was increasing. There were new meetings with Joan's supporters, suggestions for changes in care and support offered as family, staff, and peers struggled with the inevitability of further change. In early 2019, Joan appeared to be showing signs of the end-stage of dementia; sleeping for approximately fourteen to eighteen hours, lethargy during waking hours and decreased engagement during mealtimes, personal care, and one-to-one interactions with staff. She was having trouble with her swallow technique and displayed signs of silent aspiration. Again, support was offered to ensure that Joan could continue living in her familiar and comfortable home. The organisation's clinical team and the community palliative care team were extremely helpful. Joan displayed many dying-phase symptoms, including persistent pain, breathlessness, anxiety, distress, delirium, respiratory secretions, nausea, and vomiting. Symptom management medications were prescribed, and care recommendations offered.

Joan was a woman of strong faith. She practised her faith regularly by attending Mass on Sunday and participating in the evening Angelus. Joan had previously attended a special religious education programme, resulting in strong friendships with both participants and the sisters who led the programme. These relationships, along with many others, her peers, old staff members, and her family were maintained despite her advancing dementia diagnosis. As Joan's memory deteriorated, she relied on staff to support her spiritual engagements. As well as meeting Joan's spiritual needs, staff and family talked about the value of upholding her identity and engaging in reminiscence. Family, relatives, friends, staff, and chaplain spent time with her, saying prayers, reciting the Rosary, reminiscing, and singing songs that were familiar to Joan as she moved through the dying phase. Joan took pride in her own bedroom and had been supported by her key workers and family to personalise it to her own taste. Her bedroom contained her personal belongings, memory box, life story book, favourite music, and photo albums. Family and staff reported that engaging with these personal belongings provided comfort and familiarity to Joan, particularly during the dying phase.

Joan's family was provided with the means to stay and sleep, so they were close to Joan as she neared the end of life. This included the niceties of comfortable chairs, access to the internet, family room, and kitchen. It was some weeks before Joan died. The family expressed that the availability of the niceties during this period helped enormously in their ability to share connectedness with Joan. When death came, family, staff, and peers were present and supported Joan and each other. The organisation's clinical team and the community palliative care team were then available to all the people important in Joan's life and supported their expressions of grief and bereavement.

*Assumed names are used in this story. This story is drawn from experiences in supporting people with Down syndrome and dementia in their last days at Avista and Muiriosa Foundation in Ireland.

The COVID-19 Pandemic and Its Implications for People Living with Dementia

This chapter has shown that supporting people with Down syndrome and advanced dementia is complex and involves responding to physical, psychological, social, and spiritual needs, discussing wishes and care preferences, and supporting peers and families before and after bereavement. The relationship-based approach described in Joan's story has not been possible during the COVID-19 pandemic. The pandemic has heightened mortality risk for individuals with intellectual disability including Down syndrome (Watkins, 2020) and dementia (Bianchetti et al., 2020), suggesting that end-of-life care during this pandemic is particularly relevant to people with Down syndrome and dementia. This is especially concerning when the clinical presentation of COVID-19 is atypical in people with dementia, which can frustrate detection. In nursing homes, physical distancing has proven impractical due to residents' dependence on healthcare workers, reflected too in living situations for people with Down syndrome and dementia. The COVID-19 pandemic also represents a particular challenge for family caregivers (Migliaccio & Buzigues, 2020), who are at increased risk of social isolation as services offering day programmes and respite have had to close or operate at greatly reduced capacity, translating to increased isolation with its own health risks (Armitage & Nellums, 2020).

As lockdowns and visiting restrictions have occurred, people with Down syndrome and dementia have often been further isolated from their families. This isolation is exacerbated when hospitalisations occur, as key workers, often with long-term relationships with the person, together with family and peers who would otherwise be key to providing person-centred and compassionate end-of-life care, are also excluded (McCarron, McCallion, Carroll, et al., 2017). The unprecedented COVID-19 situation has resulted in high levels of moral distress for staff working in intellectual disability services and for family caregivers (see also Chapter 9), potentially translating to complicated grief and trauma (Breen et al., 2017; Embregts et al., 2021).

The group living nature of residential facilities and group homes has increased risk for infection for people with intellectual disability. Staff and family caregivers have reported the additional stress they themselves feel in protecting and supporting people living in out-of-home placements (McCarron et al., 2020). People with Down syndrome and dementia, who often experience multiple co-morbidities, are already at increased infection risk (Dard et al., 2020), and therefore at increased risk for the negative outcomes associated with COVID-19. The stress of social distancing regulations, visiting restrictions, and reduced social contacts with family and friends increases anxiety, agitation, and depression (McCarron et al., 2020). For people living with dementia, social distancing regulations and caregivers wearing personal protective equipment (PPE) may be poorly understood, frightening and confusing, leading to feelings of being abandoned by those they love. For example, the person with Down syndrome and dementia may not recognise a usually familiar staff person who is wearing PPE. The need to implement crisis standards of care during the pandemic means that front line staff in intellectual disability services have been required to implement protocols that don't sit comfortably in a home-like environment, for example preventing family and friend visits, and discouraging touch. Mary's story illustrates the departure from usual care practices, necessitated by urgent responses to the pandemic.

Mary's* story

Mary had Down syndrome and a moderate degree of intellectual disability. Aged 56 years, she had been diagnosed three years earlier with Alzheimer's dementia. She had been supported to live in her usual home in the community that she shared with three other people with intellectual disability. Her two sisters were very involved in her life. In the year prior to her death, Mary required increasing support with activities of daily living but continued to participate in a wide range of valued life pursuits, supported by staff who knew her well over many years. Mary had a personal support plan about how she wanted to live her life, and the things that were important to her, including if she was unwell.

In early July 2020, three months into the COVID-19 pandemic, Mary developed a cough and fever. She was moved to an isolation house for symptom monitoring. Staff from her usual home offered remote virtual support using technology such as facetime and zoom to minimise distress and assist the new staff to get to know Mary. Her sisters, who had not been allowed to visit since March, were particularly distressed about the move, given this meant Mary had no in-person contact with family, staff or peers who knew her. Despite testing as COVID-19 negative, Mary's health continued to deteriorate, and she was moved again, this time to the acute hospital. Here too, given COVID-19 restrictions, it was not possible for her family or staff to accompany her.

Mary's health rapidly deteriorated and she died within a week of hospitalisation. In her final hours, one of her sisters was finally permitted to be present, dressed in full PPE clothing. Mary did not recognise her. Her other sister, peers, and staff never got to say goodbye. Staff and family members talked of their final sorrow that after death, all the usual rituals were stripped away.

*Assumed names are used in this story. This story is drawn from experiences in supporting people with Down syndrome and dementia in their last days at Avista and Muiriosa Foundation in Ireland.

Prolonged COVID-19 restrictions have also meant the usual personal care conversations with the person with dementia and their family have been compromised. There is increasing concern about insufficient time and opportunity for the conversations about preferred end-of-life care, as people face urgent and ethically challenging decisions necessitated by the COVID illness. The COVID-19 pandemic has robbed people with Down syndrome and dementia of connectedness with loved ones as they die. The experience is often typified by limits to the presence of loved ones, and the requirement that those present wear PPE, resulting in lack of intimate touch or embrace. Lockdowns have meant the inability to celebrate the dead person's life with funeral rituals, particularly challenging for the bereaved. The use of technology has meant that staff in intellectual disability services and family caregivers have been enabled some opportunity for collective mourning, however, the experience is compromised compared to in-person presence.

For people with Down syndrome and dementia who also may have COVID-19, patterns in their end of life may be different and more accelerated. COVID-19 has shone a light, highlighting the importance of advance care planning given the rapid deterioration in some patients with COVID-19. It is particularly pressing for people with Down syndrome and dementia, as many may lack the capacity to make important life decisions for themselves. In Ireland, similar to many other countries, providing intensive education and facilitation of advance care planning implementation has been a central component of government response to the COVID-19 pandemic. For society at large during the COVID-19 pandemic, awareness around the importance of end-of-life planning has been recognised. Historically, people with intellectual disability, including those with Down syndrome, have had limited opportunity to discuss how they want to live their lives, let alone discuss how they want to die. There has been limited opportunity to discuss end-of-life wishes and preferences pre COVID-19, meaning people with Down syndrome, and their families/caregivers are often not prepared to have these conversations during a pandemic when there is pressed time to reflect and discuss wishes and preferences.

Joan's story illustrates the importance of physical presence, touch, social interaction, and engagement in valued life pursuits as reflective of quality dementia care. In dealing with the COVID-19 pandemic, public health guidelines have resulted in significantly compromised ability to deliver care in this way, resulting in major challenges for staff and family caregivers. In addition to the final-days care principles previously summarised in Table 7.3, additional care considerations during the COVID-19 pandemic include:

- Increased attention to infection control: supporting the person with infection control measures such as handwashing and social distancing.
- Technology: for example, using online video calls to maintain relationships during lockdowns or restricted visitor numbers.
- PPE: caregivers may be able to wear transparent items, or a tag with photograph of self to assist recognition.

Achieving quality end of life care during COVID-19 has required innovative approaches to support people living and dying with dementia. This chapter has highlighted that conversations and advance care planning are critical components of care. Pre-pandemic, families as proxy decision-makers have perhaps relied on front-line staff and medical teams to support difficult and ethically challenging decisions, and this reliance has increased during COVID-19 experience. For example, the difficult decision to transfer a patient with Down syndrome and dementia to hospital is often fraught at the best of times, with the knowledge that comorbid health conditions are more common, and transfer to hospital may mean a poor outcome. The importance of relief from psychological, spiritual, and physical distress during this period cannot be over-emphasised, particularly so in the conditions necessitated by COVID protocols in hospital. Ann's story below articulates the complex interplay of issues, and how they can be resolved with creative end-of-life care approaches involving a range of supportive caregivers.

Ann's* story

Ann was a 46-year-old woman with Down syndrome and mid-stage dementia. She had multiple health co-morbidities, including obesity, diabetes, respiratory disease, and recent onset epilepsy. Ann was described as a fun-loving woman with a full life and enjoyed living in her group home in the community that she shared with two other people with intellectual disability. Her sister and brother were very involved in her life. Ann's father was devoted to her, but due to his own failing health, he had not been able to visit for several years but was kept in touch. In July 2020, Ann had a severe seizure and was hospitalised. She tested negative for COVID-19 on admission. Staff and family were unable to visit due to public health restrictions and this was a great source of distress. Ann's condition deteriorated and despite the clinical team's best efforts she did not respond to treatment. Staff in the group-home service worked closely with the hospital staff and the family, and a rapid response was activated to transfer Ann back to the intellectual disability service and an isolation house staffed with well-prepared intellectual disability nurses with strong links to community palliative care. Staff prepared for the transition by having her own bed in place, her CD collection and TV. Ann was surrounded by family photos and her collection of handbags. Family, friends, and staff who had worked with Ann were supported to make socially distant visits. Ann's bedroom had a large window with the bed placed close by and she was able to see and hear her family and friends through the glass.

In her final days, video calls with family and visits by the window were part of the person-centred support offered to Ann. However, Ann was getting distressed, constantly calling for her sister. Her sister, wearing full PPE, was then supported to be regularly present with Ann during her final days. This was of immense comfort to her father and brother.

Ann passed away peacefully. Her sister spoke of the specialness of being there with staff and with Ann's brother and friends gathered by the window. While limited numbers could attend the funeral, a small guard of honour was formed. Family, friends, and staff joined the funeral service by video conference.

*Assumed names are used in this story. This story is drawn from experiences in supporting people with Down syndrome and dementia in their last days at Avista and Muiriosa Foundation in Ireland.

Conclusion

People with intellectual disability are more likely to develop dementia than the general population and do so at a younger age. Although most people with intellectual disability will have an existing level of support, the onset and progression of dementia is likely to require very careful attention to care and support needs. People with intellectual disability should be supported in engaging with decision-making around the end of their life as much as possible, and the circle of support in dying should be as expansive as possible. The stories presented in this chapter demonstrate the importance of relationships between people with intellectual disability and their family, support staff and peers. They also illustrate that the COVID-19 pandemic has frustrated efforts to provide what would usually be expected in end-of-life care. Yet families, staff, and peers have adapted to the challenges of the pandemic with creative innovations. Relationships have been successfully sustained despite these barriers and highlight where service development must be targeted for the future. For too long it has been suggested that supports at end of life are not as possible for people with intellectual disability. Differences in opportunities for support by the dying person with disability and dementia must no longer be accepted. Instead, innovation and creativity means positive solutions.

Reflection

1. Think about what you would hope someone's last days would be like despite intellectual disability and advancing dementia?
2. If you have supported a person with intellectual disability and dementia in their last days, did you recognise that the last days had arrived? What went well, and are there things you would do differently in the future?
3. How would you respect a person's own wishes and ensure that family, staff, and peers are able to support their dying person, and have their own support needs met when COVID-19 means so many restrictions on visiting, spending time and supporting rituals that comfort?

Resources

1. *Glancing Back, Planning Forward: Facilitating end of life conversations with persons with intellectual disability: A guide for carers.* Findings from the end-of-life interviews of the Intellectual Disability Supplement to The Irish Longitudinal Study on Ageing (IDS-TILDA). https://www.tcd.ie/tcaid/accessibleinformation/carersguide.pdf
2. *"He'd mind you, you mind him": Experiences of end-of-life care for people with intellectual disability as perceived by staff carers.* Findings from the end-of-life interviews of the Intellectual Disability Supplement to The Irish Longitudinal Study on Ageing (IDS-TILDA). https://www.tcd.ie/tcaid/research/endoflifepublications/ReportExperiencesofendoflife.pdf

3. *People living with Dementia Supports: Trinity Centre for Ageing and Intellectual Disability.* A range of resources and links that may be of assistance to health and social care professionals during the COVID-19 outbreak: https://www.tcd.ie/tcaid/about/dementia.php
4. *The National Intellectual Disability Memory Service.* Trinity Centre for Ageing and Disability site describing a dementia assessment service and easy-read support materials about dementia. https://www.tcd.ie/tcaid/research/NIDMS.php
5. *Understand Together.* A campaign designed for the general population by the Health Service Executive, Ireland aimed at combatting stigma around dementia. https://www.understandtogether.ie/

References

Adams, N., & Grieder, D. M. (2005). *Treatment planning for person-centred care*. Elsevier.

Ali, A., Hassiotis, A., Strydom, A., & King, M. (2012). Self-stigma in people with intellectual disabilities and courtesy stigma in family carers: A systematic review. *Research in Developmental Disabilities, 33*(6), 2122–2140. https://doi.org/10.1016/j.ridd.2012.06.013

Armitage, R., & Nellums, L. B. (2020). COVID-19 and the consequences of isolating the elderly. *The Lancet Public Health, 5*(5), e256. https://doi.org/10.1016/S2468-2667(20)30061-X

Benbow, S. M., & Jolley, D. (2012). Dementia: Stigma and its effects. *Neurodegenerative Disease Management, 2*(2), 165–172. https://doi.org/10.2217/nmt.12.7

Bianchetti, A., Rozzini, R., Guerini, F., Boffelli, S., Ranieri, P., Minelli, G., Bianchetti, L., & Trabucchi, M. (2020). Clinical presentation of COVID-19 in dementia patients. *The Journal of Nutrition, Health & Aging, 24*, 560–562. https://doi.org/10.1007/s12603-020-1389-1

Biernacki, C., & Barratt, J. (2001). Improving the nutritional status of people with dementia. *Journal of Nursing, 10*, 1104–1114. https://doi.org/10.12968/bjon.2001.10.17.9949

Breen, L. J., Hall, C. W., & Bryant, R. A. (2017). A clinician's quick guide of evidence-based approaches: Prolonged grief disorder. *Clinical Psychologist, 21*, 153–154. https://doi.org/10.1111/cp.12124

Burke, E., O'Dwyer, C., Ryan, K., McCallion, P., & McCarron M. (2017). *Glancing back, planning forward: Facilitating end of life conversations with persons with intellectual disability: A guide for carers.* Intellectual Disability Supplement to the Irish Longitudinal Study on Ageing. Trinity College Dublin. http://www.professionalpalliativehub.com/sites/default/files/Carers%20Guide.pdf

Clark, D. (2007). From margins to centre: A review of the history of palliative care in cancer. *The Lancet Oncology, 8*(5), 430–438. https://doi.org/10.1016/S1470-2045(07)70138-9

Coppus, A. M., Evenhuis, H. M., Verberne, G. J., Visser, F. E., Oostra, B. A., Eikelenboom, P., Van Gool, W. A., & Van Duijn, C. M. (2008). Survival in elderly persons with Down syndrome. *Journal of the American Geriatrics Society, 56*(12), 2311–2316. https://doi.org/10.1111/j.1532-5415.2008.01999.x

Cosgrave, M. P., Tyrrell, J., McCarron, M., Gill, M., & Lawlor, B. A. (2000). A five year follow-up study of dementia in persons with Down's syndrome: Early symptoms and patterns of deterioration. *Irish Journal of Psychological Medicine, 17*(1), 5–11. https://doi.org/10.1002/gps.502

Dard, R., Janel, N., & Vialard, F. (2020). COVID-19 and Down's syndrome: Are we heading for a disaster? *European Journal of Human Genetics, 28*(11), 1477–1478. https://doi.org/10.1038/s41431-020-0696-7

Doyle, A., O'Sullivan, M., Craig, S., & McConkey, R. (2021). People with intellectual disability in Ireland are still dying young. *Journal of Applied Research in Intellectual Disabilities, 34*, 1057–1065. https://doi.org/10.1111/jar.1285

Embregts, P. J. C. M., Tournier, T., & Frielink, N. (2021). Experiences and needs of direct support staff working with people with intellectual disabilities during the COVID-19 pandemic: A thematic analysis. *Journal of Applied Research in Intellectual Disabilities, 34*(2), 480–490. https://doi-org.ezproxy.uws.edu.au/10.1111/jar.12812

European Association for Palliative Care. (2015). *Consensus norms for palliative care of people with intellectual disabilities in Europe: EAPC White Paper.* www.learningdisabilityanddementia.org/uploads/1/1/5/8/11581920/eapc-white-paper-id_full-version_april2015.pdf

Fahey-McCarthy, E., McCallion, P., Connaire, K., & McCarron, M. (2008). *Supporting persons with intellectual disability and advanced dementia: Fusing the horizons of intellectual disability, palliative and person-centred dementia care*. Trinity College Dublin. https://nursing-midwifery.tcd.ie/assets/publications/pdf/fusing-horizons-ofcare.pdf

Finucane, T. E., Christmas, C., & Travis, K. (1999). Tube feeding in patients with advanced dementia: A review of the evidence. *Journal of the American Medical Association, 282*, 1365–1370. https://doi.org/10.1001/jama.282.14.1365

Good, P., Richard, R., Syrmis, W., Jenkins-Marsh, S., & Stephens, J. (2014, April). Medically assisted hydration for adult palliative care patients. *Cochrane Database of Systematic Reviews, 23*(4), CD006273. https://doi.org/10.1002/14651858.CD006273.pub3

Jokinen, N., Janicki, M. P., Keller, S., McCallion, P., & Force, L. T. (2013). Guidelines for structuring community care and supports for people with intellectual disabilities affected by dementia. *Journal of Policy & Practice in Intellectual Disabilities, 10*(1), 1–24. https://doi.org/10.1111/jppi.12016

Kingdon, A., Spathis, A., Brodrick, R., Clarke, G., Kuhn, I., & Barclay, S. (2021). What is the impact of clinically assisted hydration in the last days of life? A systematic literature review and narrative synthesis. *BMJ Supportive & Palliative Care, 11*, 68–74. https://doi.org/10.1136/bmjspcare-2020-002600

Kirkendall, A. M., Waldrop, D., & Moone, R. P. (2012). Caring for people with intellectual disabilities and life-limiting illness: Merging person-centred planning and patient-centered, family-focused care. *Journal of Social Work in End-of-Life and Palliative Care, 8*(2), 135–150. https://doi.org/10.1080/15524256.2012.685440

Kitwood, T. M. (1997). *Dementia reconsidered: The person comes first*. Open University Press.

Margallo-Lana, M. L., Moore, P. B., Kay, D. W. K., Perry, R. H., Reid, B. E., Berney, T. P., & Tyrer, S. P. (2007). Fifteen-year follow-up of 92 hospitalized adults with Down's syndrome: Incidence of cognitive decline, its relationship to age and neuropathology. *Journal of Intellectual Disability Research, 51*(6), 463–477. https://doi.org/10.1111/j.1365-2788.2006.00902.x

McCallion, P. (1999). Maintaining communication. In M. P. Janicki & A. J. Dalton (Eds.), *Mental retardation, aging, and dementia: A handbook* (pp. 261–277). Taylor & Francis.

McCallion, P., & Ferretti, L. A. (2021). Psychosocial concerns among aging family caregivers. In V. P. Prasher, P. Davidson, & F. H. Santos (Eds.), *Mental health, intellectual disabilities and the aging process* (pp. 249–259). Blackwell. https://doi.org/10.1007/978-3-030-56934-1_15

McCallion, P., Ferretti, L., Beange, H., & McCarron, M. (2019). Epidemiological issues. In V. P. Prasher & M. P. Janicki (Eds.), *Physical health of adults with intellectual disabilities* (pp. 9–26). Wiley.

McCallion, P., Hogan, M., Santos, F. H., McCarron, M., Service, K., Stemp, S., Keller, S., Fortea, J., Bishop, K., Watchman, K., Janicki, M. P., & The Working Group of the International Summit on Intellectual Disability and Dementia. (2017). Consensus statement of the International Summit on Intellectual Disability and Dementia related to end-of-life care in advanced dementia. *Journal of Applied Research in Intellectual Disabilities, 30*(6), 1160–1164. https://doi.org/10.1111/jar.12349

McCallion, P., McCarron, M., Fahey-McCarthy, E., & Connaire, K. (2012). Meeting the end of life needs of older adults with intellectual disabilities. In E. Chang & A. Johnson (Eds.), *Contemporary and innovative practice in palliative care* (pp. 255–267). Intech. https://doi.org/10.57772/29438

McCarron, M. (2009). Dementia (in people with intellectual disability). In O. Barr & B. Gates (Eds.), *Oxford handbook of learning and intellectual disability nursing* (pp. 236–237). Oxford University Press.

McCarron, M., Carroll, R., Kelly, C., & McCallion, P. (2015). Mortality rates in the general Irish population compared to those with intellectual disability from 2003 to 2012. *Journal of Applied Research in Intellectual Disabilities, 28*(5), 406–413. https://doi.org/10.1111/jar.12194

McCarron, M., Gill, M., McCallion, P., & Begley, C. (2005). Health comorbidities in ageing persons with Down syndrome and Alzheimer's dementia. *Journal of Intellectual Disability Research, 49*(7), 560–566. https://doi.org/10.1111/j.1365-2788.2005.00704.x

McCarron, M., McCallion, P., Carroll, R., Burke, E., McGlinchey, E., O'Donovan, M., McCausland, D., Sheerin, F., O'Dwyer, M., Foran, S., MacGiolla, P., Mulryan, N., O'Connell, J., & Ryan, J. (2017). Health, wellbeing and social inclusion: Ageing with intellectual disability in Ireland. *Evidence from the first ten years of The Intellectual Disability Supplement to The Irish Longitudinal Study on Ageing (IDS-TILDA)*. IDS-TILDA, Trinity College Dublin. https://www.tcd.ie/tcaid/assets/pdf/wave3report.pdf

McCarron, M., McCallion, P., Fahey-McCarthy, E., & Connaire, K. (2011). The role and timing of palliative care in supporting persons with intellectual disability and advanced dementia. *Journal of Applied Research in Intellectual Disabilities, 24*(3), 189–198. https://doi.org/10.1111/j.1468-3148.2010.00592.x

McCarron, M., McCallion, P., Reilly, E., Dunne, P., Carroll, R., & Mulryan, N. (2017). A prospective 20-year longitudinal follow-up of dementia in persons with Down syndrome. *Journal of Intellectual Disability Research, 61*(9), 843–852. https://doi.org/10.1111/jir.12390

McCarron, M., McCallion, P., Reilly, E., & Mulryan, N. (2014). A prospective 14-year longitudinal follow-up of dementia in persons with Down syndrome. *Journal of Intellectual Disability Research, 58*(1), 61–70. https://doi.org/10.1111/jir.12074

McCarron, M., McCausland, D., Allen, A. P., Luus, R., Sheerin, F., Burke, E., McGlinchey, E., Flannery, F., & McCallion, P. (2020). *The impact of COVID-19 on people ageing with intellectual disability in Ireland.* IDS-TILDA, Trinity College Dublin. https://www.tcd.ie/tcaid/assets/pdf/wave4idstildareport.pdf

McCarron, M., Sheerin, F., Roche, L., Ryan, A. M., Griffiths, C., Keenan, P., Doody, O., D'Eath, M., Burke, E., & McCallion, P. (2018). *Shaping the future of intellectual disability nursing in Ireland*. Health Services Executive. https://www.inmo.ie/tempDocs/shaping-the-future-of-intellectual-disability-nursing-in-ireland-2018.pdf

McCausland, D., McCallion, P., Brennan, D., & McCarron, M. (2018). Interpersonal relationships of older adults with intellectual disability in Ireland. *Journal of Applied Research in Intellectual Disabilities, 31*(1), e140–e153. https://doi.org/10.1111/jar.12352

McKenzie, N., Mirfin-Veitch, B., Conder, J., & Brandford, S. (2017). "I'm still here": Exploring what matters to people with intellectual disability during advance care planning. *Journal of Applied Research in Intellectual Disabilities, 30*(6), 1089–1098. https://doi.org/10.1111/jar.12355

Migliaccio, R., & Bouzigues, A. (2020). Dementia and COVID-19 lockdown: More than a double blow for patients and caregivers. *Journal of Alzheimer's Disease Reports, 4*, 231–235. https://doi.org/10.3233/ADR-200193

Mitchell, S. (2015). Advanced dementia. *New England Journal of Medicine, 372*(26), 2533–2540. https://doi.org/10.1056/NEJMcp1412652

O'Leary, L., Cooper, S. A., & Hughes-McCormack, L. (2018). Early death and causes of death of people with intellectual disabilities: A systematic review. *Journal of Applied Research in Intellectual Disabilities, 31*(3), 325–342. https://doi.org/10.1111/jar.12417

Prasher, V. P. (2005). *Alzheimer's disease and dementia in Down syndrome and intellectual disabilities*. Radcliffe Publishing.

Scior, K., Hamid, A., Hastings, R., Werner, S., Belton, C., Laniyan, A., Patel, M., Groce, N., & Kett, M. (2016). Consigned to the margins: A call for global action to challenge intellectual disability stigma. *The*

Lancet Global Health, 4(5), e294–e295. https://doi.org/10.1016/S2214-109X(16)00060-7

Sheehan, R., Ali, A., & Hassiotis, A. (2014). Dementia in intellectual disability. *Current Opinion in Psychiatry, 27*(2), 143–148. https://doi.org/10.1097/YCO.0000000000000032

Sinai, A., Mokrysz, C., Bernal, J., Bohnen, I., Bonell, S., Courtenay, K., Dodd, K., Gazizova, D., Hassiotis, A., Hillier, R., McBrien, J., McCarthy, J., Mukherji, K., Naeem, A., Perez-Achiaga, N., Rantell, K., Sharma, V., Thomas, D., Walker, Z., … Strydom, A. (2018). Predictors of age of diagnosis and survival of Alzheimer's disease in Down syndrome. *Journal of Alzheimer's Disease, 61*(2), 717–728. https://doi.org/10.3233/JAD-170624

Stancliffe, R. J., Wiese, M. Y., Read, S., Jeltes, G., & Clayton, J. M. (2017). Assessing knowledge and attitudes about end of life: Evaluation of three instruments designed for adults with intellectual disability. *Journal of Applied Research in Intellectual Disabilities, 30*(6), 1076–1088. https://doi.org/10.1111/jar.12358

Strydom, A., Chan, T., King, M., Hassiotis, A., & Livingston, G. (2013). Incidence of dementia in older adults with intellectual disabilities. *Research in Developmental Disabilities, 34*(6), 1881–1885. https://doi.org/10.1016/j.ridd.2013.02.021

Strydom, A., Livingston, G., King, M., & Hassiotis, A. (2007). Prevalence of dementia in intellectual disability using different diagnostic criteria. *The British Journal of Psychiatry, 191*(2), 150–157. https://doi.org/10.1192/bjp.bp.106.028845

Strydom, A., Shooshtari, S., Lee, L., Raykar, V., Torr, J., Tsiouris, J., Jokinen, N., Courtenay, K., Bass, N., Sinnema, M., & Maaskant, M. (2010). Dementia in older adults with intellectual disabilities—Epidemiology, presentation, and diagnosis. *Journal of Policy and Practice in Intellectual Disabilities, 7*(2), 96–110. https://doi.org/10.1111/j.1741-1130.2010.00253.x

Swaffer, K. (2014). Dementia: Stigma, language, and dementia-friendly. *Dementia, 13*(6), 709–716. https://doi.org/10.1177/1471301214548143

Todd, S., Bernal, J., Shearn, J., Worth, R., Jones, E., Lowe, K., Madden, P., Barr, O., Forrester Jones, R., Jarvis, P., Kroll, T., McCarron, M., Read, S., & Hunt, K. (2020). Last months of life of people with intellectual disabilities: A UK population-based study of death and dying in intellectual disability community services. *Journal of Applied Research in Intellectual Disabilities.* https://doi.org/10.1111/jar.12744

Tuffrey-Wijne, I., Rose, T., Grant, R., & Wijne, A. (2017). Communicating about death and dying: Developing training for staff working in services for people with intellectual disabilities. *Journal of Applied Research in Intellectual Disabilities, 30*(6), 1099–1110. https://doi.org/10.1111/jar.12382

Tuffrey-Wijne, I., Wicki, M., Heslop, P., McCarron, M., Todd, S., Oliver, D., de Veer, A., Ahlström, G., Schäper, S., Hynes, G., O'Farrell, J., Adler, J., Riese, F., & Curfs, L. (2016). Developing research priorities for palliative care of people with intellectual disabilities in Europe: A consultation process using nominal group technique. *BMC Palliative Care, 15*, 36. https://doi.org/10.1186/s12904-016-0108-5

Visser, F. E., Aldenkamp, A. P., Van Huffelen, A. C., & Kuilman, M. (1997). Prospective study of the prevalence of Alzheimer-type dementia in institutionalized individuals with Down syndrome. *American Journal on Mental Retardation, 101*(4), 400–412. PMID: 9017086.

Voss, H., Vogel, A., Wagemans, A. M., Francke, A. L., Metsemakers, J. F., Courtens, A. M., & de Veer, A. J. (2017). Advance care planning in palliative care for people with intellectual disabilities: A systematic review. *Journal of Pain and Symptom Management, 54*(6), 938–960. https://doi.org/10.1016/j.jpainsymman.2017.04.016

Watkins, A. (2020). *COVID-19-related deaths in Wales amongst people with learning disabilities from 1st March to 26th May 2020*. Improvement Cymru. https://phw.nhs.wales/publications/publications1/covid-19-related-deaths-in-wales-amongst-people-with-learning-disabilities/

Wilkinson, H., Kerr, D., & Cunningham, C. (2005). Equipping staff to support people with intellectual disability and dementia in care home settings. *Dementia, 4*(3), 387–400. https://doi.org/10.1177/1471301205055029

8

Building Shared End-of-Life Supports and Cross-Training for Hospice/Palliative and Intellectual Disability Services Providers

Lisa A. Ferretti, Mary McCarron, and Philip McCallion

Hospice and palliative care providers seeking to support people with intellectual disability, their families, and their staff caregivers will benefit from cross-training in intellectual disability from intellectual disability service providers. Similarly, intellectual disability provider staff will

L. A. Ferretti (✉)
School of Social Work, College of Public Health, Temple University, Philadelphia, PA, USA
e-mail: lisa.ferretti@temple.edu

M. McCarron
Trinity Centre for Ageing and Intellectual Disability, Trinity College Dublin, Dublin, Ireland
e-mail: mccarrm@tcd.ie

P. McCallion
School of Social Work, Temple University, Philadelphia, PA, USA
e-mail: philip.mccallion@temple.edu

R. J. Stancliffe et al. (eds.), *End of Life and People with Intellectual and Developmental Disability*, https://doi.org/10.1007/978-3-030-98697-1_8

benefit from insights, training, and supports from hospice and palliative care providers. Largely unaddressed to date is the training needed so that hospice and palliative care providers may better support people with intellectual disability who have lived independently or with family, whether those services are offered in a hospice or other setting or delivered in the person's own home. Equally, in some jurisdictions, there is growing provision by hospitals themselves of palliative and even hospice services independent of formal hospice and palliative care providers; they too have training needs. This chapter reviews evolving issues in palliative care provision and formal cross-training packages provided in several different countries. Guided by the WHO definition of palliative care (WHO, 2016) and other frameworks, the chapter also identifies critical issues to be addressed in such trainings and how training should be modified and delivered to deal with the growing community-based and in-home provision of end-of-life care.

John's Story*

John was in his last days and was being cared for in the group home where he had lived for 23 years. He was surrounded by staff who had worked with him for many years and his peers who often sat with him and the assigned care worker when it was clear that no additional medical procedures would be initiated. His sister, who lived on a different coast, returned to be there in his last days. She was not well known to staff and peers other than through her phone calls and occasional gifts and did not seem to engage with John when she came to visit. She thanked staff and peers for their support of her brother.

Some months after John had died, the manager of the group home received a letter from John's sister and she again thanked everyone who had helped her brother but did say that when she came to the group home that she felt like an unwelcome intruder, that there was no effort to include her in John's care. Since his death she had struggled with her involvement or lack of involvement with John and the many feelings she was having about how her parents had accepted the advice of professionals that John should be placed and how she had not stepped in and taken over responsibility although she was not sure how she could have done this as she had her own special needs child who she was determined would not enter care as John had. Nevertheless she felt judged and a failure by what happened in John's last days and thought she would share that.

At a staff meeting, the manager shared the letter and asked what they all might have done differently. There were mixed opinions. Some felt sorry for the sister and regretted that they appeared not welcoming, others questioned if she was trying to transfer her own guilt onto them and there were several statements that the group home had become John's home and they as staff and his peers were really his family in his last days and that she was in fact a visitor who they had tried to be kind to, but their focus was on John.

The manager reminded staff that their home included several other individuals who had chronic conditions and that this issue might come up again as there were family members who would likely become more involved in the person's last days and asked how might they approach the situation differently. After much discussion it was decided to invite a staff member from the local hospice to come to the group home to talk about death and including families, as well as the needs of the person dying.

*Assumed names are used in this story.

As noted by Bolognesi et al. (2013) a key challenge in end-of-life care is that while there is a commitment to address human and social needs when people are in the end stages of incurable disease, there is considerable disarray in the availability and content of training programmes, a need for agreement on what should be addressed in such training and a tailoring of any training for different categories of healthcare professionals and other participants in care. They particularly note the barrier of a lack of comprehensive end-of-life training in general healthcare education for physicians, nurses, social workers,and others. The Center to Advance Palliative Care's clinical training recommendations (https://www.capc.org/clinical-training-recommendations-for-all-clinicians-caring-for-patients-with-serious-illness/) argue the issues to be addressed in training are: (1) assessing the individual's needs and concerns, (2) strengthening the clinician-patient relationship, (3) understanding care goals, (4) managing pain and symptoms, (5) preventing crises, and (6) helping individuals plan ahead. Bolognesi et al. (2013) would agree with these priorities and add the need to approach each with an eye to protect individual dignity and autonomy, and to provide both intervention content and delivery approaches that are customised to the individual and which support quality of life to the end. Others have argued that another key concern in end-of-life care is the management and support of grief and bereavement in all of its forms (Moon, 2013) and that the critical concerns that must be addressed in training (McCallion et al., 2012) include:

- Understandings of end-of-life care, palliative care, and advance planning
- Movement from a medical to a social care/quality of life focus
- Areas for training
- Mechanisms for training
- Development of a collaborative ethos.

Understandings

End-of-life care. Such care addresses the support and medical care delivered at the end of life often formally beginning with a decision that death is imminent and that continued curative care is futile, including when a decision is made to forgo further procedures. The focus is upon providing the person with comfort care in their last days addressing pain, breathing problems, skin irritation, digestive problems, temperature sensitivity, and fatigue. There is also recognition that for the person dying there are emotional, mental health and spiritual needs to be addressed. Ideally end-of-life care desires have been expressed by the individual either during this time or previously, in a preferably written statement and there is also recognition that during this time family/caregivers also have needs that should be discussed and addressed (https://www.nia.nih.gov/health/providing-comfort-end-life) (see also Chapters 4 and 14). The practical and formal ways in which end-of-life care is provided are captured in hospice care, palliative care and advance planning and their implementation is often guided by person-centred principles and consideration of professional and family/caregiver roles. There are not always shared understandings of these concepts or the same levels of implementation experience, and therein lies a major cross-training need and opportunity.

Hospice care has been defined as care focuses on comfort, and quality of life when a person has a serious illness and is approaching the end of life. It is seen today as an approach to care that may be offered at home or in a nursing home, hospital, or separate hospice centre. Previously it was seen as care in a separate dedicated centre without capacity to offer curative services in order to emphasise that care in one's last days is different and focused on quality of life (https://www.nia.nih.gov/health/what-are-palliative-care-and-hospice-care#hospice). When end-of-life care is delivered as hospice there is much greater agreement about what the type of care to be delivered, a clearer understanding of the professional individuals to be involved, and a more agreed set of services to be delivered. It brings together a specific and trained team with special skills—nurses, doctors, social workers, spiritual advisors, and trained volunteers. This group then work together with the person who is dying, the caregiver, and/or family and they offer medical, emotional, and spiritual support. As well as training in their specific professions, the individuals involved have received training within the hospice itself and benefit from dedicated attention to end-of-life care issues and the learnings that emerge from their daily work and experiences. Over time, while out of home dedicated hospice care continues, there has been considerable growth of in-hospital hospice care and also the extension of care to one's own home. In the at-home scenario, the emphasis is on the hospice team offering support with an individual's family and friends providing the day-to-day care. For people with intellectual and developmental disabilities, there has been growth in hospital-based and even nursing-home-based delivery of hospice care. There has also been the development of specialised group homes for persons with intellectual and developmental disabilities who are medically frail and/or who have dementia that may then offer the opportunity to spend last days in this same or in an adjacent home with hospice-like services (McCarron et al., 2018). To the extent that services providers have their own medical and nursing services, professionals and other paid staff are likely to deliver end-of-life care in the last days.

Palliative care is specialised medical care for people living with a serious illness focused upon management of their symptoms, comfort and quality of life for the individual and their family/caregivers and may include treatments still intended to cure their serious illness. The WHO defines palliative care as "an approach that improves the quality of life of patients and their families facing the problem associated with life-threatening illness, through the prevention and relief of suffering by means of early identification and impeccable assessment and treatment of pain and other problems, physical, psychosocial and spiritual" (WHO, 2016). This definition raises the need for continued medical intervention, particularly for pain management. The team supporting the person varies depending upon their needs but usually involves palliative care specialist doctors and nurses, social workers, nutritionists, and chaplains. Together they support the person and family/caregivers offering medical, social, emotional, and practical support in hospitals, nursing homes, outpatient palliative care clinics, other specialised clinics or at home (https://www.nia.nih.gov/health/what-are-palliative-care-and-hospice-care#palliative). The plan that palliative care begins earlier than hospice care in the treatment journey means that the focus is not exclusively on end of life, and it begins with a much more medical orientation although the intention is to enhance a person's current care by focusing on quality of life for the individual and their family/caregivers. Again, there is related training in the team's individual professions and "palliative" is in many cases a specialised training and certification for those involved. Current accreditation standards in the United States, for example, require a team consisting of at least one physician, an advanced practice, or other registered nurse, a social worker, and a chaplain, but ideally this should also include pharmacists, rehabilitation therapists, direct care workers, and other clinical and nonclinical specialists. Survey findings, however, are that many hospitals rely upon advanced practice nurses with physician support, meaning fewer individuals are served and that broader areas of support, particularly social, emotional, and spiritual are not well addressed, gaps which then may carry over into linked hospice programmes and care at the end of life (Spetz et al., 2016).

Advance planning (advance care planning) involves thinking and talking about one's wishes for how to be cared for in the final months of life, including identifying treatments and approaches that are not desired during that time, and possibly identifying someone to speak on one's behalf and/or make care-related decisions when the person dying is no longer able to (Voss et al., 2017). Many jurisdictions have specific legislation that codifies what is and is not possible in advance planning, how the person's wishes may be stated and memorialised, and what are the legal obligations of practitioners and families in implementing advance plans or directives. As noted in other chapters, there are training programmes and supportive materials that may assist many individuals with intellectual and developmental disabilities in pursuing advance planning (see Chapter 17).

Movement from a Medical to a Social Care/Quality of Life Focus

End-of-life care approaches remain largely medically focused and driven by healthcare professionals. Coyle and Ferrell (2015) reject this view as inadequate and argue that our approach should be social care rather than be medical care driven.

To deliver social care, end-of-life care must be focused upon the individual who is dying and the family/caregivers (however constituted) with a need for training for family/caregivers and for all involved staff to address: (1) *Transition from living with to now dying from….* (2) *Shifting from what used to be to what is now* so as to avoid feelings of frustration, anger, loneliness, and abandonment. (3) *Understanding and validation of the "reciprocity of suffering"* where family members and other caregivers experience severe physical and emotional distress as they deal with the impending death and often changes in their roles as caregivers. (4) *Burdening*: reflecting that while there is physical burden for caregivers such as exhaustion, sleeplessness, and deterioration of one's own health, there is also the social burden of consuming end-of-life care for one person, meaning little time for self-care and a growing sense of isolation. Yet both contrast with a continuing sense of caregiving satisfaction in being with the person in these last days. (5) *Offering support to family and caregivers* where each decides what will actually be helpful with professionals understanding that "support" may mean knowing there are resources and supports available, even if they do not make use of such resources. (6) *Struggling with the paradox* that the person is both living and dying and may believe and seek to survive yet know they will not, and with families and caregivers with this same struggle and their own paradox, wanting to care for and spend time with the individual, but also wanting a "normal" life often called for by their other responsibilities. (7) *Contending with changes in relationships and roles* with different meanings and consequences among family members and other caregivers, yet coping by attempting to keep everything as "normal" as possible. (8) *Searching for meaning* perhaps through personal growth, or simply tolerating the situation, or finding new spiritual insights or rigidly adhering to their religion and its tenets. (9) *Living day to day* by "making the best of it" and enjoying the time left with or just enduring the situation. (10) *Preparing for death* by taking concrete actions to prepare for when the person has passed. (11) *Location of care*: making decisions about and seeking supports for the location that offers the family, other caregivers, and the person themselves greatest satisfaction for symptom management and emotional support for all. Decisions to rely on hospital or other non-usual locations generally mean that there is a lack of confidence that formal services will support the decisions of the family, other caregivers, and the person themselves. (12) *Caregiving at a distance*: in today's societies many family members live at a distance from other family yet provide important emotional and instrumental support and feel connected to the person who is dying although they do not visit often. Stress related to the distant caregiving is quite common, but most services and care are focused upon local caregivers. Outreach, such as increased telephone communication and consultation on key decisions are important sources of support.

Although not necessarily its original intention, in one jurisdiction, the nine-year Project on Death in America funded by the Soros Foundation was developed to address some of these challenges by exploring the meanings of death in American culture and highlighting what works, what does not work and what is needed in care at the end of life (Clark, 2013). Activities included building the evidence base for palliative care, addressing the needs of underserved communities (including people with intellectual and developmental disabilities), and resolving barriers to improving care including legal and ethical challenges. The goal was to transform cultures in end-of-life care and promote organisational change and the empowerment of individuals. Despite its initial funding overwhelmingly supporting medical and nursing professionals and researchers, there was a change as it moved forward to extending to other disciplines, particularly social work, in order to build capacity for service-delivery models for the dying and their network of family and friends that better reflected that most end-of-life care was becoming home and hospital delivered and was focused upon quality of life, the spiritual, responsive services, and personal decision-making rather than health-mediated palliative intervention (Clark, 2013). Although not a primary focus, addressing unique issues for persons with intellectual and developmental disabilities was addressed in several projects and responsive end-of-life care for this group has benefited from the expansion of end of care foci and related training that has resulted.

Areas for Training

There are several areas that need to be addressed through training and preparation. Many involve better communication between staff in hospice/palliative care including hospital-based resources, but work is also needed on addressing communication and genuine involvement of the person who is dying, peers, and family members. Specific training needs are in the areas of (1) person-centred care, relationship-centred care, (2) caring for a person with an intellectual and developmental disability, (3) palliative and hospice care, (4) nutrition, hydration and pain concerns, (5) facilitating support of grief and loss, (6) expanding the care team, and (7) reimagining end-of-life care.

Person-centred care. The person-centred challenge for end-of-life care lies in finding agreement amongst the individual, family, caregiver, and palliative/hospice team on expected outcomes in relief from distressing symptoms, the easing of pain, and/or the enhancement of quality of life. In intellectual and developmental disability services, there are well-established systematic processes for the discovery of an individual's gifts, capacities, experiences, core beliefs, and dreams. The collaborative development of plans to realise those dreams is enhanced by the commitment to the individual by valued persons who help to implement the plan. The exercise of choice by the person, the use of the person's abilities, the fostering of the expression of feelings and enabling the person to live in the context of their relationships are also critical to their care (McCallion et al., 2012).

There are differing resources focused on person-centred care in the intellectual and developmental disability services, palliative and hospice care, and dementia care areas. A useful training exercise would be to have groups of trainees compare and contrast and then find commonality in their approaches. For some families, the ideas of person-centredness may be new or where they have little experience. Staff in all settings must be trained to explain the approaches and their implications for decision-making and care implementation, but also to deliver this information in a collaborative and respectful way. In trainings there should be discussion of how disagreements about care may be managed, when, for example, through an advance directive or in the absence of a formal advance directive a family may see themselves as the decision-maker.

Relationship-centred care. Critiques of person-centred care acknowledge its far-reaching impact on care, but argue that it is not enough to consider the individual without considering their relationships. Relationship-centred care occurs where relationships have developed over long periods of time such as in intellectual and developmental disability care settings and it recognises that care involves the needs of the person being cared for, the family, caregivers, peers, and health professionals and subsequently the palliative/hospice care team (see also Chapter 7).

Moss and Moss (2002) previously highlighted that in nursing homes and hospitals there has been an emphasis on maintaining professional distance and for staff not to perceive themselves as family, although there may be intensity to the relationships that do form in caring for someone with a chronic condition and dying, they are often not of long duration. This is not as true in intellectual and developmental disability services, where caregiving for some may have been of long duration and there was a determined effort to create a family-like environment. Also, the existence of such relationships can be challenging for family members struggling with how placement in a setting disrupted and shaped their own relationship with the person. For hospice, palliative, and hospital staff, it is important that they understand and find ways to recognise the "relationship" between intellectual and developmental disability services staff and the person who is dying, as well as the family relationship (McKibben et al., 2021), and also how to incorporate the peers with whom the person may have lived with for an extended period. Among intellectual and developmental disability services staff and family members, activities and information must support each recognising their different but important relationships and how this effects the support each offers and the roles they play as the person is dying. Most available training materials identify the need to recognise relationship-centred care, but there is an absence of recommended approaches to its successful incorporation in care, care decisions, and communication.

Knowledge and skills of caring for a person with an intellectual and developmental disability. A lack of knowledge of the needs of persons with intellectual and developmental disabilities by staff in hospital settings has long been identified as leading to unexpected and unexplained deaths (McCallion et al., 2012). Key challenges for hospital and for hospice/palliative care staff include being able to (1) interpret non-verbal and alternative communication strategies, (2) appropriately assess and manage pain, and (3) recognise and respect the person's right and ability to be self determining despite prior cognitive impairment (McCallion & Ferretti, 2017). Hospice, palliative care, and hospital staff must also be equipped to understand the care provided by staff and families to persons with intellectual and developmental disabilities and the respect owed to their ability to understand and express the person's wishes and desires. For staff in intellectual disability settings, there remains a need for training and education about the ageing of persons with intellectual disability and their experience of terminal illness (Jokinen et al.,

2013; McCarron et al., 2018; McCallion, Hogan, et al., 2017; McCallion, Jokinen, et al., 2017) (see Resources at the end of this chapter for some available training materials).

Knowledge and skills of palliative care. The key areas to be addressed in knowledge and skills of palliative care as articulated by the World Health Organization (2016) are relief from pain, affirmation of life, seeing dying as a normal process, ensuring to neither hasten nor postpone death, integrate psychological and spiritual aspects, support individuals to live as actively as possible until death, help families/caregivers to cope with illness, death and their own bereavement, use a team approach to addressing needs, enhance quality of life, and be a positive influence on the course of illness. Staff in intellectual and developmental disability services are not generally prepared in these palliative care skills and often do not feel they have the expertise in practice. They are also often unprepared for their own concerns and for family dynamics around caring for someone with a terminal illness and the emotions involved.

Whittaker et al. (2007) argued that family/caregivers determine the quality of care delivered to people who are dying and there is a need for training in the pharmacological and non-pharmacological management of pain and other symptoms to improve palliation. These findings for care of the general population have been confirmed as needs too for staff within intellectual and developmental disability services (Fahey-McCarthy et al., 2009). Collaboration training with hospice/palliative care would support an exchange of expertise, help in building understanding of what to expect in the progression of the disease/illness and how to interpret important changes that might for example be indicative of pain. Training should also address what intellectual and developmental disability services staff and family caregivers should be reporting to palliative care teams to gain the support they need (McCallion et al., 2012), what are the additional services and supports that hospice/palliative care can and will offer, and how those services and supports may supplement rather than replace the good care already being provided (see Resources at the end of this chapter for some available training materials).

Nutrition, hydration, and pain concerns. Feeding difficulties and challenges experienced as part of end-stage dementia care by persons with intellectual and developmental disabilities mirror difficulties described in the general care literature in persons with Alzheimer's dementia. Difficulty in eating, of holding food in one's own mouth, chewing swallowing, spitting, and food inhalation/aspiration all culminate in stress for the person, the family, and staff caregivers (McCarron et al., 2018) (see also Chapter 14). There are similar issues when there are significant breathing difficulties around decisions to utilise ventilators, which have been highlighted during the COVID-19 pandemic. Staff and family caregivers are often attracted by what they perceive as the benefits of artificial tools such as ventilators and feeding tubes. End-of-life decision-making is an emotional and value-laden time. Cultural and religious values, emotions, relationship bonds and conflicts, limited ability to know and understand the wishes of the person, and a lack of undisputed outcome data make related decisions formidable clinical and care challenges. In intellectual and developmental disabilities care arenas, there is a history of successful use of tube feeding among children and adults with intellectual disability that confounds the discussion (McCallion et al., 2012), meaning there is a need for further informed and respectful discussion of use of these measures at end of life. Training materials in the resources section will assist in addressing some of these knowledge and skill issues but issues around the use of ventilators are not covered.

Assessment of pain in persons with intellectual and developmental disabilities has also been problematic and the assessment instruments used within hospice and palliative care services for pain are rarely helpful for these individuals (McCallion et al., 2012; McCarron et al., 2018). Here staff caregivers in intellectual and developmental disability services and families may have guidance for hospice and palliative care staff on the verbal and non-verbal pain indicators for the person they have cared for. Equally staff caregivers in intellectual and developmental disability services and families would benefit from additional guidance from hospice and palliative care services on pain management (see Resources at the end of this chapter for some available training materials).

Knowledge and skills of facilitating support for grief and loss. Loss, physical loss, and psychosocial loss are often expressed in grief, that is, the psychological, behavioural, and social reactions to the experience of the loss. Mourning is the public display of grief, which can be uncomplicated, complicated, or unresolved. During end-of-life care, individuals may also experience anticipatory grief, particularly if disease symptoms mean that the person known no longer seems to be there. There is also disenfranchised grief, defined by Doka (1989, p. 4) as "the grief that persons experience when they incur a loss that is not or cannot be openly acknowledged, publicly mourned or socially supported". This is of particular relevance for staff caregivers in intellectual and developmental disability services but may also be relevant for family members who previously have been at a distance and may be viewed as uninvolved.

We know much more about the participation, the frustrations, the ways in which formal services, particularly hospital-based service exclude staff caregivers from decision-making and the lack of bereavement support for staff. The traditional reliance in intellectual and developmental disability service systems in many countries on out-of-family-home placements often meant geographic relocation that effectively severed relationships with family members (Lightfoot & McCallion, 2016; McCallion et al., 2012). Where there is staffing stability, the emphasis then on staff creating family-like living environments means an opportunity for staff caregivers to function and care as family members. This reality has been recognised and documented by researchers and is part of the experience of many service systems, intellectual and developmental disability as well as hospice and palliative. Yet the majority of people with intellectual and developmental disabilities live their lives with family members, and there are many families who maintain contacts even when their family member lives somewhere else. Death has been noted as a time of reconnections and reconciliation among family members in the general population (Renz et al., 2020), and there is no reason why this may not also be the case for families of people with intellectual and developmental disabilities. Long distance and estranged family members feel justified in reclaiming a primary role, sometimes to the exclusion of staff caregivers; a role that hospital systems in particular may support in their protocols around notification and consultation. There are complex issues that are currently poorly understood.

Training across service systems must first address building awareness of this complexity and then offer strategies for respectful management. Training for families on what to expect in end of life must do the same. Next staff and administrators in the various services systems must become aware of the grief issues for staff caregivers and explore the provision of supportive services. For families caring at home there is likely not to be grief support from intellectual and developmental disabilities services systems, but for their family member many receive or did receive case management and/or day programming/employment services. The staff involved in these programmes are likely to have the most trusting relationships with families and need training on reconnecting and offering grief and bereavement-related supports. In the absence of such initiatives then it will fall to staff in hospice/palliative care and hospital programmes to provide such support. Additional training may be needed to address when a family has provided a life-time of care-giving; the sense of loss and its consequences are likely to be profound and perhaps longer lasting in its acute phase than may be true for other families. Training also needs to focus on the support of grieving peers with intellectual and developmental disabilities. Symptoms of normal grief as defined by ICD-11 usually occur within one month of the bereavement and do not exceed 6 months duration. For persons with intellectual and developmental disabilities, later onset and longer duration of grief symptoms are more likely (McCallion et al., 2012), perhaps with increased levels of depression, anxiety, and distress. Grief reactions may also include manifestations superficially perceived as behavioural difficulties. More specialised approaches will be needed by intellectual and developmental disabilities services and here support from hospice /palliative care services may assist in developing appropriate training for staff. Some of the resources at the end of this chapter will also be helpful.

Expanding the Care Team

It has already been noted that there can be an over-reliance on the perspectives of physicians and nurses with a risk for a more medically oriented approach that misses the important additions of social and spiritual care. This challenge is being taken up by palliative care in general but may be more pressing for end-of-life care for persons with intellectual and developmental disabilities, their families, and staff caregivers. An important training strategy may be the convening of focus and task groups that involve the recommended spectrum of physicians, nurses, social workers, and chaplains supplemented by pharmacists, rehabilitation therapists, direct care workers, and other clinical and nonclinical specialists, as well as by case managers and day programme/employment staff who are in contact with families caring at home. Building greater understanding of what each does and is able to contribute is an excellent step that may lead to new insights, invitations to offer more specific trainings and the building of supportive relationships. The circle of training and consultation support that will be created should be expanded to include the voices of families and of people with intellectual and developmental disabilities.

Particular attention should be paid to the role of chaplains. There is perhaps an assumption that chaplains emerge from their training as experts in end of life and in support of grief, bereavement and the attainment of spiritual meaning. Like other groups, this role is but one of many for chaplains. Success in supporting spiritual needs appears to require awareness of a variety of spiritual care practices and a willingness to support individuals in building and utilising spiritual competencies (Gijsberts et al., 2019) (see also Chapter 11). Success is also dependent upon the level of visibility of spirituality and spiritual care in health care. Addressing these issues successfully will require collaboration across professions, and in the case of people with intellectual and developmental disabilities, their families and staff caregivers will further benefit from more specific preparation of the chaplain to work with them. An example of a supportive resource is the Spiritual Support Certification programme offered by the American Association on Intellectual and Developmental Disabilities (AAIDD) Religion and Spirituality Interest Network (see Resources at the end of this chapter).

For families caring at home, there are likely to be many challenges in maintaining the household, dealing with anticipatory and other grief issues, and continuing to provide care. In some jurisdictions and in some hospice programmes, there is a role for volunteers to help both with managing the other aspects of a caregiver's life, offering emotional support, and in being present for the person who is dying. Training and supervision for the volunteer are important, and there are examples of successful training programmes that may be adapted from general population focused hospice/palliative care (Vanderstichelen et al., 2018).

Mechanisms for Training. As noted earlier, there are already several resources available as well as some areas where training needs to develop further. As contributors to a number of the resources cited, we also acknowledge that, being developed a decade ago, they lack sufficient attention to families and their issues and tend to "fit" people with intellectual and developmental disabilities, families, and staff caregivers into the prevalent forms of support of end-of-life care. In such an approach staff in any system may think the task is to know more about the other system and not to change ways of thinking and acting to reflect that today more and more families are providing significant care. The extent therefore to which there can be joint training enterprises between intellectual and developmental disabilities services and hospice/palliative care and hospital services—as has happened for example in Ireland (Fahey-McCarthy et al., 2009) and New York State (Botsford & Force, 2004; McCallion, 2006)—increases the likelihood of joint learning and mutual respect. As noted, this approach still does not address including the voices of families and indeed people with intellectual and developmental disabilities themselves. There has also been work in improving dialogue between disciplines around end-of-life care issues to better understand what each can contribute, to expand beyond medical/health interventions and to better emphasise social and spiritual care (https://www.health.ny.gov/professionals/palliative_care/docs/pcetc_recommendations.pdf).

Such large-scale efforts must be followed with the building of local coalitions and provider-level joint trainings. Resources for such provider-level training and for group and individual training for families and for people with intellectual and developmental disabilities are also needed. The growth of utilising online training means that joint training may be more easily delivered and encourages the development of video and other resources that may be accessed asynchronously. Finally, as Bolognesi et al. (2013) noted, the academic preparation of all disciplines (health and social care) is lacking in terms of readiness to offer end-of-life care in general, and the authors would note, to persons with intellectual and developmental disabilities in particular. At a minimum there is a need for a greater comprehensiveness in the scope of practice addressed and an inter-professional-, family-, and individual-focused ethos in the preparation of all disciplines around end-of-life issues.

Reimagining End-of-Life Care: Participating in Collaborative Practice

Efforts around inter-professional education have increasingly focused on creating positive care experiences for individuals and their families, with an emphasis on compassionate and individual- and-family-centred care (Pfaff & Markaki, 2017). A compassionate, collaborative-care model and framework have emerged emphasising (1) experiencing and acting upon one's compassion, (2) collaborating, communicating, and partnering with individuals and family members to the extent they need and desire, (3) commitments by all who provide and support health care to communicate and collaborate with each other, and (4) a belief in the resilience and well-being of professional and family caregivers (https://www.theschwartzcenter.org/media/FINAL-CCC-Model-and-Framework.pdf). The goal in training to support end-of-life care for people with intellectual and developmental disabilities and their families and staff caregivers is to achieve such innovative collaborative practice.

Reflection

Denise's Story*

Denise lived at home with her parents for many years and attended a day programme that she really enjoyed. As her parents aged, her brother David took over many of the responsibilities and in particular was a regular visitor to the day programme stopping on his way home from work. He helped out when he visited and would always help with any fundraising activities and talked with other participants. As a result, there were strong relationships between Denise, David, the day programme staff, and the other participants in the day programme. As Denise got older there were a number of chronic health conditions. She had several falls despite some environmental modifications made to the home including a second stair rail and soon she only occasionally participated in the day programme, but then stopped going and was largely cared for at home. After another fall and the onset of pneumonia, Denise was largely bed-ridden but David ensured she was able to stay at home as this was what her mother wanted and previously Denise had always indicated she never wanted to move out of the family home. The level of care was increasing and David was struggling to balance the care with his work responsibilities and eventually reduced his hours and was focused upon Denise. This presented a financial hardship exacerbated by David also trying to support his ageing parents with their own health needs. He reached out to the day programme asking for help or suggestions for help. There was little the day programme could do but one staff member did have the phone number and a contact name for the local hospice. David continued to provide care largely alone although staff from the day programme did check in and tried to provide moral support and to be a sounding board as he struggled with the responsibilities and the realisation that Denise would soon be gone. They were fond of Denise and of him and they were struggling themselves with the thought that someone they had supported for many years might soon die.

*Assumed names are used in this story.

1. For this family caregiver caring at home what services might he have been provided with and by whom?
2. Given that the only service many families caring at home for person with intellectual and developmental disabilities may access is a current or past day programme, what training should be provided to day programme staff that they may be better able to support and inform families facing end-of-life care?
3. What additional could and should have been done to better understand Denise's wishes as the level of care needs increased.

Resources

1. Fahey-McCarthy, E., McCallion, P., Connaire, K., & McCarron, M. (2008). *Supporting persons with intellectual disability and advanced dementia: Fusing the horizons of care. An introductory education and training programme*. Trainer's Manual. Dublin: Trinity College Dublin.

   Contact: https://nursing-midwifery.tcd.ie/

   The curriculum emerged from an effort to understand staff experiences in supporting persons with intellectual disability and advanced dementia. Intellectual disability service providers and a specialist palliative care provider in Ireland were involved. Their experiences were interpreted to gain an understanding of their education and training needs. This information was then the basis for a pilot educational intervention which was designed, delivered and evaluated with these services. A partnership approach which involved the Trinity College School of Nursing and Midwifery research team, intellectual disability service providers and a specialist palliative care service was crucial to success.
2. *End of Life Care: A guide for supporting older person with intellectual disabilities and their families* (Botsford & Force, 2004; McCallion, 2006).

   Contact: http://www.nysarc.org

   A collaborative report involving both intellectual disability and hospice/palliative professionals providing guidance for staff and families on the policy context and day-to-day management strategies when persons with intellectual disability approach the end of life. The guide is supported with a CD-ROM supplement also available through NYSARC, Inc.
3. *Let's Talk about Death*. A booklet about death and funerals for people who have an intellectual disability available from Down syndrome Scotland. This booklet may be used both to prepare individuals with an intellectual disability to participate in decisions about their end-of-life care and to support peers of individuals who are dying. http://www.dsscotland.org.uk/resources/shop/talkaboutdeath

4. *Spiritual Support Professional Certification.* A certification process that seeks to ensure that individuals whose work focuses on providing and/or facilitating ministry with people with intellectual and developmental disabilities is of a competent standard. Run by the American Association on Intellectual and Developmental Disabilities (AAIDD) Religion and Spirituality Interest Network (https://aaiddreligion.org/certification/).

References

Bolognesi, D., Brighi, N., Muciarelli, P.-A., & Biasco, G. (2013). Palliative care training and research: The development in Europe and the Bologna experience. *Indian Journal of Palliative Care, 19*(1), 20–26. https://doi.org/10.4103/0973-1075.110221

Botsford, A., & Force, L. (2004). *End of life care: A guide for supporting older person with intellectual disabilities and their families.* NYSARC & NYSDDPC.

Clark, D. (2013). *Transforming the culture of dying? The work of the Project on Death in America.* Oxford University Press.

Coyle, N., & Ferrell, B. R. (2015). Social Aspects of care. In R. Steele & B. Davies (Eds.), *Supporting families in palliative care* (pp. 1–26). Oxford University Press. https://doi.org/10.1093/med/9780190244132.001.0003

Doka, K. J. (1989). *Disenfranchised grief: Recognising hidden sorrow.* Lexington Books.

Fahey-McCarthy, E., McCarron, M., Connaire, K., & McCallion, P. (2009). Developing an education intervention for staff supporting persons with intellectual disability and advanced dementia. *Journal of Policy and Practice in Intellectual Disabilities, 6*(4), 267–275. https://doi.org/10.1111/j.1741-1130.2009.00231.x

Gijsberts, M.-J. H. E., Liebroer, A. I., Otten, R., & Olsman, E. (2019). Spiritual care in palliative care. *Medical Science, 7*(2), 25. https://doi.org/10.3390/medsci7020025

Jokinen, N., Janicki, M. P., Keller, S., McCallion, P., & Force, L. T. (2013). Guidelines for structuring community care and supports for people with

intellectual disabilities affected by dementia. *Journal of Policy & Practice in Intellectual Disabilities, 10*(1), 1–24. https://doi.org/10.1111/jppi.12016
Lightfoot, E., & McCallion, P. (2016). Older adults and developmental disabilities. In B. Berkman & D. Kaplan (Eds.), *Handbook of social work in health & aging* (2nd ed., pp. 489–499). Oxford University Press.
McCallion, P. (2006). *End of Life Care: Supporting older people with intellectual disabilities and their families*. A cd-rom curriculum supplement. NYSARC & NYSDDPC.
McCallion, P., & Ferretti, L. A. (2017). Understanding, supporting and safeguarding self-determination as we age. In M. L. Wehmeyer, K. A. Shogen, T. D. Little, & S. J. Lopez (Eds.), *Development of self-determination through the life-course* (pp.145–158). Springer. https://doi.org/10.1007/978-94-024-1042-6_11
McCallion, P., Hogan, M., Santos, F. H., McCarron, M., Service, K., Stemp, S., Keller, S., Fortea, J., Bishop, K., Watchman, K., & Janicki, M. P. (2017). Consensus statement of the international summit on intellectual disability and dementia related to end-of-life care in advanced dementia. *Journal of Applied Research in Intellectual Disability, 30*(6), 1160–1164. https://doi.org/10.1111/jar.12349
McCallion, P., Jokinen, N., & Janicki, M. P. (2017). Aging. In M. L. Wehmeyer, I. Brown, M. Percey, K. A. Shogren, & M. Fung (Eds.), *A comprehensive guide to intellectual and developmental disabilities* (pp. 639–654). Paul Brookes Press.
McCallion, P., McCarron, M., Fahey-McCarthy, E., & Connaire, K. (2012). Meeting the end of life needs of older adults with intellectual disabilities. In E. Chang & A. Johnson (Eds.), *Contemporary and innovative practice in palliative care* (pp. 255–267). Intech.
McCarron, M., McCallion, P., Coppus, A., Fortea, J., Stemp, S., Janicki, M., & Watchman, K. (2018). Supporting advanced dementia in people with Down syndrome and other intellectual disability: Consensus statement of the International Summit on Intellectual Disability and Dementia. *Journal of Intellectual Disabilities Research, 62*(7), 617–624. https://doi.org/10.1111/jir.12500
McKibben, L., Brazil, K., McLaughlin, D., & Hudson, P. (2021). Determining the informational needs of family caregivers of people with intellectual disability who require palliative care: A qualitative study. *Palliative Support Care, 19*(4), 405–414. https://doi.org/10.1017/S1478951520001157
Moon, P. J. (2013). Grief and palliative care: Mutuality. *Palliative Care and Social Practice, 7*, 19–24. https://doi.org/10.4137/PCRT.S10890

Moss, S. Z., & Moss, M. S. (2002). Nursing home staff reactions to resident deaths. In K. Doka (Ed.), *Disenfranchised grief: New directions, challenges and strategies for practice*. Research Press.

Pfaff, K., & Markaki, A. (2017). Compassionate collaborative care: An integrative review of quality indicators in end of life care. *BMC Palliative Care, 16*(65), 1–24. https://doi.org/10.1186/s12904-017-0246-4

Renz, M., Bueche, D., Reichmuth, O., Schuett Mao, M., Renz, U., Siebenrock, R., & Strasser, F. (2020). Forgiveness and reconciliation processes in dying patients with cancer. *American Journal of Hospice and Palliative Care, 37*(3), 222–234. https://doi.org/10.1177/1049909119867675

Spetz, J., Dudley, N., Trupin, L., Rogers, M., Meier, D. E., & Dumanovsky, T. (2016). Few hospital palliative care programs meet national staffing recommendations. *Health Affairs, 35*(9). https://doi.org/10.1377/hlthaff.2016.0113

Vanderstichelen, S., Houttekier, D., Cohen, J., Van Wesemail, Y., Deliens, L., & Chambaere, K. (2018). Palliative care volunteerism across the healthcare system: A survey study. *Palliative Medicine, 32*(7), 1233–1245. https://doi.org/10.1177/0269216318772263

Voss, H., Vogel, A., Wagemans, A. M. A., Francke, A. L., Metsemakers, J. F. M., Courtens, A. E., & de Veer, A. J. E. (2017). Advance care planning in palliative care for people with intellectual disabilities: A systematic review. *Journal of Pain and System Management, 54*(6), 938–960. https://doi.org/10.1016/j.jpainsymman.2017.04.016

Whittaker, E. W., Kernohan, G., Hasson, F., Howard, V., & McLaughlin, D. (2007). Palliative care in nursing homes; exploring care assistant's knowledge. *International Journal of Older People Nursing, 2*, 36–44. https://doi.org/10.1111/j.1748-3743.2007.00038.x

WHO. (2016). *WHO definition of palliative care*. Author. http://www.who.int/cancer/palliative/definition/en/

9

Supporting People with Intellectual Disability at End of Life: Moral Distress Among Staff Caregivers During COVID-19

Mary McCarron, Kathyan Kelly, Philip McCallion, Éilish Burke, Margaret Haigh, and Andrew Wormald

Introduction

Whether paid or unpaid, carers develop ethical and moral values about caring for people with intellectual disability. As people approach the end of their life, carers' decisions are more acute and challenges to carers' values and expectations may cause moral distress to the carer as they may feel unable to give quality care in a manner which is consistent with their

M. McCarron (✉) · K. Kelly · É. Burke · M. Haigh · A. Wormald
Trinity Centre for Ageing and Intellectual Disability, Trinity College Dublin, Dublin, Ireland
e-mail: mccarrm@tcd.ie

É. Burke
e-mail: eburke7@tcd.ie

P. McCallion
School of Social Work, Temple University, Philadelphia, PA, USA
e-mail: philip.mccallion@temple.edu

R. J. Stancliffe et al. (eds.), *End of Life and People with Intellectual and Developmental Disability*, https://doi.org/10.1007/978-3-030-98697-1_9

values and beliefs. Traditionally, in Ireland, carers would stay with the person they care for during hospital visits and at their death, ensuring the dying received the optimal care and that their expressed wishes were carried out. The COVID-19 pandemic and its related public health guidelines and restrictions, which included no visitors, full personal protective equipment (PPE), and social distancing, resulted in people often dying alone and without human touch. For carers, this unsupported and lonely situation has led to a conflict of values, where carers could not make a choice that both preserved their moral values and adhered to the public health guidelines. This conflict has been reported to lead to distress for staff and family carers that may have physical, psychological, and behavioural consequences. This chapter looks at what moral distress is among carers of people with intellectual disabilities; how it develops and manifests, and how it was overcome during the COVID-19 pandemic.

The first part of the chapter describes moral distress often drawing on research evidence from outside the field of intellectual disabilities. The second section utilises findings from the Intellectual Disability Supplement to The Irish Longitudinal Study on Ageing (IDS-TILDA) to better describe moral distress among staff caregivers of persons with intellectual disabilities who died and illustrates the issues and key points with personal stories about people with an intellectual disability.

The emergence of the COVID-19 pandemic has resulted in many carers providing care in situations akin to that provided in disaster scenarios where supplies are scarce, resources are triaged, and the needs of the individual are countermanded for the good of the many (Hossain & Clatty, 2021).

The concept of moral distress in healthcare delivery is not new, with the term first being coined in the 1980s to describe the adverse psychological effects experienced by nurses in care provision when faced with having to act in a manner contrary to their core belief system or moral code (Jameton, 1984). Originally defined through nursing, this concept has been expanded to include all health care professionals (Kälvemark et al., 2004) and unpaid informal carers (Ullrich et al., 2020; Weigel, 2019).

Caregivers in the emerging literature during COVID-19 have generally been allotted to one of three categories: carers who are providing care for those with COVID-19 (front-line), professional staff who provide health-related services to those not diagnosed with COVID-19 (non-front-line) (Li et al., 2020), and unpaid family members and other unpaid carers who may effectively be front-line in care but who are not staff (Willner et al., 2020).

Carers working with people with intellectual disability have experienced changes in care provision during the pandemic under circumstances not previously encountered. These changes include the wearing of PPE, the need for social distancing, the reduction in visitors, and the lack of respite care. Where people have been providing end-of-life care, there have been enforced changes in death and dying rituals such as one family member or no one sitting with the person in hospital in the last days and very limited attendance permitted at funerals that have impacted on the dying, their families and their carers (Courtenay, 2020; Mazza et al., 2020; Willner et al., 2020).

Moral Distress

Moral distress arises when there is a disconnect between how professional and personal judgement dictates what should be done and where healthcare systems impose limitations to act on those judgements. Moral distress can be current or retrospective, where on reflection people feel distress because of later changes in their own moral code or because changing circumstances have brought a renewed and negative focus on previous decisions (Crane et al., 2013). As such, it is associated with powerlessness (Sheather & Fidler, 2021). Moral distress is best described when there are difficulties navigating care within the ethical dimensions of practice, while maintaining personal and professional values, duties, and responsibilities (Epstein & Hamric, 2009; Pauly et al., 2009).

There is much debate about the most appropriate term to use to describe the outcome of this inner struggle, with much research referring to moral distress and more still referring to moral injury as a more appropriate term. It is generally accepted that given their day-to-day contact in the stressed last days, the term moral injury is more likely to be used for front-line hospital staff and represents a deeper, more durable psychological scarring that can be brought about by the repeated exposure to moral distress or a single appalling event (Kok et al., 2020; Williams et al., 2020).

First coined in the mid-1980s, moral distress was defined as the involvement of structural factors which constrain or disallow nurses from acting within their own moral code for what they viewed as optimal care, suggesting that the individual had no agency (Jameton, 1993). Subsequent research explored the issues from a personal perspective and a notion of self-agency or self-care (Kälvemark et al., 2004; Lützén et al., 2003; Webster & Baylis, 2000) which can act as a barrier to the more damaging impact of moral distress. Generally, though, it is accepted from the literature that moral distress occurs because of one or more of three factors—external influences, site-specific systems, and the personal traits of individuals (Burston & Tuckett, 2013).

The COVID-19 pandemic has highlighted these three interdependent factors—external influences (COVID-19 pandemic and related public health restrictions), site-specific issues (arising within organisational responses to adhere to public health guidelines) and individual or personal approaches to adhering to public health guidelines and individual experiences including length of time working with the individual with intellectual disability and the relationship that has developed. All three factors have the potential to result in moral distress for those supporting people with intellectual disability at end of life.

A further concern is that many health and social care systems across the world had been in crisis even prior to the pandemic, with consistent underfunding in healthcare creating ideal conditions for moral distress, placing healthcare workers in "unending ethical conflict" (Sheather & Fidler, 2021, p. 1). Equitable access to good quality health, including end-of-life and palliative care, is a human right, and the emergence of the pandemic then heightened challenges to providing high-quality end-of-life care (Lapid et al., 2020).

End-of-Life Care for People with Intellectual Disability

End-of-life care for people with intellectual disability can be particularly challenging due to communication challenges, high levels of multi-morbidity, and earlier mortality, often further confounded by a lack of advanced care planning and crisis decision-making. Difficulties can arise around diagnostic uncertainty, communicating end-of-life concerns, pain management, and the relief of other distressing symptoms (Bekkema et al., 2014; Stein & Kerwin, 2010; Tuffrey-Wijne et al., 2007; Vrijmoeth et al., 2016). Challenges and unmet needs of people with intellectual disability centre on where they die, personal desires and networks, medical care, and decision-making (McNamara et al., 2020). The bonds and relationships between the individual with intellectual disability and staff are built over years and possibly decades and are often very close and family like. The severing of these relationships at end of life can be a great source of distress for staff caregivers (McCarron et al., 2010; McCarron, Burke, et al., 2017; McCarron, Haigh, et al., 2017; Willems et al., 2010) (see also Chapter 16).

An additional dilemma is that often end-of-life care and support are not discussed until people are in an advanced stage of illness and decline. This lack of planning has led to calls for earlier advanced care planning for end-of-life care (McCallion et al., 2017). Early planning is particularly relevant for adults experiencing dementia, which is often present at much earlier ages for people with intellectual disability, especially those with Down syndrome, than is found in the general population (Hartley et al., 2015; McCarron, Burke, et al., 2017; McCarron, Haigh, et al., 2017) (see also Chapter 7).

The United Nations Convention on the Rights of Persons with Disability (UNCRPD) (UN Department of Economic & Social Affairs, 2008), as an international human rights treaty, specifically recognises that people with intellectual disability have a right to autonomy, and this extends to decision-making in all aspects of life. The UNCRPD challenges the notion of substitute decision-making—the formal (legal) or informal appointment of one person to make decisions on behalf of another who may be considered to lack decision-making capacity.

Instead, it urges that appropriate measures be taken to provide access to the support needed in exercising an individual's legal capacity. Supported decision-making has been recognised as working well in this cohort where there is relational closeness to the individual with intellectual disability and an awareness and inclusion of that person's life story (Watson et al., 2017) (see also Chapters 4 and 13). However, many with intellectual disabilities are denied their right to autonomy, driven by assumptions about compromised ability to make decisions (Shogren et al., 2017; Ward & Stewart, 2008). This is most true of those with profound intellectual and multiple disability, with common misperceptions around communication abilities becoming a perception of incompetence around decision-making (Watson et al., 2019).

During the COVID-19 pandemic, the ability to self-determine was further eroded for people with intellectual disability. At the time of writing, a UK review was examining the possibility that some care home residents may have died because of blanket do not attempt cardiopulmonary resuscitation orders amid concern over an increasingly burdened National Health System (NHS) (Amnesty International, 2020), leading to potentially avoidable deaths (see also Chapters 6 and 15).

In professions where people are caring for those who are ageing, there are set codes of conduct with agreed norms and values guiding the actions of those working in face-to-face care (Frennert, 2020). This is especially true in end-of-life care where best practice has an emphasis on family involvement, symptom control, and multidisciplinary team collaboration (Lapid et al., 2020; McCarron et al., 2018). The COVID-19 pandemic exacerbated the challenges to good end-of-life care for those in long-term care. With limitations being placed on entry to intensive care units, family support for the dying and bereavement support for families have been largely absent as a result of restrictions deemed necessary for the control of the virus (Lapid et al., 2020). This has also meant restrictions on access to long-standing paid carers.

Moral distress has not been confined to the nursing or medical professions but has been experienced across a variety of care contexts, including palliative carers, physicians, social workers, and therapists (Brazil et al., 2010; McMillan, 2020; Morley et al., 2020) and by families. Carers during COVID-19 have had to endure risk to personal safety and in many instances witness an overwhelming number of deaths, with patients isolated and dying alone (Hossain & Clatty, 2021; Morley et al., 2020; Sheather & Fidler, 2021). For carers of people with intellectual disabilities in particular, research carried out during strict lockdown in the UK found that the measures implemented to manage COVID-19 had a severe impact on the psychological health of carers (Willner et al., 2020). Findings of an Irish study following the first wave of the COVID-19 restrictions found that while rigid adherence to public health strategies was working in terms of protecting people with intellectual disability from the disease (and subsequently serious illness or death), the prolonged restriction was having a major impact on the mental health and well-being of many older people with intellectual disability (McCarron et al., 2020). Additional impact on the carers themselves is likely.

Moral Distress and Carers—Staff and Families

As a result of their work/caring roles, the psychological needs of carers of person with intellectual disabilities can differ greatly to those of the general population, as identified in a number of studies (Shechter et al., 2020; Spoorthy et al., 2020; Tan et al., 2020; Tomlin et al., 2020; Wu et al., 2020). During COVID-19, these psychological needs may have been exacerbated by several additional challenges. In the general population, a qualitative study explored the psychological distress of nurses caring for people with COVID-19 and found stresses over and above those in normal healthcare delivery. They included anxiety about death, the nature of the illness itself, and the emotional distress of delivering bad news. There was also evidence of a moral conflict between self-care and patient care, which created a duality between fear and conscience (Galehdar et al., 2020). The reduction or absence of visitation (the denial of individual access to usual support systems) meant that individuals were more than ever reliant on carers to provide emotional support, placing an additional burden on carers (Morley et al., 2020). Such effects are likely to be even more exaggerated in the field of intellectual disability where the duration of both family and staff caregiving extends over a long period of time.

Outside of hospital settings, many older people, especially those with dementia or with intellectual disabilities, are cared for by unpaid carers, often family, and the ability to access social care and support services is key to their well-being. The sudden closure of most day and respite services because of COVID-19 restrictions resulted in prolonged isolation for many unpaid carers, leading to significant negative impacts on their psychological health (Giebel et al., 2020). Family carers of adults and children with intellectual disabilities, bereft of usual social care supports as well as any potential respite, have reported significantly higher levels of defeat/entrapment, anxiety, and depression. In one study there was a five-fold increase in rates of severe anxiety and a four-to-ten-fold increase in rates of major depression in this population (Willner et al., 2020). They are likely therefore to enter end-of-life care situations with even less resilience and capacity to meet the changed circumstances of care and last days brought about by COVID-19.

For both families and staff, supporting death or dying requires high levels of emotional support. Grief has been defined both as normal (where grief is a common human emotion) and pathological or complicated grief with the potential to lead to psychological harm, such as post-traumatic stress disorder (PTSD), depression, and prolonged grief disorder (Lichtenthal et al., 2020; Lundorff et al., 2017). Incomplete grief is exacerbated during the COVID-19 pandemic from a sense of powerlessness in the loss of last farewells, mourning rituals and the loss of human contact or visitation near the point of death (Rabow et al., 2021), producing increased stress and moral distress.

Death and dying is individual and highly personal, with no universal agreed definition of what constitutes a good death. However, both in the general and intellectual disability literature attributes associated with the concept of dying well include presence of family, friends, and loved ones, communication and touch, freedom from pain, and religious, spiritual belief, and funeral ritual desires being met (Cithambarm et al., 2021; McCarron et al., 2010). During the COVID-19 pandemic, public health restrictions (no visiting, wearing masks and full PPE, two-metre distances, etc.) posed serious challenges to optimal end-of-life care for people with intellectual disability, with heightened stress for staff caregivers not able to meet these attributes associated with dying well. The related sense of powerlessness is likely a key driver of moral distress among staff caregivers.

Theoretical Parameters and Frameworks for the Concept of Moral Distress

Moral distress can be viewed as a natural response to the more morally difficult decisions made during COVID-19 care in general and in particular, end-of-life care. It can be viewed as an inner response to a perceived threat to an objective good (Burston & Tuckett, 2013), or because carers have to become directive in allocating resources and time, counter to their ethical values (Hossain & Clatty, 2021). The phenomenon can be categorised into specific types of distress (Morley et al., 2020). These include, but are not exclusive to:

- Moral constraint: where one is constrained from what they think is the ethically appropriate action
- Moral uncertainty: where one is uncertain if this is the right thing to do
- Moral dilemma: where one is unable to choose between two or more ethically supportable options
- Moral conflict: where there is conflict about the more appropriate ethical action
- Moral tension: where one feels one cannot share beliefs with others.

Barlem and Ramos (2015) have proposed a conceptual model of moral distress, that describes the process of moral distress caused by the lack of power to successfully carry out advocacy for a person in care (Fig. 9.1).

The Barlem and Ramos (2015) model argues that challenges to the carers' moral sensitivity causes carers to experience moral strangeness. The carer can adjust their moral sensitivity through a process of moral deliberation and advocacy. However, where there is an obstruction to the

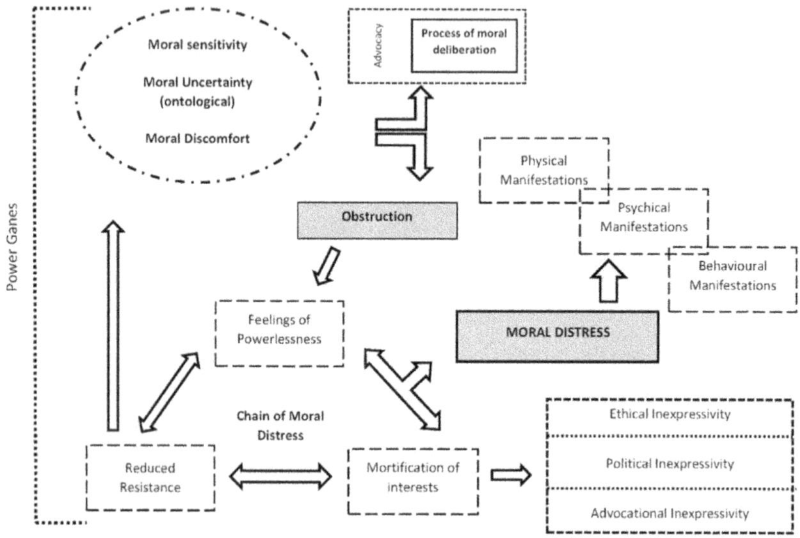

Fig. 9.1 The conceptual model of moral distress

deliberation brought about by feelings of powerlessness, the carer enters the chain of moral distress resulting in two paths:

- The carer does not consciously notice the obstruction. They reduce their resistance to the obstruction and stagnate, leading to behavioural and physical consequences.
- The carer identifies the moral distress, and despite the obstruction, perceives the moral value of the situation and can revert to a situation of moral strangeness and moral deliberation.

General powerlessness in the face of COVID-19 related restrictions is further heightened in end-of-life care given the high likelihood that the person cared for will die and will die soon. Welborn (2019) contended that when advocating for an individual, carers develop a moral sensitivity to the optimal care they should be providing. COVID-19 restrictions mean that carers are prevented from providing such care. In addition, Lutzen and Ewalds-Kvist (2013) argued that any situation where carers feel powerless in advocacy means they are subject to moral distress and lose meaning and resilience in their work.

Within the pandemic, healthcare staff around the world have reported feelings of powerlessness (Liu et al., 2020), particularly noted in community healthcare workers (Lotta et al., 2020) and among those caring for people with intellectual disabilities (Patel et al., 2021). The Barlem and Ramos (2015) model would suggest an increase in moral distress in staff at end of life caused by the COVID-19 related obstructions to care and powerlessness will lead to mortification of interests and reduced ethical, political, and advocacy resistance. Physical, behavioural, and psychic consequences result for the carer. At end of life the extent of powerlessness may rise to the level of moral distress, as evidenced in the stories outlined later in this chapter.

Escaping and Mitigating Moral Distress

Moral distress, as a natural response to morally difficult experiences, cannot be eradicated. Rather, its effects can be mitigated (Morley et al., 2020). According to Lutzen and Ewalds-Kvist (2013), escaping moral distress occurs through sharing feelings with colleagues and by respecting the person's and/or family's wishes. Such sharing and respect adds meaning to an event such as supporting someone in their last days, aids in moral deliberation and moral agency and assists carers escaping the chain of moral distress and re-adjusting their moral sensitivity. However, residual distress may be deeply troubling to the carer or intense, with knock-on effects in other morally challenging situations (Welborn, 2019). Governments and health systems need to minimise mental ill-health of carers by providing timely psychological counselling and educational supports (Galehdar et al., 2020).

Interventions

Organisational Interventions. Interventions focus on both individualistic and collaborative approaches within organisations (Burston & Tuckett, 2013). In the case of nurses, individual solutions lie with engagement in education and communication strategies, and nurses are encouraged to actively seek support or engage in personal growth as a self-improvement strategy. This is likely to be effective for other carers also. At a collaborative level, in intellectual disability services agencies, palliative care providers, and community hospitals, the focus should be on interdisciplinary education and dialogue and mentorship within a supportive organisational culture (Burston & Tuckett, 2013).

The research shows that carers in each of these organisations supporting people with intellectual disability wanted clear assurance that their organisation would offer support by listening to their concerns about COVID-19 and their ability to provide end-of-life care during the pandemic. Assurances centred around five distinct requests (Shanafelt et al., 2020):

1. Hear me: listen to and act on carers' perspectives
2. Protect me: reduce the risk of infection and subsequent transmission to family
3. Prepare me: provide training and support to enable high-quality care
4. Support me: acknowledge human limitations around work hours and family responsibilities
5. Care for me: provide holistic support if quarantine is required.

Addressing carers' concerns helps increase resilience and overcome distress, including moral distress, and increases the ability to address end-of-life care needs.

During the COVID-19 pandemic, the Health Service Executive in Ireland which supports intellectual disability services agencies, palliative care providers, and community hospitals, in common with health services in other countries, put in place accessible resources for staff, including information on staff protections, preparation for working during the pandemic, psychological support and how to keep healthy. Also, in Ireland, the All Ireland Institute of Hospice and Palliative Care (AIIHPC) collaborated with the Trinity Centre for Ageing with an Intellectual Disability to produce webinars and resource materials to adapt generic materials and provide more specific content to better support staff and the delivery of end-of-life care to people with intellectual disabilities during the pandemic (see Resources at the end of this chapter).

Individual Interventions. For those in professional care roles, self-care strategies are needed including self-stewardship, ethics education, and the development of moral resilience (Hossain & Clatty, 2021). A four-step framework for managing moral distress, first introduced in 2010 in the USA, focuses on the four "As" to Rise Above Moral Distress (McCue, 2010):

- Ask appropriate questions
- Affirm your distress and your commitment to self-care
- Assess/identify sources of distress
- Act or take action.

These ideas are also being incorporated in the supports for staff carers for persons with intellectual disabilities at end of life. The mental health of informal and family carers has not typically focussed on intellectual disability support services and therefore may not be tied into a referral route to mental health services, and the development of mental health workshops for family caregivers (Bourke-Taylor et al., 2021; Reid et al., 2016; Wei et al., 2012; Willner et al., 2020). Here too, the experiences with COVID-19 and the challenges of end-of-life care have been breaking down traditional separations between supporting staff and family caregivers.

What IDS-TILDA Data Tells Us About Moral Distress in Staff Caregivers

End-of-life interviews were conducted by researchers from the Intellectual Disability Supplement to The Irish Longitudinal Study on Ageing (IDS-TILDA) which is a longitudinal study researching ageing in Ireland among people with intellectual disability. This study is the first of its kind in Europe, and the only study able to directly compare the ageing of people with intellectual disability with the general ageing population (McCarron et al., 2020; McCarron, Burke, et al., 2017; McCarron, Haigh, et al., 2017). Across the four waves of data collection and encompassing the period 2009 to 2020, a total of 208 participants (27.6%) in the IDS-TILDA study had died. Carers of 137 deceased participants have so far supplied information about the end-of-life for those people, 102 have undertaken an in-depth interview and 8 of these were about deaths occurring during the COVID-19 pandemic. Seven of the eight pandemic-related interviews completed in 2020 were with staff who had worked with the deceased, and one was with a family member. Five of the deceased were female and three were male with ages ranging from 50 to 73 years (mean 63.75). Two of the deaths were due to COVID-19, one from cancer, two pneumonia, one dementia, and two were described as a general decline. Three of the deceased had been ill for over one year, three had been ill between one month and one year, and two had been ill for less than a week. On average the staff carers had known the deceased person for 8.4 years.

The interview findings suggested that end-of-life support for people with intellectual disabilities was severely disrupted during the COVID-19 pandemic, changing how end-of-life care was delivered during this period and exposing staff to situations that created moral distress. The stories below are an amalgam of what was reported by carers and do not represent any specific individual.

Michael's* story—Dying prior to the pandemic

Michael died in the months prior to the first COVID-19 lockdown. Michael was 63 and lived in a community home which included nursing support. He had lived nine years in the house and considered it his home. He had been very active until he was around 61 and then staff had noticed him "*slow down*". Michael had an end-of-life plan, which was updated annually. He previously discussed his desire to die at home, have a home wake, a large funeral and to be buried with his parents. Michael became ill in June and received a diagnosis of terminal cancer. A multi-disciplinary team meeting concluded that the additional care workload and lack of nursing expertise among staff was a source of concern and that a place for Michael should be sourced in a local nursing home. However, the staff who worked directly with Michael disagreed and requested that his wishes be met, despite this meaning additional work, and they came to an agreement with the service management to get additional nursing and other supports. Michael remained in his home supported by his carers and a community palliative care team called on Michael daily and were available any time by phone. Michael passed away seven months later surrounded by family, friends, and peers. Because this happened prior to the COVID-19 lockdowns, he had the traditional Irish wake in his home. Michael's funeral and burial met his wishes and was a large inclusive affair with attendees including family, friends, colleagues, and staff who had worked with Michael. Staff supported each other during this process and even though they all grieved the loss of Michael they all said they were pleased to be part of the effort that had brought Michael's wishes into reality. Staff and peers with intellectual disability arranged a month's memory service for Michael, which happened prior to the first lockdown, and continue to talk about him with his friends and each other. Michael and his passing are warmly remembered.

*Assumed names are used in this story.

Understanding Michael's story through Barlem and Ramos's (2015) conceptual model of moral distress, we can see that the carers demonstrated a developed moral sensitivity about providing the optimal care and responding to the dying person's wishes. When their moral sensitivity was challenged, they experienced moral discomfort. When they experienced moral discomfort, they advocated for the person they cared for and managed to remove the discomfort and did not enter the chain of moral distress, and therefore did not experience moral distress. Having successfully advocated for Michael, carers may in future situations adjust their moral code to take in these experiences.

Deaths During COVID-19 Pandemic Lockdown

Where people died during COVID-19 lockdowns, there was evidence that the restrictions and regulations imposed during COVID-19 caused obstructions, and carers' moral sensitivities were challenged leading to feelings of powerlessness and moral distress.

Jean's* story—Dying during the pandemic

Jean (72) had lived in an intellectual disability campus setting since she was 5 years old. Many of the carers who worked with Jean had known her for at least 20 years and considered her part of their family. Jean was also visited by her family members. While Jean experienced recurring chest infections, these were well-managed. During Jean's previous stays in hospital, staff had been assigned to stay with Jean around the clock. Jean did not have an end-of-life plan, but carers felt that if Jean were to become very ill she would want to die in her home and have the usual funeral rites.

Jean became sick in April 2020 and displayed signs of COVID-19 but only recorded a temperature of 37.5 degrees and was denied a test. Given COVID restrictions, Jean's GP and an out-of-hours doctor service would only provide phone assistance. Jean's symptoms worsened in the days following, but she could not be moved to her service provider's designated COVID-19 care unit without a positive test. Jean's carers became very concerned for her health and attempted to contact her GP and the out-of-hours doctor service. On failing to get either and not yet having a COVID-19 test result they called for an ambulance to hospitalise Jean.

Jean's carers felt torn hospitalising Jean as they knew they may never see her again and would be compromised in their ability to advocate for Jean's interest.

Jean was admitted to hospital, was diagnosed with COVID-19, and quickly assigned to palliative care. Carers phoned the hospital regularly to try and fight for Jean's interest in terms of her stated wishes, checking that staff understood how Jane communicated pain and other symptoms of distress, the supports she needed with eating and drinking, how important her rosary beads were to her, favourite music that helped her relax, and the likely distress for her in not seeing familiar staff. However, only her sister was allowed to visit. Jean died four days later. Jean's other family members had a video call with Jean in her last hours. Family members confided in staff that they were happy Jean went into hospital as they feared Jean may have been having another bout of pneumonia and if that were so, she was in the best place. Given restrictions, only Jean's immediate family attended the funeral, and staff carers were quarantined because of the confirmed COVID-19 and were not allowed to attend despite their wishes. Carers have held informal memory services for Jean with her friends and colleagues and have consoled each other on the phone. Carers feel disconnected from the reality of Jean's death and feel aggrieved at the way Jean was treated in her final days and now believe that even with the best efforts the system cannot be overcome.

*Assumed names are used in this story.

In Barlem and Ramos's (2015) terms, this story shows how differing moral sensitivities, in this case between family and staff carers and the intervention of other non-front-line carers, results in differing interpretations and actions and can lead to obstructions for each party. In this case the carers had developed moral sensitivities around how care such be conducted, including (a) testing Jean even though she did not fit into the criteria for a test, (b) GPs should do house calls to sick patients and (c) staff should support the people they are caring for when they are hospitalised. Initially, they failed to secure a test for Jean and despite their attempts at advocacy they were blocked. Once within the chain of moral distress the carers experienced moral constraint distress, when the system prevented them from getting the service they required for Jean, and moral conflict distress as they were conflicted by the decision to send Jean into the hospital. The outcome for the carers was stagnation of advocacy as they concluded the system could not be overcome and there was no forum to address their own post-death grief.

Family moral distress was also important in this story. The family wanted their sibling to survive and as Jean had a previous history of chest infections, had their own moral sensitivity about how best to care for their sister. Their view conflicted with the views of the staff and perhaps they felt the carers were the obstruction to their moral deliberations. This story suggests that by advocating for their sibling the family were able to maintain their moral sensitivity.

Reflection

1. In the absence of a COVID-19 test and no confirmed plan for end-of-life care whose decisions should have guided care for Jean?
2. How might a dialogue among staff carers and the family have been structured so that both may have felt respected in decision-making and better able to manage their moral sensitivities?
3. What aspects of COVID-19 responses aggravated the potential for moral distress for both staff carers and family members?
4. How might administrative, GP and other structures have been included in ways that would have reduced everyone's moral distress?

There have also been situations where long-term staff carers have been further excluded when the individual has been placed in a nursing home, thereby increasing distress. This is illustrated in John's story below.

John's* story—Dying during the pandemic
John had lived in a group home with a disability service provider for 15 years, and his only living relative was a cousin in Australia. He did not mix easily with other residents and would only trust familiar staff, who he considered his friends. John was diagnosed at age 51 years with

dementia and had been transferred to a dementia-specific nursing home in December 2019. Before John moved, his carers made sure John could continue to attend his regular day service and social activities. In March 2020 due to emergency COVID-19 restrictions, John's day service was closed and all his social activities suspended. Shortly thereafter a peer nursing home resident was diagnosed with COVID-19 and all nursing home residents were isolated in their rooms. John did not have the skills to use video or telephone calls without staff assistance and the nursing home did not have the staff available to assist John. Any staff John had become familiar with were quarantined and the nursing home relied on agency staff to care for residents. John was diagnosed with COVID-19 and died within a week of diagnosis. Carers from his prior home were informed by John's cousin about his death a day later. During lockdown John's former carers had repeatedly attempted to make direct contact with John and expressed their fears that nursing home staff were not trained to support with people with intellectual disabilities. Carers were upset at John's death and have expressed guilt at the decision to move John. Carers have declined the use of the employee assistance programme and have tried to support each other by phone.

*Assumed names are used in this story.

In John's story, carers experienced retrospective moral distress and regretted sending John to a nursing home. Consistent with Barlem and Ramos's (2015) model, the carers had laid out their moral code in the plan for John going forward in that John should receive an ongoing service that was as close to his familiar day routines as possible. It may be that John's carers were carrying moral residue from the decision to move John to a nursing home and were therefore more susceptible to experiencing moral distress. When the COVID-19 restrictions were put in place the carers felt powerless to see that John received the care he needed, and feelings of guilt about John's death remain. The carers also displayed evidence of moral uncertainty distress around the decisions they had made for John in the past. The carers in this story declined the use of the formal helplines and indeed in all the stories carers have preferred to use informal support networks, indicating an absence of valued forums to process grief.

Conclusion

Moral distress is a serious issue with behavioural and physical manifestations, potentially leading to a carer abandoning their ideals and failing to advocate for those in their care (Sheather & Fidler, 2021). During the COVID-19 crisis, paid and unpaid carers of people with intellectual disability have been subject to situations that have left them feeling powerless to provide optimal care and maintain duties and responsibilities within their ethical frameworks (Epstein & Hamric, 2009).

When carers experience moral distress, it threatens their professional values, patient care, and moral resilience (Bartone, 2009; Daubman et al., 2020; Morley & Horsburgh, 2021). Increases in moral distress and a lack of resilience in staff may mean carers lose their ideals and sense of meaning in their work. Carers experiencing moral distress may fail to advocate for people with intellectual disabilities in medical situations, be unsure of how to work to the standards they previously set for themselves, or become disconnected from the work and service. The COVID-19 pandemic with its restrictions on access, particularly when people with intellectual disability move to hospitals or other settings, has increased the potential for moral distress and reduced the effectiveness of prior coping and reframing strategies. Supportive workplaces provide training and support encourage self-help, open discussions, and increase resilience (Hines et al., 2021). Experiences in COVID-19 have highlighted that further work is needed in offering organisational/structural supports and interventions to augment previous approaches to managing such stress.

In all three stories there was family, with varying levels of involvement in the prior and current life of the person with intellectual disability. Family presence is not always typical for people with intellectual disabilities, particularly among those who are older or who moved out-of-family home care early in life. In every story the hospitals, nursing homes and GPs all accorded the family primary position in being informed and in some cases made decisions which did not reflect or consider the often long history of care provided by staff, the family-like relationships they had established with the individual and the level of knowledge they had of the person's wishes. An absence of connectedness between family and staff carers on these issues and struggles around who should lead decision-making is a recurrent feature of end-of-life care for people with intellectual disabilities and often reflects the way moral distress is experienced differently by families and staff. The additional restrictions present during the COVID-19 pandemic further isolated both families and the staff carers when transfers in care occurred. The pandemic experience has also highlighted that in restricted access situations, the needs and roles of staff carers are paid even less attention, exacerbating the moral distress staff experience when a death was not as they believe the person wished, or that care role preferences were not followed.

Reflection

1. Think about your own moral sensitivities around supporting a person at the end-of-life. What kind of obstructions could you encounter to their fulfilment? How would you overcome these obstructions?
2. If you have supported a person with intellectual disability in their last days, did you find any obstructions to them receiving their wishes or receiving optimal care? Did you advocate for them to help them receive this care?
3. Have you ever experienced moral distress? What kind of supports did you use to assist you in overcoming that distress? If you have not experienced moral distress what supports are available to you if you were to encounter a morally stressful situation?

Resources

1. *Family Carers' Experiences of Caring for a Person with Intellectual Disability.* Report by the National Disability Authority, Ireland. Looks at coping distress and caring in family carers of people with an intellectual disability available at: http://nda.ie/nda-files/family-carers%E2%80%99-experiences-of-caring-for-a-person-with-intellectual-disability.pdf

2. *Glancing Back, Planning Forward: Facilitating end of life conversations with persons with an intellectual disability: A guide for carers.* Findings from the end of life interviews of the Intellectual Disability Supplement to the Irish Longitudinal Study on Ageing (IDS-TILDA). http://www.professionalpalliativehub.com/sites/default/files/Carers%20Guide.pdf
3. *Health Services Executive, Ireland.* A range of accessible resources for care staff, including information on staff protections, preparation for working during the pandemic, psychological support and how to keep healthy. https://healthservice.hse.ie/staff/coronavirus/?source=banner-hse-staff
4. *Moral Distress in the Health Professions*—Book edited by Connie Ulrich and Christine Grady. This book addresses moral distress and dealing with moral distress in the workplace. ISBN 978-3-319-64626-8
5. *Moral Resilience: Transforming moral suffering in healthcare*—Book edited by Cynda Hylton Rushton. This book looks at how moral resilience can be cultivated in individuals and argues for a new architecture to support ethical practice. https://doi.org/10.1093/med/9780190619268.001.0001
6. *Trinity Centre for Ageing and Intellectual Disability (TCAID)*—Information and reports about the ageing process for people with an intellectual disability https://www.tcd.ie/tcaid/
7. *Trinity Centre for Ageing and Intellectual Disability (TCAID)*—How to support people with an intellectual disability with a diagnosis of COVID-19. https://www.tcd.ie/tcaid/about/webinars.php

References

AIIHPC—All Ireland Institute of Hospice and Palliative Care. https://aiihpc.org

Amnesty International. (2020). *As if expendable: The UK government's failure to protect older people in care homes during the Covid-19 pandemic*. https://www.amnesty.org.uk/files/2020-10/Care%20Homes%20Report.pdf?kd5Z8eWzj8Q6ryzHkcaUnxfCtqe5Ddg6=

Barlem, E. L., & Ramos, F. R. (2015). Constructing a theoretical model of moral distress. *Nursing Ethics, 22*(5), 608–615. https://doi.org/10.1177/0969733014551595

Bartone, P. T. (2009). Resilience under military operational stress: Can leaders influence hardiness? *Military Psychology, 18*(sup1), S131–S148. https://doi.org/10.1207/s15327876mp1803s_10

Bekkema, N., de Veer, A. J. E., Albers, G., Hertogh, Cees M. P. M., Onwuteaka-Philipsen, B. D., & Francke, A. L. (2014). Training needs of nurses and social workers in the end-of-life care for people with intellectual disabilities: A national survey. *Nurse Education Today, 34*(4), 494–500. https://doi.org/10.1016/j.nedt.2013.07.018

Bourke-Taylor, H. M., Joyce, K. S., Grzegorczyn, S., & Tirlea, L. (2021). Mental health and health behaviour changes for mothers of children with a disability: Effectiveness of a health and wellbeing workshop. *Journal of Autism and Developmental Disorders*. https://doi.org/10.1007/s10803-021-04956-3

Brazil, K., Kassalainen, S., Ploeg, J., & Marshall, D. (2010). Moral distress experienced by health care professionals who provide home-based palliative care. *Social Science & Medicine, 71*(9), 1687–1691. https://doi.org/10.1016/j.socscimed.2010.07.032

Burston, A. S., & Tuckett, A. G. (2013). Moral distress in nursing: Contributing factors, outcomes and interventions. *Nursing Ethics, 20*(3), 312–324. https://doi.org/10.1177%2F0969733012462049

Cithambarm, K., Duffy, M., & Courtney, E. (2021). What constitutes good quality end-of-life care? Perspectives of people with intellectual disabilities and their families. *Journal of Policy and Practice in Intellectual Disabilities*. https://doi.org/10.1111/jppi.12376

Courtenay, K. (2020). Covid-19: Challenges for people with intellectual disability. *BMJ, 369*, m1609. https://doi.org/10.1136/bmj.m1609

Crane, M. F., Bayl-Smith, P., & Cartmill, J. (2013). A recommendation for expanding the definition of moral distress experienced in the workplace. *Australasian Journal of Organisational Psychology, 6*. https://doi.org/10.1017/orp.2013.1

Daubman, B. R., Black, L., & Goodman, A. (2020). Recognizing moral distress in the COVID-19 pandemic: Lessons from global disaster response.

Journal of Hospital Medicine, 15(11), 696–698. https://doi.org/10.12788/jhm.3499

Epstein, E. G., & Hamric, A. B. (2009). Moral distress, moral residue, and the crescendo effect. *The Journal of Clinical Ethics, 20*(4), 330–342. PMID: 20120853.

Frennert, S. (2020). Moral distress and ethical decision-making of eldercare professionals involved in digital service transformation. *Disability and Rehabilitation: Assistive Technology*, 1–10. https://doi.org/10.1080/17483107.2020.1839579

Galehdar, N., Kamran, A., Toulabi, T., & Heydari, H. (2020). Exploring nurses' experiences of psychological distress during care of patients with COVID-19: A qualitative study. *BMC Psychiatry, 20*(1), 1–9. https://doi.org/10.1186/s12888-020-02898-1

Giebel, C., Cannon, J., Hanna, K., Butchard, S., Eley, R., Gaughan, A., Komuravelli, A., Shenton, J., Callaghan, S., & Tetlow, H. (2020). Impact of COVID-19 related social support service closures on people with dementia and unpaid carers: A qualitative study. *Aging & Mental Health*, 1–8. https://doi.org/10.1080/13607863.2020.1822292

Hartley, D., Blumenthal, T., Carrillo, M., DiPaolo, G., Esralew, L., Gardiner, K., Granholm, A.-C., Iqbal, K., Krams, M., & Lemere, C. (2015). Down syndrome and Alzheimer's disease: Common pathways, common goals. *Alzheimer's & Dementia, 11*(6), 700–709. https://doi.org/10.1016/j.jalz.2014.10.007

Hines, S. E., Chin, K. H., Glick, D. R., & Wickwire, E. M. (2021). Trends in moral injury, distress, and resilience factors among healthcare workers at the beginning of the COVID-19 pandemic. *International Journal of Environmental Research and Public Health, 18*(2). https://doi.org/10.3390/ijerph18020488

Hossain, F., & Clatty, A. (2021). Self-care strategies in response to nurses' moral injury during COVID-19 pandemic. *Nursing Ethics, 28*(1), 23–32. https://doi.org/10.1177/2F0969733020961825

Intellectual Disability Supplement to the Irish Longitudinal Study on Ageing (IDS-TILDA). https://www.idstilda.tcd.ie

Jameton, A. (1984). *Nursing practice: The ethical issues*. Prentice-Hall. http://hdl.handle.net/10822/800986

Jameton, A. (1993). Dilemmas of moral distress: Moral responsibility and nursing practice. *Awhonn's Clinical Issues in Perinatal and Women's Health Nursing, 4*(4), 542–551. PMID: 8220368.

Kälvemark, S., Höglund, A. T., Hansson, M. G., Westerholm, P., & Arnetz, B. (2004). Living with conflicts-ethical dilemmas and moral distress in the health care system. *Social Science & Medicine, 58*(6), 1075–1084. https://doi.org/10.1016/s0277-9536(03)00279-x

Kok, N., Hoedemaekers, A., van der Hoeven, H., Zegers, M., & van Gurp, J. (2020). Recognizing and supporting morally injured ICU professionals during the COVID-19 pandemic. *Intensive Care Medicine, 46*, 1653–1654. https://doi.org/10.1007/s00134-020-06121-3

Lapid, M. I., Koopmans, R., Sampson, E. L., Van den Block, L., & Peisah, C. (2020). Providing quality end-of-life care to older people in the era of COVID-19: Perspectives from five countries. *International Psychogeriatrics, 32*(11), 1345–1352. https://doi.org/10.1017/S1041610220000836

Li, Z., Ge, J., Yang, M., Feng, J., Qiao, M., Jiang, R., Bi, J., Zhan, G., Xu, X., Wang, L., Zhou, Q., Zhou, C., Pan, Y., Liu, S., Zhang, H., Yang, J., Zhu, B., Hu, Y., Hashimoto, K., … Yang, C. (2020). Vicarious traumatization in the general public, members, and non-members of medical teams aiding in COVID-19 control. *Brain, Behavior, and Immunity, 88*, 916–919. https://doi.org/10.1016/j.bbi.2020.03.007

Lichtenthal, W. G., Roberts, K. E., & Prigerson, H. G. (2020). Bereavement care in the wake of COVID-19: Offering condolences and referrals. *Anals of Internal Medicine, 173*(10), 833–835. https://doi.org/10.7326/M20-2526

Liu, Q., Luo, D., Haase, J. E., Guo, Q., Wang, X. Q., Liu, S., Xia, L., Liu, Z., Yang, J., & Yang, B. X. (2020). The experiences of health-care providers during the COVID-19 crisis in China: A qualitative study. *The Lancet Global Health, 8*(6), e790–e798. https://doi.org/10.1016/s2214-109x(20)30204-7

Lotta, G., Coelho, V. S. P., & Brage, E. (2020). How COVID-19 has affected frontline workers in Brazil: A comparative analysis of nurses and community health workers. *Journal of Comparative Policy Analysis: Research and Practice, 23*(1), 63–73. https://doi.org/10.1080/13876988.2020.1834857

Lundorff, M., Holmgren, H., Zachariae, R., Farver-Vestergaard, I., & O'Connor, M. (2017). Prevalence of prolonged grief disorder in adult bereavement: A systematic review and meta-analysis. *Journal of Affective Disorders, 212*, 138–149. https://doi.org/10.1016/j.jad.2017.01.030

Lützén, K., Cronqvist, A., Magnusson, A., & Andersson, L. (2003). Moral stress: Synthesis of a concept. *Nursing Ethics, 10*(3), 312–322. https://doi.org/10.1191%2F0969733003ne608oa

Lutzen, K., & Ewalds-Kvist, B. (2013). Moral distress and its interconnection with moral sensitivity and moral resilience: Viewed from the philosophy

of Viktor E Frankl. *Bioethical Inquiry, 10*(3), 317–324. https://doi.org/10.1007/s11673-013-9469-0
Mazza, M., Attanasio, M., Pino, M. C., Masedu, F., Tiberti, S., Sarlo, M., & Valenti, M. (2020). Moral decision-making, stress, and social cognition in frontline workers vs. population groups during the COVID-19 pandemic: An explorative study. *Frontiers in Psychology, 11*, 588159. https://doi.org/10.3389/fpsyg.2020.588159
McCallion, P., Hogan, M., Santos, F. H., McCarron, M., Service, K., Stemp, S., Keller, S., Fortea, J., Bishop, K., & Watchman, K. (2017). Consensus statement of the International Summit on Intellectual Disability and Dementia related to end-of-life care in advanced dementia. *Journal of Applied Research in Intellectual Disabilities, 30*(6), 1160–1164. https://doi.org/10.1111/jar.12349
McCarron, M., Burke, E. M., White, P., O'Dwyer, C., Ryan, K., O Farrell, J., & McCallion, P. (2017). *"He'd mind you, you mind him", Experiences of end of life care for people with an intellectual disability as perceived by staff carers: Findings from the End of Life Interviews of the Intellectual Disability Supplement to the Irish Longitudinal Study on Ageing (IDS-TILDA).* https://www.tcd.ie/tcaid/research/endoflifepublications/ReportExperiencesofendoflife.pdf
McCarron, M., Haigh, M., McCallion, P., McCallion, P., Carroll, R., Burke, E., McGlinchey, E., O'Donovan, M. A., McCausland, D., Sheerin, F., & O'Dwyer, M. (2017). *Health, wellbeing and social inclusion: Ageing with an intellectual disability in Ireland. Evidence from the first ten years of The Intellectual Disability Supplement to The Irish Longitudinal Study on Ageing (IDS-TILDA).* https://www.tcd.ie/tcaid/assets/pdf/wave3report.pdf
McCarron, M., McCallion, P., Coppus, A., Fortea, J., Stemp, S., Janicki, M., & Wtachman, K. (2018). Supporting advanced dementia in people with Down syndrome and other intellectual disability: Consensus statement of the International Summit on Intellectual Disability and Dementia. *Journal of Intellectual Disability Research, 62*(7), 617–624. https://doi.org/10.1111/jir.12500
McCarron, M., McCallion, P., Fahey-McCarthy, E., & Connaire, K. (2010). Staff perceptions of essential prerequisites underpinning end-of-life care for persons with intellectual disability and advanced dementia. *Journal of Policy and Practice in Intellectual Disabilities, 7*(2), 143–152. https://doi.org/10.1111/j.1741-1130.2010.00257.x
McCarron, M., McCausland, D., Allen, A. P., Luus, R., Sheerin, F., Burke, E., McGlinchey, E., Flannery, F., & McCallion, P. (2020). *Understanding the*

impact of COVID-19 on the health and well-being of older adults with an intellectual disability in Ireland: Findings from Wave 4 of The Intellectual Disability Supplement to The Irish Longitudinal Study on Ageing (IDS-TILDA). https://idstilda.tcd.ie/assets/docs/wave4idstildareport.pdf

McCue, C. (2010). Using the AACN framework to alleviate moral distress. *Online Journal of Issues in Nursing, 16*(1), C1. https://doi.org/10.3912/OJIN.Vol16No01PPT02

McMillan, N. (2020). Moral distress in residential child care. *Ethics and Social Welfare, 14*(1), 52–64. https://doi.org/10.1080/17496535.2019.1709878

McNamara, B., Same, A., & Rosenwax, L. (2020). Creating person-centred support for people with intellectual disabilities at the end of life: An Australian qualitative study of unmet needs and strategies. *Journal of Intellectual Disabilities, 24*(4), 543–558. https://doi.org/10.1177%2F1744629518823887

Morley, G., & Horsburgh, C. C. (2021). Reflective debriefs as a response to moral distress: Two case study examples. *HEC Forum.* https://doi.org/10.1007/s10730-021-09441-z

Morley, G., Sese, D., Rajendram, P., & Horsburgh, C. C. (2020). Addressing caregiver moral distress during the COVID-19 pandemic. *Cleveland Clinic Journal of Medicine.* https://doi.org/10.3949/ccjm.87a.ccc047

Patel, V., Perez-Olivas, G., Kroese, B. S., Rogers, G., Rose, J., Murphy, G., Cooper, V., Langdon, P. E., Hiles, S., Clifford, C., & Willner, P. (2021). The experiences of carers of adults with intellectual disabilities during the first COVID-19 lockdown period. *Journal of Policy and Practice in Intellectual Disabilities.* https://doi.org/10.1111/jppi.12382

Pauly, B., Varcoe, C., Storch, J., & Newton, L. (2009). Registered nurses' perceptions of moral distress and ethical climate. *Nursing Ethics, 16*(5), 561–573. https://doi.org/10.1177%2F0969733009106649

Rabow, M. W., Huang, C.-H. S., White-Hammond, G. E., & Tucker, R. O. (2021). Witnesses and victims both: Healthcare workers and grief in the time of COVID-19. *Journal of Pain and Symptom Management.* https://doi.org/10.1016/j.jpainsymman.2021.01.139

Reid, C., Gill, F., Gore, N., & Brady, S. (2016). New ways of seeing and being: Evaluating an acceptance and mindfulness group for parents of young people with intellectual disabilities who display challenging behaviour. *Journal of Intellectual Disabilities, 20*(1), 5–17. https://doi.org/10.1177/1744629515584868

Shanafelt, T., Ripp, J., & Trockel, M. (2020). Understanding and addressing sources of anxiety among health care professionals during the COVID-19 pandemic. *Jama, 323*(21), 2133–2134. https://doi.org/10.1001/jama.2020.5893

Sheather, J., & Fidler, H. (2021). Covid-19 has amplified moral distress in medicine. *BMJ, 372*, n28. https://doi.org/10.1136/bmj.n28

Shechter, A., Diaz, F., Moise, N., Anstey, D. E., Ye, S., Agarwal, S., Birk, J. L., Brodie, D., Cannone, D. E., & Chang, B. (2020). Psychological distress, coping behaviors, and preferences for support among New York healthcare workers during the COVID-19 pandemic. *General Hospital Psychiatry, 66*, 1–8. https://doi.org/10.1016/j.genhosppsych.2020.06.007

Shogren, K. A., Wehmeyer, M. L., Uyanik, H., & Heidrich, M. (2017). Development of the supported decision making inventory system. *Intellectual and Developmental Disabilities, 55*(6), 432–439. https://doi.org/10.1352/1934-9556-55.6.432

Spoorthy, M. S., Pratapa, S. K., & Mahant, S. (2020). Mental health problems faced by healthcare workers due to the COVID-19 pandemic—A review. *Asian Journal of Psychiatry, 51*, 102119. https://doi.org/10.1016/j.ajp.2020.102119

Stein, G. L., & Kerwin, J. (2010). Disability perspectives on health care planning and decision-making. *Journal of Palliative Medicine, 13*(9), 1059–1064. https://doi.org/10.1089/jpm.2010.0159

Tan, B. Y., Chew, N. W., Lee, G. K., Jing, M., Goh, Y., Yeo, L. L., Zhang, K., Chin, H.-K., Ahmad, A., & Khan, F. A. (2020). Psychological impact of the COVID-19 pandemic on health care workers in Singapore. *Annals of Internal Medicine, 173*(4), 317–320. https://doi.org/10.7326/M20-1083

Tomlin, J., Dalgleish-Warburton, B., & Lamph, G. (2020). Psychosocial support for healthcare workers during the COVID-19 pandemic. *Frontiers in Psychology, 11*, 1960. https://doi.org/10.3389/fpsyg.2020.01960

Trinity Centre for Ageing and Intellectual Disability. Trinity College University of Dublin. https://www.tcd.ie/tcaid/

Tuffrey-Wijne, I., Bernal, J., Butler, G., Hollins, S., & Curfs, L. (2007). Using nominal group technique to investigate the views of people with intellectual disabilities on end-of-life care provision. *Journal of Advanced Nursing, 58*(1), 80–89. https://doi.org/10.1111/j.1365-2648.2007.04227.x

Ullrich, A., Theochari, M., Bergelt, C., Marx, G., Woellert, K., Bokemeyer, C., & Oechsle, K. (2020). Ethical challenges in family caregivers of patients with advanced cancer—A qualitative study. *BMC Palliative Care, 19*(1), 70. https://doi.org/10.1186/s12904-020-00573-6

UN Department of Economic and Social Affairs. (2008). *The UN Convention on the rights of persons with Disabilities*. https://www.un.org/development/desa/disabilities/convention-on-the-rights-of-persons-with-disabilities.html

Vrijmoeth, C., Barten, P., Assendelft, W. J., Christians, M. G., Festen, D. A., Tonino, M., Vissers, K. C., & Groot, M. (2016). Physicians' identification of the need for palliative care in people with intellectual disabilities. *Research in Developmental Disabilities, 59*, 55–64. https://doi.org/10.1016/j.ridd.2016.07.008

Ward, T., & Stewart, C. (2008). Putting human rights into practice with people with an intellectual disability. *Journal of Developmental and Physical Disabilities, 20*(3), 297–311. https://doi.org/10.1007/s10882-008-9098-4

Watson, J., Voss, H., & Bloomer, M. J. (2019). Placing the preferences of people with profound intellectual and multiple disabilities at the center of end-of-life decision making through storytelling. *Research and Practice for Persons with Severe Disabilities, 44*(4), 267–279. https://doi.org/10.1177%2F1540796919879701

Watson, J., Wilson, E., & Hagiliassis, N. (2017). Supporting end of life decision making: Case studies of relational closeness in supported decision making for people with severe or profound intellectual disability. *Journal of Applied Research in Intellectual Disabilities, 30*(6), 1022–1034. https://doi.org/10.1111/jar.12393

Webster, G. C., & Baylis, F. E. (2000). Moral residue. In S. B. Rubin & L. Zoloth (Eds.), *Margin of Error: The Ethics of Mistakes in the Practice of Medicine* (pp. 217–230). University Publishing Group. http://hdl.handle.net/10822/931017

Wei, Y. S., Chu, H., Chen, C. H., Hsueh, Y. J., Chang, Y. S., Chang, L. I., & Chou, K. R. (2012). Support groups for caregivers of intellectually disabled family members: Effects on physical-psychological health and social support. *Journal of Clinical Nursing, 21*(11–12), 1666–1677. https://doi.org/10.1111/j.1365-2702.2011.04006.x

Weigel, C. (2019). Caregiving and moral distress for family caregivers during early-stage Alzheimer's Disease. *International Journal of Feminist Approaches to Bioethics, 12*(2), 74–91. https://doi.org/10.3138/ijfab.12.2.05

Welborn, A. (2019). Moral distress of nurses surrounding neonatal abstinence syndrome: Application of a theoretical framework. *Nursing Forum, 54*(4), 499–504. https://doi.org/10.1111/nuf.12362

Willems, A. P., Embregts, P. J., Stams, G. J., & Moonen, X. M. (2010). The relation between intrapersonal and interpersonal staff behaviour towards

clients with ID and challenging behaviour: A validation study of the Staff-Client Interactive Behaviour Inventory. *Journal of Intellectual Disability Research, 54*(1), 40–51. https://doi.org/10.1111/j.1365-2788.2009.01226.x
Williams, R. D., Brundage, J. A., & Williams, E. B. (2020). Moral injury in times of COVID-19. *Journal of Health Service Psychology, 46*(2), 65–69. https://doi.org/10.1007/s42843-020-00011-4
Willner, P., Rose, J., Stenfert Kroese, B., Murphy, G. H., Langdon, P. E., Clifford, C., Hutchings, H., Watkins, A., Hiles, S., & Cooper, V. (2020). Effect of the COVID-19 pandemic on the mental health of carers of people with intellectual disabilities. *Journal of Applied Research in Intellectual Disabilities, 33*(6), 1523–1533. https://doi.org/10.1111/jar.12811
Wu, P. E., Styra, R., & Gold, W. L. (2020). Mitigating the psychological effects of COVID-19 on health care workers. *Canadian Medical Association Journal, 192*(17), E459–E460. https://doi.org/10.1503/cmaj.200519

10

Accessible Funerals and People with Intellectual Disability

Rachel Forrester-Jones, Philip McCallion, Mary McCarron, Roger J. Stancliffe, and Michele Y. Wiese

Contemporary Approaches to Funerals

Funerals are an enduring ritual, meeting social, religious/spiritual and emotional needs. O'Rourke and et al. (2011, p. 731), noted that "The term funeral refers to a ceremony in which a body or remains are present, in contrast to a memorial service, where the body or remains are generally absent". Although this chapter is predominantly about funerals, we

R. Forrester-Jones (✉)
Western University, London, ON, Canada
e-mail: rforre@uwo.ca

Department of Social and Policy Sciences, University of Bath, Bath, UK

P. McCallion
Temple University, Philadelphia, PA, USA

M. McCarron
Trinity Centre for Ageing and Intellectual Disability, Trinity College Dublin, Dublin, Ireland

R. J. Stancliffe et al. (eds.), *End of Life and People with Intellectual and Developmental Disability*, https://doi.org/10.1007/978-3-030-98697-1_10

also consider other Western funeral-related rituals such as wakes. We also focus on recent trends towards personalisation and participation at funerals.

Brief History of Western Funeral Practices

Funeral preparations, such as in the U.S. until the mid-nineteenth century, were a family responsibility, following death at home, with the funeral held in a place of worship and presided over by a religious leader (Garces-Foley & Holcomb, 2006). The twentieth century saw the rise of the funeral industry, higher funeral costs and a marked growth in dying in hospitals or nursing homes. Increasingly, funeral directors took control of the body, arranged the ceremony and saw to legal requirements. Held in a place of worship or funeral-home chapel, funerals were typically religiously focussed, with only a brief eulogy (Garces-Foley & Holcomb, 2006).

From the late twentieth century, some Americans (not all) became dissatisfied with what they saw as predictable, impersonal, clergy-led traditional funerals. They sought a more personalised, participatory and authentic way of saying goodbye and adapted the traditional funeral with a strong focus on the eulogy, intended to tell the deceased's life story. Garces-Foley and Holcomb (2006) noted that the funeral industry has accommodated these changing preferences.

Elsewhere, Holloway et al. (2013) studied 46 funerals in northern England and concluded that while funerals continue to be significant events, here too there is a trend towards "personalisation, secularisation, consumer choice, and individual stories" (p. 30). They found that many contemporary funerals are personally tailored celebrations of the deceased's life with active participation by mourners, through delivering

R. J. Stancliffe
University of Sydney, Sydney, NSW, Australia

M. Y. Wiese
Western Sydney University, Penrith, NSW, Australia

a tribute, presenting a reading, approaching and touching the coffin, lighting a candle, playing music or scattering items at the graveside. Whether religious or secular, all funerals included a eulogy that celebrated the life and told the story of the deceased. Appropriate dress was important to show respect and sometimes included symbolic items, for example, a colour theme reflecting the dead person's preference or a symbol of a favourite sporting team.

In many Western countries, non-religious civil celebrants may officiate at funerals. These ceremonies are often jointly constructed by the family and celebrant to have personal meaning. The deceased's known/perceived wishes are central—"the family took great satisfaction from feeling that the deceased would have liked it" (Holloway et al., 2013, p. 43). In Ireland, the wake, held in the days before the funeral, has been a major tradition around death though its popularity has been waning over the past century (Lysaght, 2003). Irish wakes and funerals are discussed further in *Funeral Rituals in Ireland* below.

Burial and Cremation

In Western countries, the usual options for respectful disposal of the body are burial or cremation, although cremation is not acceptable to all religions (e.g. Islam). The Cremation Society (n.d.) reported that in the UK in 2017, 77.2% of deaths involved cremation, representing a dramatic rise from 0.3% (1920) to 34.7% (1960), and 71.5% (2000). This large change is partly because cremation costs less than burial. While widespread, there are differences in cremation rates between Western countries and variability within countries—in the UK, rates have been persistently lower in Scotland (68.2% in 2017) and Northern Ireland (21.0% in 2017).

Online Options

Contemporary options include attending a funeral via online streaming, online tribute sites to which mourners can add photos or reminiscences, and stored video of the ceremony (Garces-Foley & Holcomb,

2006). These developments allow participation by mourners unable to attend the ceremony in person. During the COVID-19 pandemic, there have been severe restrictions in many countries on numbers attending funerals and being at the graveside. Travel restrictions and self-isolation have further limited mourners' ability to join funerals in person. Consequently, use of online streaming has increased to try to compensate. However, these restrictions have been experienced as onerous by many (Selman et al., 2021).

Finding Meaning

Followers of a religion often have a prescribed way of conducting a funeral whereas non-religious people may accept varied funeral options. Meaning-making is central to funerals, be they religious or secular: "active participation by all actors in the co-creation and enacting of the funeral as a meaningful event … was at the heart of the funerals we studied" (Holloway et al., 2013, p. 50). Likewise, in the U.S., Garces-Foley and Holcomb (2006) characterised personalised funerals as reflecting the wish for meaning through authentic, personal ceremonies, although they emphasised that many Americans find meaning in traditional religious funerals. O'Rourke et al. (2011) found that religiosity affects the form and meaning of U.S. funerals.

Finding meaning in a funeral is commonly reflected through participation. Activities might include helping to organise and make the arrangements, giving a eulogy, singing, presenting a reading or simply attending the ceremony. O'Rourke et al. (2011, p. 746), described participation in "sharing stories, expressing condolences, crying, and otherwise interacting with others at the funeral", interpreted as "the social negotiation of emotion, the confirmation of one another's emotions …the most important feature of funeral satisfaction". O'Rourke et al. also concluded that funerals are social events where participants give and receive social support. Emotional expressions of loss and grief are common—crying in public is acceptable at a funeral. This collective mourning is thought to facilitate acceptance of the loss. Funerals are also occasions for social relationships among mourners to be renewed.

As noted, COVID-19 restrictions have greatly limited in-person attendance at funerals. In addition, practising social distancing at the ceremony has meant that physical expressions of condolence and emotional support have not been possible. These restrictions may well have reduced opportunities for all mourners to find meaning in the person's death. Sadly, this situation has reflected the exclusionary funeral experiences of many people with intellectual and developmental disability long before the pandemic.

Online streaming of funerals has provided a workaround for people to participate remotely, but research is yet to determine the extent to which collective mourning is truly enabled in an online environment. In response to these unknowns, some funeral directors recommend that families hold a small private funeral, with a memorial service later when mourners could attend in person (Reardon, 2020). Alternative in-person practices have developed, such as mourners lining the road for the hearse's journey to the burial site, while maintaining sufficient distance from others.

Implications for People with Intellectual Disability

For accuracy, throughout the rest of this chapter, we use the term *intellectual disability* rather than the more inclusive term *intellectual and developmental disability*, because all the disability research on funerals we have cited only involves people described as having intellectual disability. Indeed, we were unable to locate any peer-reviewed studies of funeral attendance by people with autism spectrum disorder or cerebral palsy and without co-occurring intellectual disability.

Two elements regarding research on contemporary funerals have important implications for people with intellectual disability. The more personalised and participatory aspects of contemporary funerals open the way for accommodations to enable meaningful funeral participation. Even more fundamentally, O'Rourke et al.'s (2011) finding that funeral satisfaction is strongly influenced by shared social and emotional participation with other mourners indicates that in-person attendance, participation in, and emotional expression at funerals are important for all mourners, including people with intellectual disability. As one man with intellectual disability said while discussing how to deal with death and grief, "If you share it with somebody it makes it a whole lot lighter than trying to deal with it yourself" (McRitchie et al., 2014, p. 181). Active participation at funerals does not necessarily require verbal skill or literacy, meaning greater accessibility for people with intellectual disability.

For participants with intellectual disability, cognitive, emotional, social, and spiritual accessibility are important for meaning making at all funeral stages: prior to the funeral, at the funeral, and subsequent remembrance activities. Celebrants who work to personalise the funeral to the preferences of the deceased and the family have a role to play in additional tailoring to meet the needs of people with intellectual disability.

Participation in Funerals by People with Intellectual Disability

In the general community, there is consensus that the shared social and emotional participation enabled by funeral rituals is therapeutic (Doka, 2006; O'Rourke et al., 2011). While it would seem intuitive that attendance would similarly offer an opportunity for meaningful participation, this does not automatically occur for all people with intellectual disability (Dodd et al., 2008; Raji et al., 2003). Their findings revealed that there was little understanding about the optimum level or type of participation that influences therapeutic gain or shared meaning making. Dodd et al. (2008) also reported complicated grief symptoms and separation distress following parental bereavement and involvement in rituals including funerals. These authors proposed that previous bereavement history and understanding the purpose of the ritual had a mediating role. Taken together, Dodd et al. (2008) concluded that participation in funerals requires an individualised approach.

Emerging research does suggest that people with intellectual disability can and do meaningfully participate in funerals. McMaugh et al.'s (2017) Australian qualitative study of funeral-related activities with four families reported, for example, that "Elizabeth was involved with the family in caring for her father as he was dying and said a eulogy with her siblings at the funeral" (p. 288), and that "Anne included her son Samuel by ensuring his attendance at the funeral so that he could grieve with his family…He was very, very upset as we all were" (p. 290). Similarly, in Young et al.'s (2017) study, a parent described the supportive presence of her daughter with profound intellectual and multiple disabilities at a funeral, "she helped me tremendously, because, sharing the grief with her and seeing how she coped with it helped me to cope with it too" (p. 1040).

Rituals around death extend beyond the funeral itself. In Ireland, when a person living in an intellectual disability care setting dies, the wake may be held in the setting, with staff carers, family and other residents involved. Carers report that this can help preserve memory of the person, allow others to pay their respects and can normalise the process of death and dying (McCarron, Burke et al., 2017). As with funerals, there may be concerns that a person with intellectual disability may not understand the context of the wake. However, wakes and funerals can bring family members of people with intellectual disability more closely together with residential services to support meaning-making and understanding.

More recent data from the Intellectual Disability Supplement to the Irish Longitudinal Study on Ageing (IDS-TILDA) Wave 4 suggests that involvement of peers with intellectual disability in funeral rituals has expanded (McCarron, Burke et al., 2017). Of 26 carers of individuals who had lived in a community disability setting prior to their death, 24 reported peers with intellectual disability attended wakes held in the residence of the deceased and paid their last respects (see *Funeral Rituals in Ireland* below for more detail).

The literature also reports mixed findings on funeral participation. Forrester-Jones's (2013) UK research noted that 60% of participants with intellectual disability had on occasions not been invited to funerals of a person they cared about. Yet, those same individuals described an understanding of the purpose of funerals, as a mark of respect to the dead person and an opportunity to say goodbye and to share in the comfort offered by the funeral ritual. Alcedo Rodríguez et al.'s (2018) research largely echoed these findings, but also showed that being informed of the death of a loved one or being able to go to the funeral was more common among those with milder intellectual disability.

The outcome of inconsistent approaches to funeral participation by people with intellectual disability too often is exclusion, which may result in disenfranchised grief (Read & Elliott, 2007), wherein sorrow is hidden, and not acknowledged or supported within a socially sanctioned mourning group (Doka, 2002). At worst, inconsistent approaches to sensitively supporting people with intellectual disability to understand and participate (to the extent they feel comfortable) in all dimensions of dying and death may lead to serious mental health issues such as persistent complex bereavement disorder (American Psychiatric Association, 2013).

The following story about Adam and Joel illustrates several important issues about death, grief, funeral participation, religion and disability services.

Adam and Joel's* story

Adam and Joel lived together in a group home. They were best friends. Joel was admitted to hospital one night with severe respiratory distress. By morning he had died. As Joel's family were Jewish and orthodox, it was important that he be buried immediately. Although the death was sudden it was not unexpected, so no autopsy was required. Joel's body was removed to a nearby funeral home and then transported to his family's home city 150 miles away and the burial completed. By the time group-home staff and Adam were informed, Joel was already in transit to his family. Adam was very distraught at not getting to say goodbye or attend Joel's funeral. Staff were also upset. The group-home administrator contacted the local rabbi and asked if it was possible to hold a memorial service. The rabbi was initially reluctant to interfere in the family's desired arrangements. He felt that the requested memorial service was to meet the needs of Christian staff and peers but without respect for Jewish traditions.

The administrator acknowledged this was not well thought through, but the request was well-intentioned and reflected concern particularly for Adam's grief. He explained that the way the hospital had managed communications (believing that since no one in the group home was next of kin, no one was entitled to be informed) had exacerbated the situation. He suggested that a memorial service might offer a learning opportunity for everyone. After some discussion, the rabbi agreed to come to the group home that evening. As instructed by the rabbi, the administrator arranged to get a candle, ensure handwashing, cover mirrors and have stools or boxes available. That evening the rabbi explained the traditions of sitting Shiva and had Adam light the candle. Adam, staff and other peers told stories about Joel as one does during Shiva. The rabbi explained that Shiva is usually for seven days, but it was held for just the one night. Staff supported Adam in lighting the candle briefly each evening for seven days, but group-home regulations prevented the candle staying lit throughout that period. Although Adam and the staff appreciated the rabbi supporting them in this way, they still did not feel it was a "proper" memorial or funeral. Adam remained upset that he did not get to say goodbye to his friend or go to his funeral.

*Assumed names are used in this story.

Barriers to Funeral Participation

Many people with intellectual disability want to go to the funerals of friends and family. For example, one of McRitchie et al.'s (2014) participants commented "No I didn't go. I wish I had. To say goodbye...No they didn't ask me. I wish they had" (p. 182). Available evidence suggests that distinct barriers compromise funeral participation for those with intellectual disability. Among caregivers, there is well-intended, but misguided protection and avoidance (Wiese et al., 2015; see also Chapter 2) fearing that the person with intellectual disability will not understand, or not cope. Paradoxically, this approach instead can exacerbate problems of understanding by robbing people with intellectual disability of needed opportunities to learn. Such protection has been witnessed even with individuals who have an understanding of death (Tuffrey-Wijne et al., 2020). Further, caregivers report that they find it difficult to explain what a funeral is, and are unsure whether any offered explanation assists understanding (Read, 2011). When the learning problems associated with intellectual disability are more severe, the desire to protect an individual from the reality of death increases (Alcedo Rodríguez et al., 2018; Tuffrey-Wijne et al., 2020).

People with intellectual disability have reported being forbidden from attending funerals. The reasons they have been given range from the funeral being too far away with no available transport to it being beyond their comprehension, or simply that they have not been invited. (Forrester-Jones, 2013; McRitchie et al., 2014; Read & Carr, 2014). Although likely informed by the protection discussed, there may also be practical issues like funeral scheduling, location, and accessibility that can affect the feasibility of funeral attendance for all mourners, including people with intellectual disability.

Given the sensitivities and social taboos that continue to shroud death, caregivers also perhaps wish to protect themselves (Forrester-Jones, 2013; Wiese et al., 2013). Staff may not feel comfortable with the topic (Forrester-Jones, 2013) and/or ill-equipped to provide support at funerals (Sango & Forrester-Jones, 2019). Parents report such trepidations about involving their adult children with intellectual disability in funeral rituals (McMaugh et al., 2017). There appears to be a particular struggle for families in understanding if and how people with intellectual disability actually experience or comprehend bereavement, especially for those with more profound intellectual and multiple disabilities (Young et al., 2017). Reporting on the experience of funeral directors, Raji et al. (2003) found they had very little understanding about how to assist people with intellectual disability to meaningfully participate in the shared ritual of funerals, but also a willingness to try. Most celebrants and funeral directors report needing guidance about how to do this. We currently know of no such resource specifically for this purpose, although resources about funerals which have people with intellectual disability as their intended audience may be helpful (see Resources at the end of this chapter).

The available research implies that if we maximise accessibility through the provision of tangible supports, then the person with intellectual disability will be able to better understand and meaningfully participate in funeral rituals, and experience shared emotions, acceptance and inclusion, which also may make healthy grieving possible. Specific strategies are presented in *Maximising Meaningful Participation* below. Young et al.'s (2017) research demonstrates clearly that people with profound intellectual and multiple disabilities (PIMD), although perhaps not intellectually understanding the funerals they attended, did demonstrate an emotional understanding. So too, did their funeral presence offer solace and support to other mourners. Their conclusions are instructive: "when one sees [PIMD]…as relational and emotional, rather than in terms of their cognitive impairments, one begins to feel challenged in our perceptions of what they are capable of understanding" (pp. 1042–1043). Perhaps then our understanding of meaningful funeral participation by people with intellectual disability should consider *both* (a) the extent to which the provision of tangible supports acts to maximise participation and (b) the extent to which the person emotionally connects with and responds to, the ritual of funeral. Understanding meaningful participation in this way may also relieve caregivers' reported fears about the person understanding and coping, and ease their turmoil about if and how to assist people with intellectual disability to attend funerals.

Maximising Meaningful Participation

The ability of funerals to be spiritually and socially inclusive of people with intellectual disability encompasses a variety of issues, many of which are only beginning to be articulated and understood. Below, we offer initial ideas about dealing with these issues, but make no claim to have covered the full range of concerns or responses to them. Our field's nascent understanding of what accessibility means for people with intellectual disability and funeral participation should mean that new topics and new approaches are identified in future as understanding grows. It is also important to note that we know of no research that has directly evaluated the effectiveness of the ideas set out below in enhancing meaningful funeral participation. Rather, our proposals represent a brief distillation of available research, resources and experience.

Doka (2006) offered three simple principles for funeral participation as a guide: *choice, support* and *do not wait*. In essence, the person should have sufficient information and support to decide if they wish to attend the funeral, and that when attending, choice about meaningful participation together with planned support throughout is critical. The principles also recommend that rather than waiting till the death of a loved one, when emotions are high, it is better to introduce learning about funeral rituals early. This learning has the added benefit of informing choice about funeral attendance and meaningful participation.

Nor should emotions be avoided. Often the reason given to explain why the person with intellectual disability was not invited to a family or peer's funeral was that the emotional impact of the funeral would likely distress the individual and that subsequent behaviour that might arise, would be challenging to other family members and too much to bear at such a time of loss. However, as noted previously, grief and bereavement and their expression are part of experiencing death for us all. Not having the opportunity to participate in death rituals may mean unresolved grief and contribute to not understanding why someone important in the person with intellectual disability's life is no longer present, which may have its own emotional consequences (McCallion et al., 2012).

In practice, meaningful participation can be maximised via a range of supports, many involving active, concrete participation. It could mean viewing the body to assist the person with intellectual disability to understand the permanence of death. As one person with intellectual disability stated about such a viewing "Well I would've felt it worse if I hadn't seen him... Because he was away" (McRitchie et al., 2014, p. 182). Especially for those less familiar with funerals, bearing in mind the social and cultural context of the individual, it is important to provide accessible and appropriate information beforehand about what occurs at a funeral, what will happen to the body, and the social expectations about what people usually wear and how they behave. Given the current widespread use of cremation, an explanation of this particular process will often be needed. There are excellent resources available that explain cremation and burial in simple, clear, direct terms aided by pictures or animations (see Resources at the end of this chapter). Pre-funeral preparation could also mean a prior visit to the funeral site so the person is more familiar with the setting. And, it could mean offering and rehearsing active roles at the funeral such as handing out orders of service, co- or part-presenting the eulogy, being a pallbearer or choosing a shared favourite song to be played. Similar active roles could be planned at the wake, and later too, at any annual remembrances or graveside visits. Importantly, any written material associated with the funeral could be revised to enable easy-read formats, with familiar pictures.

During the funeral, sitting with a trusted support person is valuable because that person can explain, interpret, console and if needed, support the individual to take a break. As described previously, the online funeral options necessitated by COVID-19 may offer a way for people with intellectual disability to experience a funeral that may be too stressful in person. These options, if individually tailored, align with the findings of Dodd et al. (2008) who reminded us of the importance of carefully preparing the individual prior to, during and after the funeral.

Addressing the emotional connection and response to funeral participation requires thinking about emotions that are typically evoked at funerals or wakes and how we engage with them. Examples might include offering condolences to chief mourners, touching, hugging, hand-shaking, reminiscing, singing and prayer. A common feature is that they are shared, and they evoke emotions typical of funerals; sadness, pain and sometimes relief, joy and pride. So, to maximise meaningful participation through emotional connection, we propose that people with intellectual disability are actively encouraged to offer condolence, to touch as others do at funerals and to join in singing, story-telling and prayer. Many have developed through experience, an intuitive understanding about the socially acceptable expression of emotion at funerals, but people with intellectual disability may not automatically understand these rules. This should not mean that expression of emotion be curbed. Instead, a combination of tangible supports and active participation, together with enablers of emotional connection, could act to maximise meaningful funeral participation.

The story of Paul's funeral is an example of an accessible personalised ceremony, with active involvement by people with intellectual disability.

Paul's* story.

Paul lived his life in disability service settings. For the last 15 years, he shared a home with his childhood friend Stephen. He had no family contact for decades, but was close to his long-term (40 years) friends, Judy and David (former staff members) and their extended families. Paul worked throughout adulthood until a dementia-related decline in his mid-forties. He died at home aged 50, having received wonderful care from his group-home staff, general practitioner (physician) and a community palliative care service.

Judy and David arranged the funeral, with collaborative input from important people in Paul's life. A number chose to take an active role in the funeral. Held in the funeral-home chapel, it was a secular ceremony with a celebrant, because Paul was not religious. The ceremony featured music from Paul's favourite singers, a slide show of people, places and events throughout his life, and a eulogy telling his life story delivered by Judy and David. Joanne (group-home team leader) gave a reading and close friend Jennifer sang a much-loved song. Paul's fellow house-mates, including Stephen, sat together in the front pew, as they were in every way Paul's "family", and each placed a flower on the coffin, supported by staff if needed. One house mate, visibly upset after placing his flower, remained by the coffin and said "I stay here". This totally understandable expression of emotion was received empathetically by other mourners and the celebrant, who were content to wait until he felt able to resume his seat.

Stephen was a pallbearer. Mourners included long-time friends, former and current staff. Everyone joined a heartfelt singalong of Paul's favourite pop song. All agreed that Paul would have loved the singing, photos and presence of friends. The ceremony was streamed online, with an online tribute site that featured photos and information.

Paul was buried in a small local cemetery that is easy to navigate and accessible. His headstone includes a favourite colour photo of him as a memento and to identify his grave for those with reading difficulties.

*Assumed names are used in this story.

Funeral Attendance by People with Intellectual Disability

Before considering funeral attendance, a preliminary issue arises. Do people with intellectual disability encounter deaths that would usually lead to funeral attendance? More than one in five respondents with intellectual disability in Ireland aged 40 and older identified the death of a friend as a major life event in the previous year alone (McCarron et al., 2017). When adults with intellectual disability were asked if they had ever experienced the death of another person, 88% (McEvoy et al., 2012) and 95% (Stancliffe et al., 2016) of respondents reported the deaths of one or more specific individuals they had known. These findings indicate that, like the rest of the community, adults with intellectual disability encounter the death of others and should have the opportunity to go to the funeral. In the sections that follow, we examine whether and when that opportunity is realised.

Qualitative Data on Funeral Attendance

Research evidence on funeral attendance by people with intellectual disability is limited. Some qualitative data, mostly from the UK, suggests that people with intellectual disability are rarely involved in funerals (Raji, et al., 2003; Read & Carr, 2014) and may be deliberately excluded due to concerns about their understanding or behaviour (Forrester-Jones, 2013; McRitchie et al., 2014; Wiese et al., 2018). However, other qualitative studies indicate that some people with intellectual disability do attend certain funerals (Forrester Jones, 2013; Read & Carr, 2014) and may even have an active role in the ceremony (McCarron et al., 2017; McMaugh et al., 2017; McRitchie et al., 2014). Importantly, there is evidence that people with intellectual disability report feeling a duty to go to the funeral out of respect for the deceased, to provide an opportunity for shared emotional expression and to say goodbye (Forrester-Jones, 2013). Those who have attended funerals showed understanding of the ritual and its meaning, and reported finding it a worthwhile experience (Forrester-Jones, 2013; McRitchie et al., 2014).

Only 3% of McEvoy et al.'s (2012) Irish participants said that they themselves chose not to go to the funeral and one commented "I didn't attend the funeral. I wanted to go but couldn't bring myself to go" (p. 198). McEvoy et al. did not report the number who did attend the funeral or other reasons for non-attendance, but made the following insightful comment:

> a small minority of individuals preferred not to participate in funeral arrangements or to talk about their loss. This serves to remind us of the importance of listening sensitively to individuals and to respect attempts to cope with bereavement in a personal, 'individualistic' manner. (p. 197)

Funeral Attendance in Australia

In Australia, Stancliffe et al. (2016, 2017), used the Concept of Death Questionnaire (CODQ) (McEvoy et al., 2012), which among other questions asks "Has someone close to you ever died?" and if so "Did you attend the funeral?" All participants with intellectual disability were able to communicate verbally and many had a good understanding of the concept of death. The Australian self-report findings for people with intellectual disability set out below are similar to those from Ireland reported by McEvoy et al. (2012) using the same instrument. Combining the previously unpublished Australian CODQ data from both studies (Stancliffe et al., 2016, 2017) showed that most of the reported deaths were close family members and other relatives (Table 10.1). There was no significant difference between staff and adults with intellectual disability

Table 10.1 Reported person who died by participant group

	Adults with ID *n* = 64		Disability staff *n* = 73	
Person who died	*n*	%	*N*	%
Parent/sibling	27	42.2	34	46.6
Other relative	25	39.1	26	35.6
Spouse	3	4.7	1	1.4
Friend	6	9.4	8	11.0
Other	3	4.7	4	5.5

concerning who they said had died. For the deceased people listed in Table 10.1, almost all would be considered as close relationships, whose funeral one would normally attend.

Overall, 80.6% of participants with mild intellectual disability reported attending the funeral, whereas only 60.0% of those with moderate intellectual disability did so. This difference did not quite attain statistical significance ($p = 0.078$). There were no participants with more severe intellectual disability, but presumably this group would be even less likely to go to a funeral. The difference in funeral attendance between people with intellectual disability (73.4%) and disability staff (83.6%) was not significant, and most adults with mild and moderate intellectual disability had attended at least one funeral. However, the CODQ does not directly ask whether the person had ever been prevented from attending a funeral. Some participants with intellectual disability may have experienced such exclusion, but the extent is unknown.

Funeral Rituals in Ireland

Two striking features of funerals in Ireland are waking the person (holding a wake) in their home and burial in the family plot. The Wake, the glorious send-off of departed loved ones, is a prominent feature of Irish funeral traditions. Held in the days before the funeral, it is a celebration where people gather to share stories of the lost loved one and have food and drinks. It is a grieving method where mourners get a chance to share a special moment with the deceased person. Irish people believe that the wake is a way to let the dead and the living bind together. The Irish wake generally takes place in the home of the deceased person or sometimes, at the place of someone who was close to the dead person. They prepare a room and gather items of significance belonging to the deceased, for example photographs, medals and other memorabilia. The departed person is dressed in their best clothes, and family and friends come to visit the deceased and pay their respects. Death and funerals create a gloomy atmosphere, but at an Irish wake, a celebration of the person's life, people will laugh and cry and share memories of fun times they had with the deceased.

Research on a cohort of ageing adults with intellectual disability in Ireland where staff caregivers of those who died were interviewed found approximately one quarter of the individuals had been ill for less than one month before they died (McCarron, Burke et al., 2017). Only a small number of the people who had died (7.1%) had had input into their funeral arrangements before their death. However, funeral arrangements had a high level of family input, with 79% involving family members in the process. In 20% of cases, the funeral organisers indicated they were following the deceased's wishes. Peers in intellectual disability residential settings were also reported to have had a lot of involvement in the funeral preparation and rituals around death, including choosing gifts, assisting with the Mass readings and participating in the choir service.

Funerals for people with intellectual disability conformed to Irish norms; 3.5% involved cremation of the deceased, in 4.7% of cases, there was no family attendance at the funeral and in only 1.2% of cases were peers of the deceased denied attendance at the funeral (McCarron, Burke et al., 2017). Most funerals were inclusive, emphasising respecting the deceased and allowing all those with a connection to them a chance to be part of a community of grief.

Having the wake in the intellectual disability residential setting meant that other residents had a chance to pay their respects, and respondents suggested this could help to normalise dying for other residents (all quotes shown below are from staff caregivers).

> I remember myself and one of the residents laid her out inside there.

> Having her wake here in the house because like it's very important for all our residents as well to be able to see her and you know understand what's going on.

> She came back to her community house and she was waked there. Probably actually two days and the people that she lived with had a chance to

> spend a bit of time with her. And then other people and lots of staff that would have known her over the years and her family of course.

In 70% of cases reviewed, burial was with other family members, usually the deceased's parents and was found to be in keeping with requests by the deceased when planning for their passing.

> She wanted to be buried in the family plot. Because she knew her mam was going to be there. So that was how that was decided.

When the family plot was not available for burial, or when the deceased was cremated, efforts were made to ensure the person reposed with or close to their family.

> the plan, to put the ashes in along with the mum.

> Yeah, he told me that could I find out where his mother and father were buried. And I did, and he said that's where he wanted to be buried with them. But unfortunately … I discovered there was … a pauper's grave and they told me that was full. … his other brother was buried in the same cemetery but when I tried to seek permission to get 'N' buried with him from another family member they refused. And then I got a grave for 'N' himself.

Friends of the deceased were actively involved in the funeral service.

> Yeah, her friends from the services and from Special Olympics … all her friends in Down syndrome, they did a guard of honour. All her friends that travelled together to Lourdes did guard of honour too.

> Taking up the gifts, all the service users in the house were involved.

COVID-19 restrictions disrupted established rituals for some with reports that grief may be unresolved (see *Maximising Meaningful Participation* above and Chapter 9).

> maybe for the family I'd say it was a huge thing because there wouldn't be the same closure. I suppose as a staff group we couldn't, you know go through that emotional process with them, both for ourselves and for them.

Alternative forms of remembrance have been developed, such as lining the streets as the funeral parade passes by and informal rituals that allow some of the grieving process to be undertaken.

> looking at pictures, lighting candles for them, that kind of thing was actually very useful. You know and it was a very informal kind of way and it was people they knew and I thought that worked quite well.

> We were allowed to go, stand outside the church, we weren't allowed in the church. And I suppose just for any family that's bereaved at the moment … the whole mourning process and grieving process is very different.

Following death in a residential service, the room in which a person lived was left unfilled for some time. The *month's mind* (Catholic Mass celebrated one month after the person's death) is still observed among many Irish adults. The month's mind may be of use in facilitating a sense of an ongoing relationship with the dead.

Conclusions About Funeral Attendance

Taken together, the qualitative and quantitative findings about funeral attendance are mixed. There is clear evidence of exclusion from at least some funerals, but other evidence of people attending funerals and related rituals, and even actively participating in the ceremony. Factors related to attendance include having milder disability, being supported by family (McMaugh et al., 2017), taking place in or near an intellectual disability facility, and the individual wanting to attend. The more recent data from Australia and Ireland set out in the previous sections could indicate a trend over time for more funeral attendance, compared to earlier reports of exclusion, but methodological limitations make such conclusions uncertain. What seems clear is that a number of adults with intellectual disability do successfully attend funerals and may find meaning by so doing. What is also evident is that exclusion from funerals by caregivers remains a barrier to funeral attendance, a barrier that needs to be addressed.

Spirituality

Despite academic framing of religion as an outmoded institution eclipsed by secularisation particularly in the global North (Dinham, 2020), the Pew Research Center (2017) claims 84% of the world's population report a religion or religious/spiritual belief. The UN Convention on the Rights of Persons with Disabilities asserts its purpose as enabling people with disabilities to enjoy "full and equal enjoyment of all human rights and fundamental freedoms" (United National General Assembly, 2006, Article 1), but spirituality/religion is not specifically referenced (Whiting & Gurbai, 2015). Similarly, a systematic review by Sango and Forrester-Jones (2014) found that spiritual care in relation to people with intellectual disability was generally absent from UK policy and regulations. Radstake (2021) reported that, despite advocacy work of ecumenical societies to build more disability-inclusive faith communities in North America, understandings of intellectual disability still rely on formal diagnosis and professional care that does not routinely include spiritual support (Webb-Mitchell, 2009).

Reinert and Koenig (2013) caution against making distinctions between *religion* and *spirituality*. In everyday discourse, the terms tend to be used interchangeably (for a detailed discussion, see Sango & Forrester-Jones, 2017), both referring to something that offers meaning and purpose to life and a sense of belonging with other like-minded individuals. Spirituality will be the term used for the remainder of this section.

The limited research on the topic (see Sango & Forrester-Jones's, 2017 systematic review) suggests that many individuals with intellectual disability have a spiritual identity (Carter, 2013; Hatton et al., 2004; McEvoy et al., 2012). Apart from mediating inclusive social support networks (Gaventa, 2005; Sango & Forrester-Jones, 2018) fostering feelings of acceptance, self-esteem and self-worth (Swinton, 2002), environments offering opportunities to practise spiritual beliefs have also been found to provide comfort and psychological support (Culliford, 2002) during difficult life events including death (Narayanasamy et al., 2002). Yet people with intellectual disability are often excluded from such support. Forrester-Jones (2013) found that all of her 15 study participants (split into 3 focus groups) with mild to moderate intellectual disability noted how spiritual rituals accompanying funerals including prayers, music and hymns provided comfort. While the aim of Forrester-Jones' (2013) study was to explore individuals' views concerning funerals, each focus group ended with an invitation to ask the researcher questions. An additional theme *spiritual matters* in relation to funerals was delineated from this data and is reported here with explanatory quotes.

Spiritual Matters

Participants (denoted as P1 etc.) firstly demonstrated their understanding and questions about what happens to the physical body post-funeral:

P1: Do we become skeletons?
P7: Umm…when they [deceased] have gone, where do they go? How far down do they go?
P9: Do they burn the coffin afterwards or it [body] just stays there? Someone …told me that all her [deceased] skin has gone off in the coffin…and…the coffin is for the next one to go in it—is that right?
P13: Basically when your body goes in the ground it stays there—it is just your body isn't it? What is cremation?
P11: Every time someone dies they make room for someone else being born.

In the following conversations, participants appeared to grapple with the question of the soul/spirit and where it goes post-death:

P1: What is a soul?
P2: You believe in Jesus [and] you go to heaven. I think that when you die… your ashes go to heaven. The soul is mainly in you.

P3: Yeah, what is a soul? Now tell me the truth, if people steal or kill, will they go to heaven or hell? What about the people who stole my guinea pig—will they go to hell—they shouldn't go to heaven should they?
P4: All I can say is we go to another world… and God gives us his respect.
P1: They [deceased] become spirits.
P2: Yes some people believe that—spirits.
P3: When we die, we go where no-one can get us, no-one can get to us again, no-one can get past the gate.

And:

P8: Your soul goes up to the sky.
P9: They go to heaven…they are asleep.
P11: Yeah, your spirit goes on.
P12: Up to heaven. You can actually see them [deceased] when you are in bed.

Finally, some talked about how they sensed that their loved ones were "still around":

P6: I have felt my dad touch me since he died. He always used to brush my hair and I can still feel his hand brush against me.
P3: I laughed and talked to her [deceased mother] and I said 'hello mum how are you'. I know they die and I know Jesus died for us but I don't have an understanding of it you see.

Understanding if and how people with intellectual disability conceptualise the spiritual aspect of funerals has been something both families and care staff struggle with (Sango & Forrester-Jones, 2019). Yet the theme above demonstrates that while their conceptualisations may not fit with some mainstream religions (Swinton, 2002) and the data here is limited to the Christian religion, religious groups, support workers, families and service providers may need to work together to help develop these understandings.

Conclusions and Recommendations

Among the many events that affect the day-to-day lives of people with intellectual disability, experience and discussion of others' funerals may influence thoughts about their own needs and desires at end of life; particularly the funerals of those to whom the person with intellectual disability is close. As is true for everyone, people with intellectual disability should be able to choose whether or not to engage in such rituals, what type of ritual they are comfortable with, and whose funeral they would like to attend.

The loss of a loved one evokes many emotions. Bereavement and grief are often experienced differently by individuals, and the length of time involved varies. The opportunity to attend a funeral, share emotions and make some meaning may be important contributors to successfully managing grief and loss. Attendance should be possible, but the person with intellectual disability should have the choice to attend rather than it be a requirement. Beyond simple presence at a funeral, we need to learn more about the nature of participation that people find comforting and satisfying, recognising that individuals' emotional, spiritual and participation preferences may differ.

The support each of us needs to navigate the experience of a funeral and the associated grief varies and is likely to be greater for some people with intellectual disability due to communication difficulties and their understanding of what death and burial mean. The addition of possible cremation and unfamiliar religious and other rituals adds further complication. Increasing opportunities to attend funerals alone is not a sufficient response. There are needs for supportive prior explanations of what a funeral is, and the specifics of the rituals. Having a trusted family member or staff person to accompany, interpret and support the person with intellectual disability during the funeral is also important. Continued availability of supports is required to help process what has occurred and to offer solace, explanation and counselling when needed, as the person with intellectual disability processes additional grief after the funeral.

Funerals are also a rite where there are religious practices and beliefs. These spiritual dimensions are important in experiencing and dealing with bereavement and grief. Particularly in staffed situations, there may be differences in such beliefs and practices between the individual with intellectual disability and their staff caregiver. Further, the service provider organisation itself may be secular and not wish to be involved in these issues. There is a challenge for providers and staff to support the individual with intellectual disability in exploring and practising the religious and spiritual aspects of their lives (Liu et al., 2014; Whiting & Gurbai, 2015), including death and burial, that require further research exploration and development of effective supportive practices.

Too often, adults with intellectual disability are viewed as a homogenous group with similar needs and similar experiences. Much of the available literature and practices are focussed upon those with mild and moderate intellectual disability. As well as a need for more work on how persons with severe and profound intellectual disability may be supported, there is also a need to consider similarities and differences for other individuals with developmental disabilities, such as those with autism (Liu et al., 2014) and those with cerebral palsy. The disability research examined in this chapter has focussed on adults with intellectual disability. These issues are also important for children with disability and deserve careful attention in future research.

We each will come to the end of our own lives and the experience of attending and being genuinely included in funerals will likely lead to thoughts about our own funeral arrangements. For people with intellectual disability, there must also be an openness to having such discussions about their own funerals and to incorporating expressed desires in the advance care plans they develop.

Reflection

Joan's story. Her mother is dying and cannot visit.
Joan's mother had previously visited every weekend, but had stopped visiting about 9 months ago. Her mother had been diagnosed with cancer, an addition to other chronic conditions, and she had become too frail to make the journey. About three months later, she was admitted to

a nursing home as her care needs had increased. Other family members visited Joan, but not as often and were reluctant to explain to her how ill her mother now was. Joan was increasingly upset that her mother was not visiting and expressed that something was wrong. She experienced increased sleeplessness and was noticeably eating less and less. The family was persuaded to tell Joan that her mother was ill, but they insisted on saying she was going to get better. Privately they told staff that death was only a matter of time. Staff raised that Joan needed to know this and asked about how her mother's last days would be managed and what role Joan might play in visiting her mother and attending her funeral. One sister accepted that Joan needed to know. The other sister and the brother insisted that she was not to be told or to be involved. It was then agreed among the family members that Joan was not to visit her mother (although she was asking to do so) but perhaps she would be able to attend the funeral.

*Assumed names are used in this story.

1. Family members were not court-appointed guardians for Joan. Was it their decision whether she could visit her mother or later attend the funeral?
2. If there was agreement that Joan might visit her mother in the nursing home and/or attend the funeral after her death, how might she be involved meaningfully in these experiences?

Resources

The resources listed below have all been developed specifically for people with intellectual and developmental disability, and people who support them.

1. *Bereavement and Loss: Learning Resource Pack and Training*. From PAMIS: Promoting a More Inclusive Society, guidance for those supporting bereaved people with profound or multiple learning

disabilities and their parents and carers. http://pamis.org.uk/resources/bereavement-and-loss/.

2. *Books Beyond Words: When Mum Died* and *When Dad Died.* UK wordless books designed for people with intellectual and developmental disability that include only pictures so the story can be tailored to the individual. These titles include depictions of funerals, including burial and cremation. https://booksbeyondwords.co.uk.
3. *Breaking Bad News*: This UK website contains information and guidelines for caregivers of people with intellectual and developmental disability about how to break bad news. http://www.breakingbadnews.org/.
4. *Glancing Back, Planning Forward, A Guide for Planning End-of-Life Are with People with Intellectual Disability.* Trinity Centre for Ageing and Intellectual Disability has developed an accessible planning tool for people with an intellectual disability to help them, their families and their carers plan ahead so that their wishes can be respected at the end of their life. https://www.tcd.ie/tcaid/accessibleinformation/index.php.
5. *Let's Talk About Death*. A free booklet from Down's Syndrome Scotland in easy-read format about death and funerals for adults with intellectual and developmental disability. https://www.dsscotland.org.uk/wp-content/uploads/2019/05/DSS_LETS-TALK-ABOUT-DEATH_WEB.pdf.
6. *Talking End of Life…with People with Intellectual Disability [TEL].* Australian website with 12 modules on end of life and people with intellectual and developmental disability. Designed for disability support workers but also helpful for families, health professionals and educators. TEL includes a module on *Funeral Wishes*. https://www.caresearch.com.au/TEL/.
7. *When I Die*. An example of a completed end-of-life plan by a person with intellectual and developmental disability, with a section on funeral wishes. http://www.pcpld.org/wp-content/uploads/when_i_die_2 0.pdf.

References

Alcedo Rodríguez, M. A., Christóbal Fernández, L., Gómez Sánchez, L. E., & Arias González, V. B. (2018). Evaluating the characteristics of the grieving process in people with intellectual and developmental disability. *Journal of Applied Research in Intellectual Disabilities, 31*(6), 999–1007. https://doi.org/10.1111/jar.12454

American Psychiatric Association. (2013). *Diagnostic and statistical manual of mental disorders* (5th ed.). American Psychiatric Association.

Carter, E. W. (2013). Supporting inclusion and flourishing in the religious and spiritual lives of people with intellectual and developmental disabilities. *Inclusion, 1*(1), 64–75. https://doi.org/10.1352/2326-6988-1.1.064

Culliford, L. (2002). Spirituality and clinical care: Spiritual values and skills are increasingly recognised as necessary aspects of clinical care. *British Medical Journal, 325*(7378), 1434–1436. https://doi.org/10.1136/bmj.325.7378.1434

Dinham, A. (2020). Religion, belief and public policy. In N. Ellison & T. Haux (Eds.), *Handbook on society and social policy* (pp. 101–110). Edward Elgar.

Dodd, P., Guerin, S., McEvoy, J., Buckley, S., Tyrrell, J., & Hillery, J. (2008). A study of complicated grief symptoms in people with intellectual disabilities. *Journal of Intellectual and Developmental Disability Research, 52*(5), 415–425. https://doi.org/10.1111/j.1365-2788.2008.01043.x

Doka, K. J. (Ed.). (2002). *Disenfranchised grief: New directions, challenges, and strategies for practice*. Research Press.

Doka, K. J. (2006). Thinking and planning for end of life: Rituals: How we cope with death—And life. *The Exceptional Parent, 36*(7), 69.

Forrester-Jones, R. (2013). The road barely taken: Funerals, and people with intellectual disabilities. *Journal of Applied Research in Intellectual Disabilities, 26*, 243–256. https://doi.org/10.1111/jar.12022

Garces-Foley, K., & Holcomb, J. (2006). Contemporary American funerals: Personalising tradition. In K. Garces-Foley (Ed.), *Death and religion in a changing world* (pp. 207–227). M. E. Sharpe.

Gaventa, W. (2005). A place for all of me and all of us: Rekindling the spirit in services and supports. *Mental Retardation, 43*(1), 48–54. https://doi.org/10.1352/0047-6765(2005)43%3c48:APFAOM%3e2.0.CO;2

Hatton, C., Turner, S., Shah, R., Rahim, N. & Stansfield, J. (2004). *Religious expression, a fundamental human right: The report of an action research project on meeting the religious needs of people with learning disabilities*. Foundation for People with Learning Disabilities, Mental Health Foundation, Glasgow. https://www.learningdisabilities.org.uk/learning-disabilities/publications/religious-expression-fundamental-human-right

Holloway, M., Adamson, S., Argyrou, V., Draper, P., & Mariau, D. (2013). "Funerals aren't nice but it couldn't have been nicer". The makings of a good funeral. *Mortality, 18*(1), 30–53. http://dx.doi.org/10.1080/13576275.2012.755505

Liu, E. X., Carter, E. W., Boehm, T. L., Annandale, N. H., & Taylor, C. E. (2014). In their own words: The place of faith in the lives of young people with autism and intellectual disability. *Intellectual and Developmental Disabilities, 52*(5), 388–404. https://doi.org/10.1352/1934-9556-52.5.388

Lysaght, P. (2003). Hospitality at wakes and funerals in Ireland from the seventeenth to the nineteenth century: Some evidence from the written record. *Folklore, 114*(3), 403–426. https://doi.org/10.1080/0015587032000145405

McCallion, P., McCarron, M., Fahey-McCarthy, E., & Connaire, K. (2012). Meeting the end of life needs of older adults with intellectual disabilities. In E. Chang & A. Johnson (Eds.), *Contemporary and innovative practice in palliative care* (pp. 255–267). Intech. http://dx.doi.org/10.5772/29438

McCarron, M., Burke, E., White, P., O'Dwyer, C., Ryan, K., O Farrell, J., McCallion, P. (2017). *"He'd mind you, you mind him": Experiences of end of life care for people with an intellectual disability as perceived by staff carers.* https://www.tcd.ie/tcaid/research/endoflifepublications/ReportExperiencesofendoflife.pdf

McCarron, M., Haigh, M., & McCallion, P. (Eds.). (2017). *Health, wellbeing and social inclusion: Ageing with an Intellectual disability in Ireland. Evidence from the first ten years of the Intellectual Disability Supplement to The Irish Longitudinal Study on Ageing (IDS-TILDA). Wave 3 IDS-TILDA.* School of Nursing and Midwifery, The University of Dublin, Trinity College. https://www.tcd.ie/tcaid/assets/pdf/wave3report.pdf

McEvoy, J., MacHale, R., & Tierney, E. (2012). Concept of death and perceptions of bereavement in adults with intellectual disabilities. *Journal of Intellectual Disability Research, 56*(2), 191–203. https://doi.org/10.1111/j.1365-2788.2011.01456.x

McMaugh, P. J., Wiese, M. Y., & Stancliffe, R. J. (2017). The experiences of parents in supporting their son or daughter with intellectual disability to learn about dying and death. *Journal of Intellectual & Developmental Disability, 42*(3), 285–294. https://doi.org/10.3109/13668250.2016.1236367

McRitchie, R., McKenzie, K., Quayle, E., Harlin, M., & Neumann, K. (2014). How adults with an intellectual disability experience bereavement and grief: A qualitative exploration. *Death Studies, 38*(3), 179–185. https://doi.org/10.1080/07481187.2012.738772

Narayanasamy, A., Gates, B., & Swinton, J. (2002). Spirituality and learning disabilities: A qualitative study. *British Journal of Nursing, 11*, 948–957. https://doi.org/10.12968/bjon.2002.11.14.10467

O'Rourke, T., Spitzberg, B. H., & Hannawa, A. F. (2011). The good funeral: Toward an understanding of funeral participation and satisfaction. *Death Studies, 35*(8), 729–750. https://doi.org/10.1080/07481187.2011.553309

Pew Research Center. (2017). *The changing global religious landscape*. Pew Research Centre Demographic Projections. https://www.pewresearch.org/fact-tank/2017/04/05/christians-remain-worlds-largest-religious-group-but-they-are-declining-in-europe/

Radstake, J. (2021). Serving all people: Pentecostal experiences in ministering to people with IDD in Ontario. *Journal of Disability & Religion, 25*(2), 83–109. https://doi.org/10.1080/23312521.2020.1776666

Raji, O., Hollins, S., & Drinnan, A. (2003). How far are people with learning disabilities involved in funeral rites? *British Journal of Learning Disabilities, 31*, 42–45. https://doi.org/10.1046/j.1468-3156.2003.00216.x

Read, S. (2011). Palliative care for the person with a learning disability. In M. Baldwin & J. Woodhouse (Eds.), *Key concepts in palliative care* (pp. 148–151). Sage.

Read, S., & Carr, M. (2014). Living with loss. In S. Read (Ed.), *Supporting people with intellectual disabilities experiencing loss and bereavement: Theory and compassionate practice* (pp. 36–45). Jessica Kingsley.

Read, S., & Elliott, D. (2007). Exploring a continuum of support for bereaved people with intellectual disabilities: A strategic approach. *Journal of Intellectual Disabilities, 11*(2), 167–181. https://doi.org/10.1177/1744629507076930

Reardon, A. (2020, March). Funeral providers adapt to coronavirus as gathering limit lowered to 10 people. *ABC News*. https://www.abc.net.au/news/2020-03-25/funeral-providers-try-to-adapt-to-new-coronavirus-restrictions/12089162

Reinert, K. G., & Koenig, H. G. (2013). Re-examining definitions of spirituality in nursing research. *Journal of Advanced Nursing, 12*, 2622–2634. https://doi.org/10.1111/jan.12152

Sango, P. N., & Forrester-Jones, R. (2014). Spirituality and learning disability: A review of UK Government guidance. *Tizard Learning Disability Review, 19*(4), 170–177. https://doi.org/10.1108/TLDR-01-2014-0005

Sango, P. N., & Forrester-Jones, R. (2017). Intellectual and developmental disabilities, spirituality and religion: A systematic review 1990–2015. *Journal of Disability & Religion, 21*(3), 280–295. https://doi.org/10.1080/23312521.2017.1317224

Sango, P. N., & Forrester-Jones, R. (2018). Spirituality and social networks of people with intellectual and developmental disability. *Journal of Intellectual & Developmental Disability, 43*(3), 274–284. https://doi.org/10.3109/13668250.2017.1310820

Sango, P. N., & Forrester-Jones, R. (2019). Spiritual care for people with intellectual and developmental disability: An exploratory study. *Journal of Intellectual & Developmental Disability, 44*(2), 150–160. https://doi.org/10.3109/13668250.2017.1350834

Selman, L. E., Sowden, R., & Borgstrom, E. (2021). 'Saying goodbye' during the COVID-19 pandemic: A document analysis of online newspapers with implications for end of life care. *Palliative Medicine, 35*(7), 1277–1287. https://doi.org/10.1177/02692163211017023

Stancliffe, R. J., Wiese, M. Y., Read, S., Jeltes, G., & Clayton, J. M. (2016). Knowing, planning for and fearing death: Do adults with intellectual disability and disability staff differ? *Research in Developmental Disabilities, 49–50*, 47–59. https://doi.org/10.1016/j.ridd.2015.11.016

Stancliffe, R. J., Wiese, M. Y., Read, S., Jeltes, G., & Clayton, J. M. (2017). Assessing knowledge and attitudes about end of life: Evaluation of three instruments designed for adults with intellectual disability. *Journal of Applied Research in Intellectual Disabilities, 30*(6), 1076–1088. https://doi.org/10.1111/jar.12358

Swinton, J. (2002). Spirituality and the lives of people with learning disabilities. *Tizard Learning Disability Review, 7*(4), 29–35. https://doi.org/10.1108/13595474200200037

The Cremation Society. (n.d.). *Progress of cremation in the British islands 1885–2018*. https://www.cremation.org.uk/progress-of-cremation-united-kingdom

Tuffrey-Wijne, I., Finlayson, J., Bernal, J., Taggart, L., Kar Kei Lam, C., & Todd, S. (2020). Communicating about death and dying with adults with intellectual disability who are terminally ill or bereaved: A UK-wide study of intellectual disability support staff. *Journal of Applied Research in Intellectual Disabilities, 33*(5), 927–938. https://doi.org/10.1111/jar.12714

United Nations General Assembly (UNGA). (2006). *Resolution 61/106 of 13 December 2006 on the Convention on the Rights of Persons with Disabilities, Optional Protocol to the Convention on the Rights of Persons with Disabilities.* A/RES/61/106. https://www.un.org/disabilities/convention/conventionfull.shtml

Webb-Mitchell, B. (2009). *Unexpected guests at God's banquet: Welcoming people with disabilities into the church.* Wipf and Stock Publishers.

Whiting, R., & Gurbai, S. (2015). Moving from the implicit to the explicit: 'Spiritual rights' and the United Nations Convention on the rights of persons with disabilities. *Canadian Journal of Disability Studies, 4*(3), 103–126. https://doi.org/10.15353/cjds.v4i3.233

Wiese, M., Dew, A., Stancliffe, R. J., Howarth, G., & Balandin, S. (2013). 'If and when?': The beliefs and experiences of community living staff in supporting older people with intellectual disability to know about dying. *Journal of Intellectual Disability Research, 57*(10), 980–992. https://doi.org/10.1111/j.1365-2788.2012.01593.x

Wiese, M., Stancliffe, R. J., Read, S., Jeltes, G., & Clayton, J. M. (2015). Learning about dying, death, and end-of-life planning: Current issues informing future actions. *Journal of Intellectual & Developmental Disability, 40*(2), 230–235. https://doi.org/10.3109/13668250.2014.998183

Wiese, M. Y., Stancliffe, R. J., Wagstaff, S., Tieman, J., Jeltes, G., & Clayton, J. (2018). *TEL talking end of life* [website]. https://www.caresearch.com.au/TEL/

Young, H., Hogg, J., & Garrard, B. (2017). Making sense of bereavement in people with profound intellectual and multiple disabilities: Carer perspectives. *Journal of Applied Research in Intellectual Disabilities, 30*(6), 1035–1044. https://doi.org/10.1111/jar.12285

11

End-of-Life Doulas and People Living with Intellectual and Developmental Disability

Caroline Ellison

Introduction

Death is the journey few people want to undertake alone. No one, whether living with disability or without, needs to die alone if that is not their wish. Changes in medical technology, vaccinations and pharmaceuticals over the last century have also resulted in significant cultural change in how many societies deal with dying. People now want greater choice and control over the end of life. Urbanisation arising from the industrial revolution combined with twentieth-century global wars saw death become an industry, with the emergence of the modern hospital, medicalised services and the funeral sector. Responsibility for and the perceived unpleasantness of death were put into the hands of paid others. There is now an opportunity to learn from cultures which have a recognised community role for a wise, caring person who supports people

C. Ellison (✉)
University of South Australia, Adelaide, SA, Australia
e-mail: Caroline.Ellison@unisa.edu.au

R. J. Stancliffe et al. (eds.), *End of Life and People with Intellectual and Developmental Disability*, https://doi.org/10.1007/978-3-030-98697-1_11

spiritually, physically and socially as they die. The end-of-life doula represents this role.

While accessing an end-of-life doula is becoming increasingly mainstream for people without disability, it is a novel approach to support those dying with disability. This chapter discusses the *disability-aware end-of-life doula* as an option to fill a much-needed gap in current supports to individuals with intellectual and developmental disability who are dying. The history and role of the doula are described, followed by a description of its unique position to offer the dying person with intellectual and developmental disability and those around them an increased understanding about death, end-of-life planning and support. The range of ways that an end-of-life doula might assist the dying of people with intellectual and developmental disability is outlined and the chapter is completed with the challenges ahead to accessing the disability-aware doula service for the dying person and their families. The chapter is written in first-person voice as I am a practising end-of-life doula and Developmental Educator. I describe the doula role through my own experience and also draw upon the known evidence from my additional role as a university researcher in the disability space.

Defining and Describing the Evolution of the Modern Doula

The term *end-of-life doula* has become an emerging concept to identify "lay people, primarily women, who provide a diversity of nonmedical supports—social, emotional, practical, and spiritual—for people nearing the end of life" (Krawczyk & Rush, 2020, p. 1). The end-of-life doula generated from the doula model where lay-trained women provided nonmedical assistance during and after pregnancy (Fukuzawa & Kondo, 2017; Krawczyk & Rush, 2020; Rawlings et al., 2019). End-of-life doulas provide informed companionship and resources before, during and after death, including after-death care of the body and funeral planning education and relevant services (Krawczyk & Rush, 2020).

The end-of-life doula is a trained, nonmedical professional who provides support, options and education, taking the focus of the person as a patient and refocusing on them as a person. The role may involve assisting a person who is dying, and those around them, to have their end of life unfold as they wish, while preserving quality of life, well-being and self-worth up to and beyond the end of life (McGuire & McEvoy, 2016). The end-of-life doula role can provide a range of supports from sitting with the dying, education about the dying process, to home funerals and advance care planning (Rawlings et al., 2019).

End-of-life doulas are not new. Oral and aural stories across many cultures indicate a long history of assisting both dying people and their significant others to make sense of the dying process, before, during and after death. For millennia, almost all cultures have supported people staying in their homes to die, looked after by their family, their community or some designated specialists (Krawczyk & Rush, 2020). In the modern Western world, there has been a trend towards medicalisation of death, the growth of professional funeral services and out-of-home institutionalised residential care for older people and people with disability such as hospitalisation or life in a hospice setting when death becomes imminent (Moro et al., 2017). This segregation of death and its devolved responsibility to others often lead to isolation, lack of choice and opportunity for all concerned, most particularly the dying person and their family members, friends and those whom they love who will be left behind. Such segregation is a function of the detachment of social and support networks from the dying process in order to avoid discomfort and fear. Inevitably in this context, death often becomes a lonely, sombre, medicalised, sterile, dehumanised, industrialised and disconnected experience. Death can also last much longer in the current century than would have been the case at virtually any other time in history, as more advanced and sophisticated medical intervention prolongs life, sometimes at the cost of quality of life (Feifel, 1959; Kübler-Ross et al., 1972). Although dying individuals and their families may know long in advance that their time is limited, prolongation of the dying process provides a strong rationale for guidance through it.

There is a large and growing number of formal end-of-life doulas practising in developed countries such as the USA, Canada, UK, Australia and Europe. However, end-of-life doulas also exist in villages and rural communities in developing countries, but are often less formal (Kelly,

2015). While end-of-life doulas are not consistently formally accredited or registered internationally, training courses, directories of practitioners and research into the role and place of end-of-life doulas in social care systems are slowly developing (Rawlings et al., 2019).

Research in Australia shows the majority of people would choose to conduct their dying at home (Burge et al., 2015; Foreman et al., 2006; Swerissen et al., 2014). However, staying at home until the end requires a great deal of coordinated care. Even for the most committed and willing family members, this may be difficult to manage while they are also dealing with their own grief. An end-of-life doula will work wherever their client happens to be, in a person's private home, an aged care facility, a hospice, a hospital or supported accommodation. There is usually a fee for service and a matching process involved to ensure there is a positive rapport between the doula and the individual and family. Because feeling at ease with an end-of-life doula is so important, most doulas will offer an initial free consultation as a chance to see whether there is likely to be a good fit for a supportive relationship.

People can work with an end-of-life doula at any time throughout the dying. It does not need to be only after a terminal diagnosis, or only when the person reaches a point where they qualify for intense palliative care or a hospice. The doula is able to devote themselves to a single person, engaging with no other agenda than to fulfil the needs of the person and those around them, to provide an active involvement rather than simply a passive presence. End-of-life doulas have particular expertise in gently challenging the notion that death is a taboo subject best left to religious ministers and spiritual leaders, or the medical and funeral professions.

Potential Role in the Disability Sector

The term intellectual and developmental disability describes lifelong conditions including impairment in physical, learning, language or behaviour that occurs prenatally or after birth as a result of factors such as injury or infection (Moro et al., 2017; Schalock et al., 2007). People with intellectual and developmental disability have challenges with independent social functioning and understanding new or complex information due to their cognitive condition (World Health Organization, 2016).

Many people with intellectual and developmental disability confront disability-specific barriers at the end of life, including underuse of palliative care services, communication difficulties, limited healthcare providers with disability-specific knowledge and experience, poorer pain and symptom management, and lack of preventive efforts to avoid conditions amenable to surveillance (Haveman et al., 2011; Marks et al., 2008; Moro et al., 2017; Webber et al., 2010). An ethnographic study about the lived experiences of 13 adults with intellectual disability experiencing cancer showed that participants were not helped to understand what was happening to them and their diagnosis or prognosis tended to not be discussed (Tuffrey-Wijne et al., 2010). People with intellectual disability are often excluded from discussions and from making decisions about their illness or end of life (Araujo de Carvalho et al., 2017; Flynn et al., 2016) (see also Chapters 5 and 13).

Death is an uncomfortable topic that it is often not planned for, neither are people educated about it. Failing to be prepared, open and informed, however, will often result in a poor end-of-life experience for a person with intellectual and developmental disability who is dying, and increased grieving and stress for those left behind (Lord et al., 2017). The best dying experience is likely to be one where there is an understanding of what is going on in the here and now, what is likely to happen in the future, and what options there are along the way (Moro et al., 2017; Tenzek & Depner, 2017). I have personally witnessed people living with intellectual and developmental disability dying with little control over their end-of-life experience, even if they had previously been living independently and engaging in choice and self-determination. Handing over control when dying was distressing for these individuals and highlighted a gap in the disability sector's capacity to facilitate a positive end of life (Kirkendall et al., 2017; Ryan et al., 2011b; Voss et al., 2019).

The research has shown, concurring with my own experiences, that hospices, despite their best intentions, have been unable to provide all of the services needed by a person with intellectual and developmental disability, and that paid disability services are often not included, deliberately excluded or left unsupported (Lindley et al., 2017; Wark et al., 2017). Hospice and palliative care can provide excellent care for the medical aspects of dying, but sometimes have limited knowledge, resources and training about the disability experience. The experience of intellectual and developmental disability means that additional care skills are needed at the end of life, such as ensuring information is understood, making reasonable adjustments and enabling time to identify and implement wishes around dying, all of which are equally as important for dying peacefully (Wark et al., 2017).

Choice and control are key to a positive end-of-life experience, whether a person lives with disability or not (Krawczyk & Rush, 2020). A key, and perhaps the most important role of the end-of-life doula, is to assist the dying person with intellectual and developmental disability to maximise their choice and control experience. This is enabled because the doula can take the necessary time to facilitate a more collaborative and transdisciplinary approach, where the person, family and their supports work in partnership with the doula to share knowledge and expertise in a way that suits the dying person and is matched to their capacity to understand (Gomes & McVilly, 2019; McGuire & McEvoy, 2016; Voss et al., 2019).

Being a Doula in the Disability Sector

There are many questions as to what an end-of-life doula could do in the disability sector, what they can offer and the value of and need for the role. The doula can offer a diverse range of supports, including being a paid companion who can use consumer-centred and individualised approaches to help circumvent fragmented health and palliative care services. Their focus is one or two dying people and their families at a time, rather than the larger numbers typically expected in the formal health care system (Rawlings et al., 2020). Table 11.1 offers a summary of the potential job description of the end-of-life doula in the disability space. Importantly, in this space, the end-of-life doula is *disability-aware*. Essentially, their job description is identical to what it would be when supporting dying people without disability; however, they are able to

Table 11.1 End-of-life doula job description in the disability space

Roles	Description
Informer	Listening and preparing, e.g. seeking to understand needs, values, preferences and goals, and exploring fears and anxieties Giving information about the journey and what might be expected Providing information about end-of-life choices, both while living and for after death
Planner	Assisting with end-of-life planning, e.g. assisting with advance care directives, appointment of a medical treatment decision-maker, vigil planning, coordinating informal support systems, after-death body care and funeral planning
Advocate	Assisting to complete legal and advance care planning and funeral planning so the dying person can have their choices known and wishes followed Familiarity with local jurisdictional requirements Advocating with and providing guidance for involved medical and care professionals, ensuring that they know about the person and their preferences Liaising as the dying person's "interpreter" with medical and other supports
Coordinator	Coordinating practical supports, meals, cleaning and other activities to support the person and those around them Legacy work, e.g. assisting with creating legacy documents, such as birthday cards or the sourcing of gifts to be given now or in the future Closure work, e.g. helping clients attend to unfinished business to facilitate contentment
Companion	Companioning and giving emotional support for all stages of end of life, from diagnosis to the point of death, and bereavement support for those left behind Supporting the person's right to choose an end of life that is meaningful and consistent with their belief system and values, including ethnic, cultural, religious/spiritual or personal Assisting the maintenance of an intimate, spiritual and peaceful space and experience for everyone involved, wherever the person is and irrespective of medical support

understand and respond to the particular disability-specific barriers the dying person with intellectual and developmental disability may face, and use their disability-aware skills to modify the conduct of their role accordingly.

Talking with the person who is terminally ill about dying, or their wish to engage in advance planning, can relieve anxiety both for people living with intellectual disability and for others involved (Moro et al., 2017). In fact, all participants in the conversation can benefit and be part of ensuring that final wishes regarding treatment at the end of life are honoured. People often feel awkward and ill at ease when faced with the opportunity for communication at the end of life (Keeley, 2017). The overarching need across end-of-life services for people living with intellectual and developmental disability, therefore, is the creation of greater awareness and knowledge based on current research regarding the importance of active listening and open discussions and communication at the end of life. This may be achieved by use of visual resources, adapted resources or individual strategies tailored specifically to the person's support need.

The end-of-life doula helps people *lean into* the pain and fear of the unknown, freeing up space and energy so they can experience the emotions that come with dying. The disability-aware end-of-life doula can work with families and professionals to assist a person living with disability through the experience, so they can stay close to the person who is dying and focus on what is most important in the time that remains. The presence of the doula can enable the person who is dying to connect more deeply with family and friends so as to enjoy the time that remains, focusing on possibilities, and on what they *can* control rather than on what they *can't*. The doula facilitates an end-of-life space for all.

Understanding How End-of-Life Doulas Can Provide Disability-Aware Support

This section illustrates different ways that the end-of-life doula may conduct their role in the disability space, with a specific focus on how disability awareness accommodates the individual support needs of the dying person with intellectual and developmental disability. Stories, based on true accounts but with details modified to protect privacy, are used to illustrate the role.

What the End-of-Life Doula Offers a Person Living with Intellectual and Developmental Disability that Health, Palliative Care, and Funeral Services Might not Provide

Frank's* story

Frank, a man in his early 50s with intellectual disability and limited verbal communication, was living in supported accommodation with two other residents. Frank's father had been diagnosed with a terminal illness (pancreatic cancer) only a few months before death was likely. Frank was thus presented with the need and opportunity to learn about the death of another person, as well as the opportunity to raise and discuss questions about his own mortality.

*Assumed names are used in this story. The stories are based on the author's doula work. Details have been changed to protect individual identity.

I began working with Frank, his housemates, disability staff, his parents and an uncle. This involved creating social stories, visuals and easy-to-read information sheets about the medical and health supports that would be involved as Frank's dad died. These included information about what treatments would be like, specific roles that Frank could play in supporting his dad (like preparing food with support staff that was easy for his dad to digest), talking about the funeral, the music Frank's dad would like to have playing and creating a legacy. Legacy items included picture collages, video clips and recorded conversations between Frank and his dad in the last few months, when they shared stories of fun times and experiences that shaped their relationship. Frank's active engagement in his dad's dying journey allowed him and his dad to build on their existing relationship, and reassured Frank's dad that he could die knowing that Frank was more aware of what was happening and would be better equipped to cope with his dad's death.

I also worked with disability staff and provided professional development through conversations about how to be present for Frank and answer his questions using language he could understand, whilst also avoiding child-like talk. Engagement with these staff also involved listening to them and acknowledging their own fears and uncertainties about death. The book by Gordon and Kacandes (2015) *Let's Talk About Death,* and the *Talking End of Life with People with Intellectual Disability* website (Wiese et al., 2018) (see Resources at the end of this chapter) provided tools, videos and modules that formed the basis of the support provided to Frank, his family and the paid support staff.

A disability-aware end-of-Life doula who understands how to navigate between the health and disability sectors, and the range of people involved, can facilitate between them, ensuring all involved are informed that information is modified and adjusted if needed, a plan is framed, and roles and expectations are expressed, described and communicated so that fear and anxiety can be reduced. The end-of-life doula can potentially commit to continuity of care for many hours over many months, according to a person's, their family's or staff wishes. Very few, if any, medical personnel can provide this intensive and consistent relationship, especially for people who are not in a hospital or hospice.

What the End-of-Life Doula Offers Families, Disability Staff and Peers Facing the Death of Someone Important to Them

David's* story

Forty-three-year-old David died of cancer, which, by the time it was discovered, had aggressively spread throughout his body. David died quickly without being fully aware of the reason for his decline. His family and community support staff were grief stricken and felt guilty for not picking up his illness earlier.

*Assumed names are used in this story. The stories are based on the author's doula work. Details have been changed to protect individual identity.

Following David's death, in a conversation over coffee with me as the end-of-life doula, there was discussion about David's extreme dislike of doctors, medical procedures and hospitals. Throughout his adult life, David had actively chosen not to engage with health professionals. In the years leading up to his death, he lived an independent, self-determined and active life. He went on holidays, worked, attended many sporting fixtures, socialised and went camping. I facilitated David's surviving family and disability staff to list on a whiteboard everything he had done in the past four or five years, the likely time he had unknowingly been living with cancer. A long list resulted, of activities and experiences David had chosen, planned and undertaken. Everyone present was then given a marker and asked to cross off everything that would not have been possible had David known he had cancer and would likely have been experiencing invasive treatment. Everything was crossed off the list. What then became clear was that David had had a great life until the end and had choice and control over it. It had been, after all, David's life. This process facilitated the capacity for those feeling guilt and blame to miss David in a celebratory way. It also provided everyone involved the opportunity to learn about the end-of-life, to explore their own mortality and to be more comfortable in supporting others in the future.

How the End-of-Life Doula Helps People to Understand Death and Plan if They Wish

Susan's* story

Susan, a woman in her late 30s with intellectual disability, lived at home with her mum and sister. Susan had recently learned that a friend at her supported employment workplace had died. She attended the friend's funeral. Afterwards, Susan expressed to her mum and staff at her workplace that she felt that the music at the funeral and a lot of what had been said had not been what she expected, and would not have matched the interests and music tastes of her friend. Susan also commented that the mood had been very sombre and that when she (Susan) died, she would like to have something more like a party.

*Assumed names are used in this story. The stories are based on the author's doula work. Details have been changed to protect individual identity.

Advance Care Planning (ACP) (see also Chapter 4) is becoming increasingly recognised as an intrinsic and crucial component of quality in end-of-life care, including for those living with disability. A consistent system where ACP is integrated within the health care of the dying person is not yet available (Watson et al., 2019), with few readily available easy-read ACP documents to help people with intellectual and developmental disability understand. Research shows, however, that support around decision making and understanding the purpose of planning improves engagement with the process of advance care planning, both for those with and without disability (Fink et al., 2019; McKenzie et al., 2017; Prince-Paul & DiFranco, 2017).

For Susan, her experience raised a lot of questions about death, what happens as we die, and what happens after death. Susan's mother was unsure how to have these conversations and how to build an advance care plan with Susan so that any expressed wishes she may have could be written down and respected. The disability staff at Susan's supported employment worksite were also inexperienced. Susan's comments happened to raise issues for them too, as they voiced awareness that they had not expressed their wishes in the event of being diagnosed with a terminal illness or dying suddenly.

As an end-of-life doula, I was engaged to work with Susan, her mum and sister, disability staff and other interested individuals. I listened to and recorded everyone's concerns, fears and questions. Susan had lots of questions like "What is dying and death?" and "Why should we plan?". The focus of my work was to ensure the use of communication styles that matched everyone's needs, and accessible documents modified to suit Susan's understanding. Using adapted resources based on the *Talking End of Life with People with Intellectual Disability* website (Wiese et al., 2018; see Resources at the end of this chapter), we initiated ACP with Susan, her family and her support personnel, with a calendar for annual review and update.

How the End-of-Life Doula Can Assist Legacy

A legacy emerges from a person with intellectual and developmental disability's recent or distant past, or the past of those who came before. It can take the form of a bequest through a will, but can also refer to life experiences, or a set of beliefs and values. Typical legacy activities facilitated by end-of-life doulas might include planning a dying person with intellectual and developmental disability's final days or moments, management of their body after death and planning for celebrations of life. End-of-life doulas can focus on creating a safe space for the dying person and those around them to do the emotional and *soul* work needed to help prepare for death. These meaningful and connecting activities can be more important for the individual, family and paid support personnel than dealing just with the physical dying.

A *soul story*, for example, is a legacy gift in the form of a written document that captures the dying person's thoughts and reflections on life. For a dying person with intellectual and developmental disability, it could be in the form of pictures, drawings, words or other symbolic representations. It is often a source of relief and peace that the person is leaving nothing unexpressed, and it is a gift of great comfort and care for loved ones left behind after the death. Different from a biography (generally a detailed description of the events of a person's life written by another person), a soul story legacy gift preserves what is precious in the dying person's reflections and is presented *in their own words or images*. Whether they include one or a combination of memories, messages, advice or instructions, an emotional legacy gift is a way of endowing those to be left behind, and generations yet to come.

Legacy gifts can include:

- Memories from when the person felt most alive,
- Hopes and dreams for loved ones,
- Advice and instructions,
- Final messages of love,
- Important things the person would like others to know or remember,
- The biggest lessons the person has learned,

- Pre-arranged cards for special upcoming events (e.g. the first baby to be born in the family, a 21st birthday card or graduation congratulations) or any other individualised future gifts.

End-of-Life Doulas Supporting Disability Staff

Professional staff in the disability sector who share a history of familiar relationship with a dying person with intellectual and developmental disability also need to feel safe and supported during this important transition. In the often emotionally laden time of dying, the focus may be the dying person and their family, and disability staff may inadvertently be forgotten.

Knowledge about the end of life is often a reflection of prior experience, and sometimes the disability staff's first ever experience with the end of life is in their professional role, when a person they support receives a terminal diagnosis. Further, many have never thought about their own mortality and inevitable end of life. If they are to maintain choice, control and dignity for those they support, therefore, they themselves may need assistance in order to minimise undue personal stress and concern (Gray & Kim, 2017).

Very few disability staff have access to education, training and professional development around end of life (Ryan et al., 2011a). Involvement of the disability-aware end-of-life doula fills this space, providing support to not only the person dying and their loved ones, but also staff who may have invested sometimes years of time and energy in relationship with the dying person.

Other Disability-Aware Doula Services

End-of-life doulas can have a powerful community education role (Grindrod & Rumbold, 2017). One internationally growing example is the Death Positive Movement, characterised by death cafés and death dinners (Lambert South & Elton, 2017). A death café is referred to as "a pop-up event where people get together to talk about death and have tea and delicious cake" (Miles & Corr, 2017, p. 152). Every word in this definition is purposeful, describing the essence of the experience (Miles & Corr, 2017). The phrase *pop-up* indicates that the events are not over-planned or regularly scheduled, and the word *people* intends to imply that “attendees are unknown to each other prior to coming together” (Miles & Corr, 2017, p. 152). The objective of the movement is “to increase awareness of death with a view to helping people make the most of their lives” (Miles & Corr, 2017, p. 153). Across the world, events like these are growing in popularity, further highlighting that the community wants to engage with dying and death, reclaiming it as a social process rather than a commercially controlled or medicalised one. The end-of-life doula as a facilitator at death cafés and dinners is uniquely positioned to bring the dialogue about death and dying to the public (Baldwin, 2017; Miles & Corr, 2017).

With respect to participation by people with intellectual and developmental disability in the Death Positive Movement, gatherings could be easily adapted to actively include people living with disability and their networks. Participants in a death café typically have two different perspectives: some have not yet experienced death in their circle of family and friends and wish to talk with others about their beliefs on death and dying, while others will have had some experience of death. Disability-aware end-of-life doulas are uniquely placed to facilitate death cafes and dinners inclusive of people with intellectual and developmental disability. At a practical level, this could mean information using minimal jargon or medical terminology, increased use of visuals, videos and other resources which can open up discussions and allow expression of thoughts and feelings in a safe way (Lambert South & Elton, 2017).

Challenges for the Doula in the Disability Space

My experience as an end-of-life doula in the disability space has identified a number of challenges, all of which will require attention in future if the disability-aware end-of-life doula role is to fully realise its helpful role.

The evidence suggests that people living with intellectual and developmental disability are viewed, often with little evidence, to have limited ability to engage with planning their death or funeral, and require protection from the stress, anxiety and discomfort that is perceived to surround the end of life (Lord et al., 2017; Tuffrey-Wijne & Rose, 2017; Wiese et al., 2015). Jenkinson (2015) challenges this protection, suggesting instead that taking an active role in one's end of life can often facilitate better coping and less suffering for those left behind. Currently unknown is the extent to which the presence and support offered by a disability-aware end-of-life doula provide a demonstrable positive influence, resulting in increased understanding, choice and decision making at the end of life, and offer an avenue for future research.

An additional challenge surrounds payment for the end-of-life doula service. For people with intellectual and developmental disability, who as a group experience increased risk of poverty (Emerson, 2007), affordability can represent a significant barrier. In Australia, the introduction of the National Disability Insurance Scheme Act (NDIS) in 2013 was designed to provide the person-centred support system and funding to better meet the individual needs of persons with disability (Australian Government Productivity Commission, 2011; Wark, 2015). However, debate continues about the extent to which ageing and dying supports should be approached within this funding scheme (Wark, 2015). This issue has implications for the service fee of an end-of-life doula, resulting in confusion for both the individual dying person and the doula profession.

I have observed the challenges that facilitating a so-called good death presents for people living with intellectual and developmental disability, challenges largely attributable to the dying person's, family, and disability staff's lack of knowledge, skills, capacity and understanding about the end of life. As I argued in the introduction of this chapter, this is partly attributable to the commercialisation and medicalisation of dying and death. Individuals in the community have lost touch with dying, and everyday knowledge and experiences have been also lost. This translates to the need for community education, not only in reclaiming dying and death as a community responsibility, but also about the role of a doula in that space. I have witnessed how little understanding there is about the assistance an end-of-life doula can offer individuals, families, disability support workers and medical, health and palliative care services.

Within the doula profession, there are also some inconsistencies about what comprises a doula service. There is general agreement that these inconsistencies need to be addressed (Rawlings et al., 2019, 2021). In a study with end-of-life doulas, Rawlings et al. (2021), concluded that there is a need to demystify the role of the doula, develop business models and create consistency in service expectations for people, both with and without disability, in order to increase understanding and collaboration and alleviate suspicion.

While there is an argument for better and consistent communication about the role of the end-of-life doula in the disability sector, this needs to be balanced with maintaining an individualised approach. When approaching end of life, a person does not always know what they want. Safe, consistent and open discourse is needed to mould wishes as they unfold. It is, after all, that person's death, their own experience, and it is only right to support the person in their choices (Ryan et al., 2011b). Rightly too, end-of-life doulas need to continue their quest to individually tailor services to the person, their family and their supports, so that their end-of-life journey may be undertaken with dignity and respect at every stage (Araujo de Carvalho et al., 2017).

Finally, there is work yet to be done to develop an international Community of Practice (COP) and to engage in co-design approaches to further develop the end-of-life doula role and to ensure it is seen as a valued addition at end of life as an available option for those with intellectual and developmental disability who choose to include it.

Conclusion

This chapter has highlighted the potential for end-of-life doulas to assist people with intellectual and developmental disability, their families and disability staff to engage openly in discussions of death and dying, to maximise choice and control, and to plan, just as one may choose to do at any other life transition. Into the future, the end-of-life doula would ideally be accessible to everyone, including those with intellectual and developmental disability. With growing understanding of the role, a positive impact of the disability-aware end-of-life doula role will be determined.

The benefit of the end-of-life doula role is in its potential for nimbleness and flexibility in responding to what is needed when it is needed. The journey to death can be a long one, much longer than most people acknowledge, and thinking about what would give the best death is better started sooner rather than later. As stated by a midwife nursing colleague, "We all need a purple folder, the one that outlines what we want, how we want it and how our life is to be celebrated at the end" (Anonymous Personal Communication, n.d.). We discuss, debate and plan for our careers, our aspirations and what we want in life, but we need to add in planning for the final chapter as well. Each of us, including those living with intellectual and other disability, can benefit from having someone walk alongside, providing support, education, options and championing our choices. It can make an enormous difference.

Reflection

1. What do you think you would need to learn about, understand and/or reflect upon to be able to work comfortably in the end-of-life space with people living with intellectual and developmental disability?
2. If you could spend time with an end-of-life doula, what questions you would ask? What are the things that are least clear or most challenging to you about dying and death?

Resources

Books

1. Butler, K. (2020). *The art of dying well: A practical guide to a good end of life*. Scribner. ISBN: 9781501135477; ISBN-10: 1501135473

2. Cochran, D. (2019). *Accompanying the dying: Practical, heart-centered wisdom for end-of-life doulas and health care advocates.* Sacred Life Publishers. ISBN: 9780989659352, ISBN-10: 0989659356
3. Glenn, A. W. (2017). *Holding space: On loving, dying, and letting go.* Parallax Press. ISBN: 9781941529799
4. Gordon, S., & Kacandes, I. (2015). *Let's talk about death: Asking questions that profoundly change the way we live and die.* Prometheus Books. ISBN 781633881129
5. Hebb, M. (2018). *Let's talk about death (over dinner): An invitation and guide to life's most important conversation.* Orion Publishing. ISBN: 9781841883014

Websites

1. *Death over dinner: Let's have dinner and talk about death:* A guide to planning a conversation about death over a meal. https://deathoverdinner.org/
2. *End of life and palliative care for people living with a disability.* Fact sheets about planning ahead and palliative care for people with disability. https://www.betterhealth.vic.gov.au/health/ServicesAndSupport/End-of-life-and-palliative-care-for-people-living-with-a-disability
3. *End of life planning*. The work of a death doula, training opportunities and resources. https://endoflifeplanning.com.au/portfolio/beth-obrien/
4. *International End-of-life Doula Association.* A professional association with training and events about the doula role. https://www.inelda.org/
5. *Talking End of Life…with People with Intellectual Disability (TEL).* Australian website with 12 modules on end of life and people with intellectual and developmental disability. Designed for disability support workers but is also helpful for others including families, health professionals and educators. https://www.caresearch.com.au/TEL/

References

Araujo de Carvalho, I., Epping-Jordan, J., Pot, A. M., Kelley, E., Toro, N., Thiyagarajan, J. A., & Beard, J. R. (2017). Organizing integrated health-care services to meet older people's needs. *Bulletin of the World Health Organization, 95*(11), 756–763. https://doi.org/10.2471/BLT.16.187617

Australian Government Productivity Commission. (2011). *Disability care and support: Public inquiry.* https://www.pc.gov.au/inquiries/completed/disability-support

Baldwin, P. K. (2017). Death cafés: Death doulas and family communication. *Behavioral Sciences, 7*(2), 26. https://doi.org/10.3390/bs7020026

Burge, F., Lawson, B., Johnston, G., Asada, Y., McIntyre, P. F., & Flowerdew, G. (2015). Preferred and actual location of death: What factors enable a preferred home death? *Journal of Palliative Medicine, 18*(12), 1054–1059. https://doi.org/10.1089/jpm.2015.0177

Emerson, E. (2007). Poverty and people with intellectual disabilities. *Mental Retardation and Developmental Disabilities Research Review, 13*(2), 107–113. https://doi.org/10.1002/mrdd.20144

Feifel, H. (Ed.). (1959). *The meaning of death.* McGraw-Hill.

Fink, R. M., Somes, E., Brackett, H., Shanbhag, P., Anderson, A. N., & Lum, H. D. (2019). Evaluation of quality improvement initiatives to improve and sustain advance care planning completion and documentation. *Journal of Hospice & Palliative Nursing, 21*(1), 71–79. https://doi.org/10.1097/NJH.0000000000000490

Flynn, S., Hulbert-Williams, N. J., Hulbert-Williams, L., & Bramwell, R. (2016). "You don't know what's wrong with you": An exploration of cancer-related experiences in people with an intellectual disability. *Psycho-Oncology, 25*(10), 1198–1205. https://doi.org/10.1002/pon.4211

Foreman, L. M., Hunt, R. W., Luke, C. G., & Roder, D. M. (2006). Factors predictive of preferred place of death in the general population of South Australia. *Palliative Medicine, 20*(4), 447–453. https://doi.org/10.1191/0269216306pm1149oa

Fukuzawa, R. K., & Kondo, K. T. (2017). A holistic view from birth to the end of life: End-of-life doulas and new developments in end-of-life care in the West. *International Journal of Palliative Nursing, 23*(12), 612–619. https://doi.org/10.12968/ijpn.2017.23.12.612

Gomes, M. F., & McVilly, K. R. (2019). The characteristics of effective staff teams in disability services. *Journal of Policy and Practice in Intellectual Disabilities, 16*(3), 191–200. https://doi.org/10.1111/jppi.12280

Gordon, S., & Kacandes, I. (2015). *Let's talk about death: Asking questions that profoundly change the way we live and die* (p. 781633881129). Prometheus Books.

Gray, J. A., & Kim, J. (2017). Direct care workers' experiences of grief and needs for support. *Journal of Applied Research in Intellectual Disabilities, 30*(6), 995–1006. https://doi.org/10.1111/jar.12339

Grindrod, A., & Rumbold, B. (2017). Providing end-of-life care in disability community living services: An organizational capacity-building model using a public health approach. *Journal of Applied Research in Intellectual Disabilities, 30*(6), 1125–1137. https://doi.org/10.1111/jar.12372

Haveman, M., Perry, J., Salvador-Carulla, L., Walsh, P. N., Kerr, M., van Schrojenstein Lantman-de Valk, H. M. J., & Weber, G. (2011). Ageing and health status in adults with intellectual disabilities: Results of the European POMONA II study. *Journal of Intellectual & Developmental Disability, 36*(1), 49–60. https://doi.org/10.3109/13668250.2010.549464

Jenkinson, S. (2015). *Die wise: A manifesto for sanity and soul*. North Atlantic Books. ISBN: 1583949747, 9781583949740.

Keeley, M. P. (2017). Family communication at the end of life. *Behavioral Sciences, 7*(3), 45. https://doi.org/10.3390/bs7030045

Kelly, S. (2015). Greening death: Reclaiming burial practices and restoring our tie to the earth. *Omega Journal of Death and Dying, 75*(4), 414–416. https://doi.org/10.1177/0030222816676023

Kirkendall, A., Linton, K., & Farris, S. (2017). Intellectual disabilities and decision making at end of life: A literature review. *Journal of Applied Research in Intellectual Disabilities, 30*(6), 982–994. https://doi.org/10.1111/jar.12270

Krawczyk, M., & Rush, M. (2020). Describing the end-of-life doula role and practices of care: Perspectives from four countries. *Palliative Care & Social Practice, 14*, 1–15. https://doi.org/10.1177/2632352420973226

Kübler-Ross, E., Wessler, S., & Avioli, L. V. (1972). On death and dying. *JAMA, 221*(2), 174–179. https://doi.org/10.1001/jama.1972.03200150040010

Lambert South, A., & Elton, J. (2017). Contradictions and promise for end-of-life communication among family and friends: Death over dinner conversations. *Behavioral Sciences, 7*(2), 24. https://doi.org/10.3390/bs7020024

Lindley, L. C., Colman, M. B., & Meadows, J. T. (2017). Children with intellectual disability and hospice utilization. *Journal of Hospice and Palliative Nursing, 19*(1), 28–33. https://doi.org/10.1097/NJH.0000000000000301

Lord, A. J., Field, S., & Smith, I. C. (2017). The experiences of staff who support people with intellectual disability on issues about death, dying and bereavement: A metasynthesis. *Journal of Applied Research in Intellectual Disabilities, 30*(6), 1007–1021. https://doi.org/10.1111/jar.12376

Marks, B., Sisirak, M. P. H., & Hsieh, K. (2008). Health services, health promotion, and health literacy: Report from the state of the science in aging with developmental disabilities conference. *Disability and Health Journal, 1*, 136–142. https://doi.org/10.1016/j.dhjo.2008.04.003

McGuire, B., & McEvoy, J. (2016). Service models, the macrosystem and professional issues in clinical practice. In A. Carr, C. Linehan, G. O'Reilly, P. N. Walsh, & J. McEvoy (Eds.), *The handbook of intellectual disability and clinical psychology practice* (2nd ed., pp. 811–826). ISBN 9781138806351

McKenzie, N., Mirfin-Veitch, B., Conder, J., & Brandford, S. (2017). "I'm still here": Exploring what matters to people with intellectual disability during advance care planning. *Journal of Applied Research in Intellectual Disabilities, 30*(6), 1089–1098. https://doi.org/10.1111/jar.12355

Miles, L., & Corr, C. A. (2017). Death café: What is it and what we can learn from it. *Journal of Death and Dying, 75*(2), 151–165. https://doi.org/10.1177/0030222815612602

Moro, T. T., Savage, T. A., & Gehlert, S. (2017). Agency, social and healthcare supports for adults with intellectual disability at the end of life in out-of-home, non-institutional community residences in Western nations: A literature review. *Journal of Applied Research in Intellectual Disabilities, 30*(6), 1045–1056. https://doi.org/10.1111/jar.12374

Prince-Paul, M., & DiFranco, E. (2017). Upstreaming and normalizing advance care planning conversations—A public health approach. *Behavioral Sciences, 7*(2), 18. https://doi.org/10.3390/bs7020018

Rawlings, D., Litster, C., Miller-Lewis, L., Tieman, J., & Swetenham, K. (2020). The voices of death doulas about their role in end-of-life care. *Health and Social Care in the Community, 28*(1), 12–21. https://doi.org/10.1111/hsc.12833

Rawlings, D., Litster, C., Miller-Lewis, L., Tieman, J., & Swetenham, K. (2021). End-of-life doulas: A qualitative analysis of interviews with Australian and international death doulas on their role. *Health and Social Care in the Community, 29*(2), 574–587. https://doi.org/10.1111/hsc.13120

Rawlings, D., Tieman, J., Miller-Lewis, L., & Swetenham, K. (2019). What role do death doulas play in end-of-life care? A systematic review. *Health and Social Care in the Community, 27*(3), e82–e94. https://doi.org/10.1111/hsc.12660

Ryan, K., Guerin, S., Dodd, P., & McEvoy, J. (2011a). Communication contexts about illness, death and dying for people with intellectual disabilities and life-limiting illness. *Palliative and Supportive Care., 9*(2), 201–208. https://doi.org/10.1017/S1478951511000137

Ryan, K., Guerin, S., Dodd, P., & McEvoy, J. (2011b). End-of-life care for people with intellectual disabilities: Paid carer perspectives. *Journal of Applied Research in Intellectual Disabilities, 24*(3), 199–207. https://doi.org/10.1111/j.1468-3148.2010.00605.x

Schalock, R. L., Gardner, J. F., & Bradley, V. J. (2007). *Quality of life for persons with intellectual and other developmental disabilities: Applications across individuals, organizations, communities, and systems.* American Association on Intellectual and Developmental Disabilities.

Swerissen, H., Duckett, S., & Farmer, J. (2014). *Dying well* (Grattan Institute Report No. 2014-10). ISBN: 978-1-925015-61*4. https://grattan.edu.au/wp-content/uploads/2014/09/815-dying-well.pdf

Tenzek, K., & Depner, R. (2017). Still searching: A meta-synthesis of a good death from the bereaved family member perspective. *Behavioral Sciences, 7*(2), 1–18. https://doi.org/10.3390/bs7020025

Tuffrey-Wijne, I., Bernal, J., & Hollins, S. (2010). Disclosure and understanding of cancer diagnosis and prognosis for people with intellectual disabilities: Findings from an ethnographic study. *European Journal of Oncology Nursing, 14*(3), 224–230. https://doi.org/10.1016/j.ejon.2010.01.021

Tuffrey-Wijne, I., & Rose, T. (2017). Investigating the factors that affect the communication of death-related bad news to people with intellectual disabilities by staff in residential and supported living services: An interview study. *Journal of Intellectual Disability Research., 61*(8), 727–736. https://doi.org/10.1111/jir.12375

Voss, H., Vogel, A., Wagemans, A. M. A., Francke, A. L., Metsemakers, J. F. M., Courtens, A. M., & Veer, A. J. E. (2019). Advance care planning in the palliative phase of people with intellectual disabilities: Analysis of medical files and interviews. *Journal of Intellectual Disability Research., 63*(10), 1262–1272. https://doi.org/10.1111/jir.12664

Wark, S. (2015). Ageing, end-of-life care, and the National Disability Insurance Scheme: What can we learn from overseas? *Journal of Intellectual and*

Developmental Disability, 40(1), 92–98. https://doi.org/10.3109/13668250.2014.969201

Wark, S., Hussain, R., Müller, A., Ryan, P., & Parmenter, T. (2017). Challenges in providing end-of-life care for people with intellectual disability: Health services access. *Journal of Applied Research in Intellectual Disabilities, 30*(6), 1151–1159. https://doi.org/10.1111/jar.12408

Watson, A., Weaver, M., Jacobs, S., & Lyon, M. E. (2019). Interdisciplinary communication: Documentation of advance care planning and end-of-life care in adolescents and young adults with cancer. *Journal of Hospice and Palliative Nursing, 21*(3), 215–222. https://doi.org/10.1097/NJH.0000000000000512

Webber, R., Bowers, B., & Bigby, C. (2010). Hospital experiences of older people with intellectual disability: Responses of group home staff and family members. *Journal of Intellectual & Developmental Disability, 35*(3), 155–164. https://doi.org/10.3109/13668250.2010.491071

Wiese, M., Stancliffe, R. J., Read, S., Jeltes, G., & Clayton, J. M. (2015). Learning about dying, death, and end-of-life planning: Current issues informing future actions. *Journal of Intellectual and Developmental Disability, 40*(2), 230–235. https://doi.org/10.3109/13668250.2014.998183

Wiese, M. Y., Stancliffe, R. J., Wagstaff, S., Tieman, J., Jeltes, G., & Clayton, J. (2018). *Talking End of Life … with people with intellectual disability (TEL)* [website]. https://www.caresearch.com.au/TEL/

World Health Organization. (2016). *Definition: Intellectual disability*. http://www.euro.who.int/en/health-topics/noncommunicable-diseases/mental-health/news/news/2010/15/childrens-right-to-family-life/definition-intellectual-disability

12

Palliative and End-of-Life Care for Children with Intellectual Disabilities

Carmel Doyle and Claire Quinn

Understanding Palliative and End-of-Life Care for Children with Intellectual Disabilities

The death of a child at any age is not frequently considered due to the profound sadness it brings to families. Often the difficulties experienced are hidden from view and families can struggle to gain support and understanding regarding their needs and those of their child. The concept of paediatric palliative care focuses on the relief of suffering of the child and family by controlling pain and other distressing symptoms, integrating psychological and spiritual aspects of care, offering a support system to allow children and their families to live as active and full a life

C. Doyle (✉)
School of Nursing & Midwifery, Trinity College, Dublin, Ireland
e-mail: doylec5@tcd.ie

C. Quinn
School of Nursing & Midwifery, NUI Galway, Galway, Ireland

R. J. Stancliffe et al. (eds.), *End of Life and People with Intellectual and Developmental Disability*, https://doi.org/10.1007/978-3-030-98697-1_12

as possible, and supporting families to help them cope into bereavement (Health Service Executive & Faculty of Paediatrics, RCPI, 2016).

Children's palliative care is an active and total approach to care, from diagnosis or recognition, throughout the child's life, death and beyond. It embraces physical, emotional, social and spiritual elements and focuses on the enhancement of quality of life for the child or young person and support for the family (Together for Short Lives, 2018). This means that children's palliative care seeks to provide a child with the best quality of life possible from the time they are diagnosed with a palliative care need until the point where specialist palliative services are no longer required or until the end of life. Palliative care provides not just physical care such as pain relief and symptom management but also emotional, social and spiritual support encompassing the family in the process (Together for Short Lives, 2018). A common focus of palliative care is the preparation for death that is anticipated (Grinyer, 2012).

The Specific Groups of Children with Intellectual Disabilities Requiring Palliative Care

Although understanding of and reference to children with intellectual disabilities may seem evident, much that is written regarding this cohort suggests a variety of potentially ambiguous terms. These include children with life-limiting conditions (Together for Short Lives, 2018), children with complex care needs (Brenner, Kidston et al., 2018), children who are technology dependent (Brenner et al., 2021), and children with life-limiting neurodevelopmental disabilities (Guerin et al., 2020). Children with intellectual disabilities may have conditions with limitations in the functioning of the brain or neuromuscular system encompassing congenital or acquired conditions, and a variety of neurological, genetic or metabolic aetiologies that can limit the lifespan (Goldstein & Reynolds, 2011; Health Service Executive & Faculty of Paediatrics, RCPI, 2016; Hudson, 2014). Many children with intellectual disabilities have no formal diagnosis, while some will have more common conditions such as chromosomal abnormalities, severe cerebral palsy and epilepsy.

In determining who might be expected to require children's palliative or end-of-life care, Together for Short Lives (2019) identified four typical groups (Table 12.1).

Group 4 includes those children likely to have intellectual disability. The vast majority of these children live at home with their families and may periodically require hospital and hospice-based care where available (Guerin et al., 2020). Many of these children will undergo frequent hospital and intensive care unit admissions, receive care from multiple subspecialty teams, and often suffer from distressing symptoms (Cohen et al., 2011; Schwantes & O'Brien, 2014; Walter et al., 2013). Care can be fraught with complexity of medical regimes, technology dependence, medication management and frequent health instability (Brenner et al., 2021; Doyle, 2020, 2021; Guerin et al., 2020). As a result, interaction and supports are sought regularly with many different healthcare providers (Courtney et al., 2018; Doyle, 2020).

Table 12.1 Four groups of children most likely to have palliative care needs by summary of condition

Group	Summary of condition
1	Life-threatening conditions for which curative treatment may be feasible but can fail. Where access to palliative care services may be necessary when treatment fails, children in long-term remission or following successful curative treatment are not included (Examples: cancer, irreversible organ failures of heart, liver, kidney)
2	Conditions where premature death is inevitable, where there may be long periods of intensive treatment aimed at prolonging life and allowing participation in normal activities (Example: cystic fibrosis)
3	Progressive conditions without curative treatment options, where treatment is exclusively palliative and may commonly extend over many years (Examples: Batten disease, mucopolysaccharidoses, muscular dystrophy)
4	Irreversible, but non-progressive conditions causing severe disability leading to susceptibility to health complications and likelihood of premature death (Examples: severe cerebral palsy, multiple disabilities such as those following brain or spinal cord insult)

While the number of children with intellectual disability can be difficult to ascertain, it is estimated that there are 45,325 children between the ages of 0–19 years with complex disabilities (prevalence of 4%) requiring services in Ireland (Health Service Executive & Faculty of Paediatrics, RCPI, 2016). Approximately 2.5% (351,000) of children in the UK are believed to have an intellectual disability (Office for National Statistics, 2019), while Zablotsky et al. (2019) determined that in the US about 1 in 6 (17%) children aged 3–17 years were diagnosed with a developmental disability, with a portion of these specifically identified as intellectual disability (1.2%).

Notably, in contrast to adult palliative care, children with cancer and malignancies constitute a minority of those referred to palliative care services. In Ireland, there are in the region of 490 childhood deaths per year, with approximately 350 deaths per year from life-limiting conditions (Department of Health & Children, 2010). Of these childhood deaths due to life-limiting conditions, the majority occur in the first year of life. Based on the most recent epidemiological data from England, the prevalence of children with life-limiting illnesses has risen over the last 17 years and is predicted to continue to rise in the future (Fraser et al., 2021). It is estimated that there are 8,311 children (0–18 years of age) in Ireland with a life-limiting condition. Of these, 19% (1,579 children) are said to be either unstable (15%), deteriorating (2.2%) or dying in any one year (Fraser et al., 2015). Approximately, 5,000 children die each year in England and Wales, 2,500 as a result of a life-limiting or life-threatening condition (Fraser et al., 2013).

Internationally, malignancies constitute approximately only 22% of referrals to paediatric palliative care, with diseases of the nervous system (39.1%), congenital conditions or illnesses originating in the perinatal period (22.1%) or other illnesses (16.7%) making up the majority (Widger et al., 2007). McMahon et al. (2018) received 83 perinatal referrals over a 4-year period, with chromosomal abnormalities accounting for 35% of diagnoses, congenital heart disease 25%, complex neurological abnormalities 11% and renal agenesis 4%. Children with intellectual disabilities may have specialist palliative care needs at different stages during the illness trajectory. Referrals are likely to increase as the care of children with complex disability improves and awareness of paediatric palliative care intensifies.

Children's Palliative Care Differs from Adult Palliative Care

While adult palliative care is well established, palliative care for children is developing. It is not yet widely recognised but gaining some momentum in research and policy in recent years. There are many differences in caring for the dying child versus the dying adult, including:

- The number of children dying is small.
- The conditions are extremely rare with diagnoses specific to childhood.
- Predicting a prognosis can be difficult.
- The palliative phase is often much longer and can be episodic and unpredictable.
- Children may experience several apparently terminal phases.
- Care embraces the whole family and uses a model of family centred care.
- Parents are typically involved in care and require adequate resources to support them with the responsibility of personal and nursing care.
- Siblings are vulnerable and parents must continue to care for them while often providing 24-hour care to a sick child.
- Conditions are sometimes familial—other children in the family may be living with, or have died from, the same condition.
- Children's ability to communicate and understand varies according to their age or stage of development.
- The provision of education and play when a child is sick is essential.
- Grief over the death of a child is known to be more intense, long lasting and complicated.
- Professionals caring for children need specific training and skills to communicate effectively with the child.
- Ethical and legal issues of self-determination (Department of Health & Children, 2010).

For children with intellectual disabilities, due to the complexity of care often arising from rare conditions and unknown trajectories, the palliative phase if often much longer and more unpredictable (Guerin et al., 2020). The provision of palliative and end-of-life care to children with intellectual disabilities can present challenges. Children born with complex care needs may access palliative care for their whole life, however long that trajectory may be with little or no emphasis on a curative phase, unlike children who might present with cancer or malignancies who undergo treatment aimed at cure. The provision of paediatric palliative care and access to adequate services can also be fragmented and inconsistent, whereas models of adult palliative care tend to be well established. This may be due to fewer children requiring these services. The profile of children requiring palliative care has also changed in recent decades due to technological advances and increased survival rates of preterm babies (Department of Health, 2008; Myrhaug et al., 2019).

Family Centred Approaches to Delivery of Palliative Care

While disability services provide most support to children with intellectual disability and their families, services tend to be chronically under resourced and, as previously noted, palliative care services are also less well developed (Guerin et al., 2020). The aim of support is to provide maximum levels of care as close to home as possible and ensure access to specialist supports when required. Optimal care should be facilitated through integrated service networks with supports from tertiary and quaternary specialties as required (Guerin et al., 2020). The preferred care setting is within the home with minimal interruption of normal routines (Carter et al., 2016; Olsen & Maslin-Prothero, 2001). While this has positive benefits for both the child and family, recognition of the care burden on parents needs to be acknowledged (McCann et al., 2015; Scott, 2000). Approximately 98% of children with intellectual disabilities live at home with their parents (Hourigan et al., 2018). It is assumed that children with disabilities and possible technology dependence will be cared for at home with little or no alternative afforded to parents

(Carnevale et al., 2008; Lindahl & Lindblad, 2011; Mendes, 2013). However, needs can shift over time with the birth of new siblings, the illness of another sibling with the same condition and other life events (Grinyer, 2012), therefore flexibility is an imperative. A range of care settings should be available for delivery of palliative and end-of-life care.

The interface between disability and palliative care services can be problematic and unclear. Universal services such as the general practitioner provide a good foundation with core palliative care services delivered mainly in the community, encompassing a range of services including community nursing and hospice care (Harrop & Edwards, 2013). Specialist palliative care services tend to be delivered through hospice and tertiary specialist paediatric services and are more concerned with symptom control (Harrop & Edwards, 2013).

Palliative care benefits children and families by providing expert symptom management, care coordination and supportive counselling, including advance care planning (Heckford & Beringer, 2014; Miller et al., 2012; Schwantes & O'Brien, 2014). However, identifying children most likely to benefit from paediatric palliative care can pose a challenge. Sometimes referrals are made too late in the illness trajectory (Harrop & Edwards, 2013). This may be due in part to healthcare professionals associating palliative care with dying and death, thereby failing to recognise its role in enhancing quality of life (Twamley et al., 2014). Early integration of paediatric palliative care is recommended by national organisations for a variety of paediatric illnesses and has been shown to be feasible and acceptable (American Academy of Paediatrics, 2013; Berlinger et al., 2013; Department of Health, 2004; Lafond et al., 2015; Mack & Wolfe, 2006). While gaining some momentum in adult palliative care, rarely is advance care planning considered for children as it requires initiation of difficult conversations. However, Liberman et al. (2016) reported that while challenging, early conversations in the outpatient setting can increase access to palliative care for children and often improve knowledge and comfort with the topic.

Approaches to deciding who requires palliative care are important, with one of these being the use of the Paediatric Palliative Screening Scale (PaPaS; Song et al., 2021) (see Resources at the end of this chapter), a tool that can be used with children and their families irrespective of the origin of their condition. Such an instrument is important in capturing the child's needs at an early stage and assists healthcare professionals to accurately identify children with palliative care needs. This type of assessment captures detail about the illness trajectory and its impact on daily activities, expected outcome of the illness and burden of treatment, symptom burden, preferences of the child and family, and lastly, life expectancy. This tool may be helpful in supporting timely referrals and ultimately be of benefit to the child and family.

The process of referral to palliative care services may initially commence as an information gathering and wayfinding exercise. Referrals may be received as early as the antenatal period where there is identification of life-limiting conditions. Usually, the groupings identified in Table 12.1 assist in categorising those requiring palliative care. However, with palliative care services under resourced, not all children receive optimal services. Therefore, providers need to focus on those children and families that are likely to benefit most from the available paediatric palliative care services, including symptom control and end-of-life care (Table 12.2).

As children with intellectual disabilities may have a long palliative trajectory with many surviving longer than possibly predicted, transition from paediatric to adult palliative care must be considered. This would ideally be seamless in nature, however, can be difficult to achieve. Often strong relationships have been formed through paediatric services meaning a daunting transition for the child and family. Brown et al. (2021) acknowledged that the experience of transition continues to be negative for people with intellectual disabilities, and parents are often forced to drive the fragmented and poorly managed transition process. A good transition requires coordination with a strategic focus but should encompass key elements: start early, offer flexibility and individually tailored to meet the needs of the child and family (Brown et al., 2021).

The need for a multidisciplinary integrative approach to paediatric palliative care is essential in the provision of seamless care. This type of approach encompasses the child and family and is centred on the needs of the individual with the family, often including wider family members and not just the child, siblings and parents. Clear, concise communication is a fundamental component of multidisciplinary working,

Table 12.2 Paediatric palliative care services and their function

Service type	Function
Respite support	Usually for a defined period /number of nights and may be delivered in the child's own home or in a specified respite facility. Nursing care is usually provided by suitably qualified/experienced staff
Acute admission to hospital or hospice facility	For the management of poorly controlled symptoms usually at short notice. Sometimes initiated for specialist treatment such as syringe drivers, buccal medication, second line opioids or patient controlled analgesia. May be able to return home with support from community health
Supported discharge from hospital	Children already known to palliative care services may be discharged earlier following elective admission for procedures. This can allow for more normality in a less medicalised environment
End-of-life care	This may involve a new referral or emergency referral when a child's status has changed. The child and family should be supported in the environment of their choice (e.g. home, hospital, hospice)
Compassionate withdrawal of life-sustaining treatment	Paediatric palliative care can often support withdrawal of life-sustaining treatment (e.g. ventilatory support) when families and the multidisciplinary team have made a decision that further treatment is futile and not in the best interests of the child
Sibling support	Specific support programmes may be run for siblings/bereaved siblings. This may involve weekend activities/camps aimed at normalising the experience of having a chronically ill sibling

(continued)

Table 12.2 (continued)

Service type	Function
Child and adult bereavement services	Specialist bereavement support may be offered as individual or group sessions. There may also be a remembrance day offered

Adapted from Harrop and Edwards (2013)

with each member holding distinct roles and sharing common goals for the child and family. Working in multidisciplinary teams requires specialist training and education (Bergsträsser et al., 2017). Paediatric palliative care approaches require agreement on the elements of caring and avoidance of conflicting information for the child and family. Each multidisciplinary team should appoint an identified key worker, often a nurse, as the liaison person irrespective of where care delivery takes place (Health Services Executive, 2020).

The composition of the multidisciplinary team varies according to the child's needs, but might include any, or all, of the following:

- The child and family
- Social worker
- Dietician
- General practitioner
- Palliative care consultant and team
- Pharmacist
- Physiotherapist
- Occupational therapist
- Psychologist
- Chaplain
- Teacher
- Play specialist
- Nurses—community/hospital/hospice
- Voluntary organisations

The provision of integrated care across Europe has been widely emphasised in recent years, culminating in the provision of principles to be adopted for effective, personalised care of children with complex care needs (Brenner, O'Shea, et al., 2018). The three main principles underpinned by a child-centred focus are (1) access to care, (2) co-creation of care with parents and (3) effective integrated governance. These guiding principles offer a benchmark for existing services which providers and policymakers can adopt and use to assess current and future service developments. More recently, Ireland has developed a national framework for clinical governance and operational arrangements for supporting a model of care for children with life-limiting conditions towards the end of life (Health Services Executive, 2020). This marks a fundamental milestone in its explicit recognition of children with intellectual disabilities and their need for paediatric palliative and end-of-life care.

Specialist paediatric palliative care is a relatively new area, with evidence suggesting the benefits of early engagement (Harrop & Edwards, 2013). Specialist roles see nurses in particular working across a range of settings delivering family centred palliative and end-of-life care. Settings include disability services, schools, respite facilities, voluntary homecare providers and tertiary hospital settings. In providing care for the family, the delicate nature of the task must be emphasised including the navigation of key ethical issues, when decisions might be required about life-sustaining treatment or provision of palliative and end-of-life care (Barone & Unguru, 2018; Furingsten et al., 2015). This makes skilled and sensitive communication with the child and family of paramount importance (Marsac et al., 2018; Sharkey et al., 2016; Ulrich et al., 2018).

Symptom Management

The assessment of symptoms related to palliative or end of life and the evaluation of the outcomes of symptom management require considerable attention to detail and dedication to the task. The single most important skill in assessment is the ability to engage fully and secure trust with the child and their family (Hain et al., 2021; McCrate Protus et al., 2014; Together for Short Lives, 2018). Without this, all other aspects of care are diluted, and the goals of palliative care difficult to achieve.

The professional cannot begin to respond to the experience of the child without meaningful engagement that facilitates a deep understanding of the impact of the illness on the child and family. This is particularly challenging for the child with an intellectual disability where communication and goal setting, in particular end-of-life care wishes, may go undisclosed, unknown or unheard (Feudtner, 2007; Quinn, in press). Parents are recognised as the key advocate for children with a significant disability and require collaborative opportunities to participate in advance care planning about the end of life (Kuo et al., 2011; Quinn, in press; Simon et al., 2010; Vollenbroich et al., 2016).

Assessment is a constant thread in care for individuals receiving palliative care and is an all-important element of care with children with intellectual disabilities. The symptom assessment process is often guided by tools which examine the child and family caregiver experience with a view to addressing current needs and anticipating potential future needs, irrespective of the length of the expected trajectory (Hain et al., 2021).

Key principles to guide symptom assessment include:

- Identify a suitable time and private space to take a history and conduct an assessment.
- Create an unrushed environment. Adapt and use available technology to assist with communication where appropriate.
- Recognise the essential importance of the therapeutic relationship in building rapport and trust with the child and family.
- Do not make assumptions or jump to conclusions. Listen, see, sense and feel all that is told to you (and not told to you) and make a clear, informed, unbiased clinical judgement.
- Ask the child/parent/caregiver to describe the symptom in detail. Parent/caregiver reporting about the child with intellectual disability is key (Jassal & Hain, 2016; McCrate Protus et al., 2014).

In assessing symptoms, it is important to consider both qualitative and quantitative aspects of the symptom. Some aspects of symptoms can only be assessed through a qualitative approach. For example, how pain impacts on the child's sleep, diet, mood etc. can be explored with the child and their parents/ caregivers, but cannot be quantified or measured. Other aspects of symptoms are quantifiable. For example, the severity of pain can be measured using pain intensity rating scales. A variety of assessment tools and computer-based mobile apps have also been developed in recent years and may assist the identification and experience of symptoms (Brock et al., 2018). When assessing symptoms that are indicative of the end of life it is important the following considerations are addressed:

- Time of onset of symptom.
- Associations with the symptom, e.g. movement, eating etc.
- Duration of symptom, e.g. is it constant or episodic?
- Frequency and severity of symptom.
- Symptom associations: What exacerbates the symptom? What relieves the symptom? Impact on quality of life, e.g. sleep, mood, function, social interaction.
- Believe the child or parent/caregivers responses about the symptom.
- What meaning do the child and family attach to the symptom.
- A full physical examination.
- Current medications/treatments noted.
- Whether any new medications or treatments have been recently started.
- Consider the use of diversional or alternative approaches to assessment e.g. through play.
- Consider 'break periods' for the child and family if required.
- Make allowances for communication challenges or non-English speaking children and families. Consider the use specific assessment tools that can assist with identification of symptoms.
- Document all findings (McCrate Protus et al., 2014).

The assessment of specific symptoms should result in the identification of possible cause and approach to treatment that can be implemented. The most common symptoms in palliative and end of life care for children with intellectual disabilities are outlined in Table 12.3, including their cause and associated treatment approach.

Table 12.3 Common symptoms by cause by palliative and end-of-life care treatment approaches

Symptom	Cause	Treatment approaches
Pain	Between 25 and 46% of children experience persistent chronic pain, incidence is likely higher in the palliative child. Numerous and multifactorial in causation	Team approach Non-pharmacological—physical, complementary and cognitive behavioural interventions may reduce the perception of pain, e.g. acupuncture, art therapy, aromatherapy, heat/cold application, humour clown therapy, storytelling oral motor stimulation (sucking) Pharmacological—wide range of approaches based on the child's understood pain level; by the clock; less invasive route; tailored to child's unique circumstances and needs
Agitation and delirium	Life-threatening illness can cause widespread cerebral dysfunction (biochemical, disease). Terminal agitation, also known as terminal restlessness, is agitated delirium at the end of life caused by impaired consciousness. Other causes are medication related, infection, organ failure or psychosocial	Pharmacological—medications to relieve anxiety Non-pharmacological—e.g. acupressure, acupuncture, massage, music/diversion therapies
Anorexia and cachexia	Dietary, disease related, gastrointestinal, iatrogenic (chemotherapy, radiation, medications, opioid induced constipation)	Team approach Family/carer education Pharmacological approaches are limited in children but may be considered, e.g. anti-emetic, appetite stimulant. Steroids may also assist in a limited capacity

(continued)

Table 12.3 (continued)

Symptom	Cause	Treatment approaches
Bowel patterns changes	Mechanical—physiological changes i.e. behavioural, dehydration, disease and medication related, metabolic, hyper/hypocalcaemia, hypothyroidism, neurologic with damage to nerve pathways, extension of disease tumours Non-mechanical—changes in normal intestinal muscle activity	Team approach requiring dietary review Pharmacological and non-pharmacological treatments may be considered
Cough/excessive secretions	Allergens, cardiac (heart failure), infection, malignancy, medications (ACE inhibitors, antibiotics, inhaled steroids, radiation therapy), pulmonary. Psychosocial (psychogenic cough), Tourettes and other (aspiration, reflux, secretions, seizures, rheumatic disease) Excessive secretions can be caused by over production of saliva, inability to retain saliva in the mouth and difficulty swallowing	Non-pharmacological treatments—e.g. fluids, honey, saline nasal washes, humidified air, lozenges (if appropriate), upright position, eliminate antigens with air filters, vapour rubs, acupressure or acupuncture, chest physiotherapy, airway clearance (with training). Fluid intake requires assessment Pharmacological approaches can assist/inhibit cough reflex or reduce sensitivity Medication to thin or dry up secretions can be considered (anticholinergic)
Gastroesophageal reflux and dyspepsia	Anatomical, diet, medications, motility, other (smoke related)	Team approach Positioning Bland diet Small frequent meals

(continued)

Table 12.3 (continued)

Symptom	Cause	Treatment approaches
Nausea and vomiting	Anxiety, anatomical, bowel obstruction, constipation, gastrostasis, infection, increased intracranial pressure, pain, medications, metabolic disorders (hypercalcemia, renal failure), radiation, reflux	Non-pharmacological—e.g. acupuncture, avoid triggers, eliminate medications if possible, play/distraction, good oral care, clear liquids, positioning, relaxation techniques Pharmacological approaches
Seizures	Nervous system anomalies (known/deterioration), comprehensive system decline (chronic or acute), metabolic deterioration (hypoglycaemia, hypocalcaemia, hyper/hyponatremia)	Team approach Assessment is key Non-pharmacological—e.g., positioning, monitoring, education and reassessment Pharmacological—e.g., advance planning for emergency medication application, titration and effect
Dysnoea	Physical, psychosocial, spiritual factors may contribute Causes are extensive and have multiple contributing factors Underlying causes may be treated successfully, others cannot be reversed and treatment is palliative	Non-pharmacological approaches can have significant success, eg., breathing strategies. Use of fans or open windows, repositioning, relaxation, minimise triggers (strong perfume, smoke) Ensure companionship Pharmacological management is aimed at decreasing the perception of dyspnoea and considers the use of benzodiazepines and opioids Oxygen therapy may be helpful but is rarely beneficial at the end of life and hinders communication and connection

(continued)

Table 12.3 (continued)

Symptom	Cause	Treatment approaches
Communication	May already have difficulty in communication and require additional supports	Ensure all staff have knowledge of and are familiar with communication method

Hain and Jassal (2010), Friedrichsdorf et al. (2015), McCrate Protus et al. (2014), National Institute for Health and Care Excellence (NICE) (2019), and Zadeh et al. (2015)

To conclude this section, it is essential symptoms are reassessed regularly. Reassessment includes attention to symptoms already assessed, responses to treatment, screening for new symptoms and medication side effects. In sum:

- Listen to both the child and their parent/carer.
- Work with the parent/carer as experts in their child's care and be open to new approaches/assessment.
- Keep approaches simple and child/family focussed.
- Pay close attention to detail.
- Work as a team.
- Keep all lines of communication open at all times.
- Know your own limits, scope of practice and ask for help.

Ethical Considerations in End-of-Life Decision-Making

A great deal of stress can be experienced by the family when discussions concerning possible need for palliative and end-of-life care occur. Decision-making where all stakeholders are in agreement is the ideal. Ethical concerns can arise concerning treatment options or withholding treatment and resuscitation, or other conflicts between parents and with health professionals (Hain et al., 2021). Families likely have significant knowledge and experiences that influence their decision-making processes (Mitchell et al., 2019), but most remain unprepared for end-of-life decision-making. Listening skills by health professionals, including awareness of both their own verbal and non-verbal communication as well as others, cannot be overemphasised (Inglin et al., 2011; Mitchell et al., 2019). Information exchange should be treated as an honest and reciprocal process, while being sensitive to the child and family needs (Weidner et al., 2011; Winger et al., 2020). While end-of-life care decision-making for children with intellectual disabilities is complex and emotionally overwhelming, preparation becomes possible if parents reach a place of acceptance (Mitchell et al., 2019). Amy's story demonstrates the importance of addressing key symptoms, with an appropriate management approach co-created and agreed with parents.

Amy's* story

As Amy's condition started to deteriorate, she began to experience several uncomfortable symptoms requiring careful management to alleviate and ensure comfort. Amy had always experienced seizures (although usually well controlled), but with the physical deterioration in her condition, the seizures became more intense. Seizure management goals were discussed with Amy's parents by the lead paediatrician and community palliative care team with a focus on Amy's comfort, dignity and quality of life during this palliative phase. Medications were reviewed and titration dosages were considered in response to anticipated disease progression and expected management, especially during the end-of-life period. Due to seizure activity, increased irritability was another concern, with Amy showing her frequent distress with a prolonged distressed cry. This was upsetting for her, as well as her parents, wider family and friends. The lead paediatrician and community palliative care team had a further discussion with Amy's parents where anxiolytics and pain management medications were explored, and a plan put in place. Over the next few days, as Amy's condition deteriorated as anticipated, cessation of feeding (diet and hydration) was discussed with Amy's family. This difficult decision had the ultimate goal to reduce an unnecessary burden on Amy's body, especially the management of secretions which became an increasing concern during the last days of life. While this was an extremely difficult discussion for Amy's parents, together everyone recognised that the delicate balancing of hydration with the challenges of maintaining a good gentle death was paramount. Amy died peacefully with her symptoms well controlled. Amy's parents and a night nurse were in attendance.

*Assumed names are used in this story.

Bereavement Support

Bereavement has been referred to as the grief experienced after a loss, especially of a loved one, and includes emotional, psychological, physical, behavioural and functional reactions (Weafer, 2014). This understanding of bereavement allows for a broader consideration of core issues for many families around loss which may have accompanied the initial diagnosis of earlier childhood disability, with the ultimate death of the child being the second loss (Zisook & Shear, 2009). The specific experiences of parents of children with intellectual disability losing their child are not well explored in the literature (Reilly et al., 2008). Grief, if prolonged, can become a pathological condition that may require psychological intervention to promote recovery (Comtesse et al., 2020).

Contrary to earlier views around grief, a bereaved parent's ability to reconstruct their relationship with their deceased child and integrate this memory into their inner and social worlds is now considered central to adjustment following bereavement (McNeil et al., 2020). Possessions and physical objects offer links to the deceased child and evoke memories, assisting to keep their child intact (Papadatou et al., 2021). These memories support families to recount their feelings, thoughts and events following a bereavement, assisting them to adjust while giving meaning to their loss and remaining connected to their loved one (Papadatou et al., 2021). The idea of continuing bonds may require professional support if parents experience difficulty talking about their deceased child to family members and the wider social world. The inability to talk about the deceased child can mean loneliness and isolation (De Clercq et al.,

2017; Papadatou et al., 2021). Healthcare professionals can support parents and families through therapeutic interventions like listening, conversational remembering and referring families to support groups of those who have suffered the same loss. Grief care is not solely allocated to after the death as it is recognised that grief can begin at illness diagnosis, known as anticipatory grief (see also Chapter 2). This is particularly prolonged for families of a child with intellectual disability whereby the trajectory is unknown.

Parental bereavement literature acknowledges that for families losing a child is a tragedy and evokes a significant emotional response (Hannan & Gibson, 2005), implying that those bereaved are at higher risk of ill health when unable to cope with their grief and loss (Johns, 2016; Keegan, 2013; Papadatou et al., 2021; Tatsuno et al., 2012). Additionally, the experience of child bereavement for both parents and other carers can present as emotional distress, reduced life expectancy or reduced overall quality of life and increased anxiety (Eilertsen et al., 2013). However, the experience can also allow for post traumatic growth for families, through creating an opportunity for them to find meaning after their great loss (Barrera et al., 2009). While not all families and caregivers will require professional support, it is recommended that supports and services are available as required. This support can range from a caring listener to requiring more structured assistance. Either way, there is a need to respect and support the family's natural coping mechanisms.

Conclusion

Appropriate healthcare services are vital for the health and wellbeing of children with intellectual disabilities, including for children who are dying. This should incorporate a family centred model of palliative and end-of-life care which responds to all stages of the palliative trajectory and extends from diagnosis to beyond death. Palliative care can be seen as a philosophy of care that stretches across a broad range of services, where the primary goal is to make a child's life as comfortable as possible while experiencing palliative symptoms and end-of-life needs. This care provides supports to the child and family and can range from emotional support to palliative care services within the home or elsewhere. Care also encompasses support for bereavement and loss and adjusting to life after losing a child. Quality care requires a cohesive, collaborative and fully integrated partnership inclusive of all professionals together with the child and their family.

Reflection

Bella's* story

Bella is 4 years old and has profound intellectual disabilities and complex health needs. She lives 3 hour away from the main children's hospital but attends a special needs playschool and is well established within the health care services and early intervention team. On a planned admission, Bella is admitted for a review of her metabolic disorder and deteriorating symptoms. Unfortunately, investigations show that she is doing less well and her condition is no longer amenable to treatment. Her organs are now demonstrating dysfunction and she is significantly more symptomatic. It is expected she will continue to deteriorate in the coming weeks. In view of further treatment not being undertaken, Bella's treating paediatrician discusses the new findings and implications with her family. Her parents are understandably devastated and wish her to be discharged home immediately. A paediatric palliative care assessment is undertaken. Bella and her parents are involved in the plan and wish to engage with the paediatric team at her nearest paediatric hospital unit, seek help from the community Specialist Palliative Care Team (SPCT) and allied colleagues to ensure they can obtain help and support. A teleconference is to be held prior to Bella's return home. Her parents are made aware of the need to seek home nursing supports including alongside Bella's current disability services (for respite and continuity with family support), as well as the (new) contribution of the SPCT and the Paediatrician in symptom management. It is agreed the overall management will be co-ordinated locally by a Clinical Nurse Co-ordinator (CNC) for children with life-limiting illness. This nurse liaises with the family directly to offer information and guide access to voluntary home nursing services, in home respite care and end-of-life nursing hours from the

Cancer Society Night Nursing Services. Bella's death is not expected in the immediate hours or days following this discharge, but it is made clear that her condition may deteriorate suddenly. The CNC, paediatrician, SPCT, other disciplinary colleagues and Bella's family engage in a teleconference to agree a plan of care and relevant documentation is shared securely via email. Communication is accelerated and Bella and her parents leave the tertiary hospital with a hard copy of their referral documentation forwarded to all teams including the general practitioner, the paediatric team in their nearest hospital unit, the community SPCT, the CNC and public health nurse. Voluntary sector service referrals are completed remotely by the CNC and submitted with parents involvement and expressed consent. A team approach to end-of-life care is expedited to ensure that Bella's requirements and the needs of her parents and family are fully met as they journey towards her end of life.

Bella lived for four more months with the coordination managed by the CNC with the focus on her quality of life and making memories, guided by the collaborative effort of both voluntary and statutory providers. This collaborative effort ensured that Bella and her family remained connected to established services and personnel (who had become familiar and valued by the family). These professionals (public health, community nurses and disability providers) had previously attended various educational programmes to ensure they had an awareness and knowledge of the principles of children's palliative care and the competencies required to deliver skilled palliative and end-of-life care. The participation and contribution of previously known professionals (including respite carers and allied health colleagues) had the benefit of consistency in care giving and family support, recognising Bella's individual needs as well as changes or new symptoms as they arose. With this full team effort and excellent interdisciplinary communication Bella was able to remain in her preferred home setting, even as her physical condition deteriorated, with providers contributing their unique skill set with the overall aim to support Bella and her family to her final days.

*Assumed names are used in this story.

Questions

1. Consider how this example of paediatric and end-of-life care offers support to the child and family.
2. An integrated family centred home-based plan of care is described. Consider the role of each person involved in the care process.
3. What are the key competencies that professionals working with children with intellectual disabilities can contribute to in the end-of-life period?

Resources

1. *Paediatric Palliative Screening Scale (PaPaS).* A tool for screening paediatric patients for palliative care needs. Useful for primary care clinicians unfamiliar with palliative care. Not purposefully designed to accommodate the individual needs of people with intellectual disability. https://bmcpalliatcare.biomedcentral.com/articles/10.1186/s12904-021-00765-8
2. *Together for short lives.* A website of resources to support children who are dying and their families. While not purposefully designed for children with intellectual disability, the information is helpful. https://www.togetherforshortlives.org.uk/

References

American Academy of Pediatrics. (2013). Pediatric palliative care and hospice care commitments, guidelines, and recommendations. *Pediatrics, 132*(5), 966–972. https://doi.org/10.1542/peds.2013-2731

Barone, S., & Unguru, Y. (2018). Ethical issues around paediatric death: Navigating consent, assent, and disagreement regarding life-sustaining medical treatment. *Child and Adolescent Psychiatrics Clinics, 27*(4), 539–550. https://doi.org/10.1016/j.chc.2018.05.009

Barrera, M., O'Connor, K., D'Agostino, N. M., Spencer, L., Nicholas, D., Jovcevska, V., Tallet, S., & Schneiderman, G. (2009). Early parental adjustment and bereavement after childhood cancer death. *Death Studies, 33*(6), 497–520. https://doi.org/10.1080/07481180902961153

Bergsträsser, E., Cignacco, E., & Luck, P. (2017). Health care professionals' experiences and needs when delivering end-of-life care to children: A qualitative study. *Palliative Care, 10*. https://doi.org/10.1177/1178224217724770

Berlinger, N., Barfield, R., & Fleischman, A. R. (2013). Facing persistent challenges in pediatric decision-making: New Hastings Center guidelines. *Pediatrics, 132*(5), 789–791. https://doi.org/10.1542/peds.2013-1378

Brenner, M., Alexander, D., Quirke, M. B., Eustace-Cook, J., Leroy, P., Berry, J., Healy, M., Doyle, C., & Masterson, K. (2021). A systematic concept analysis of 'technology dependent': Challenging the terminology. *European Journal of Pediatrics, 180*(1), 1–12. https://doi.org/10.1007/s00431-020-03737-x

Brenner, M., Kidston, C., Hilliard, C., Coyne, I., Eustace-Cook, J., Doyle, C., Begley, T., & Barrett, M. J. (2018). Children's complex care needs: A systematic concept analysis of multidisciplinary language. *European Journal of Pediatrics, 11*, 1641–1652. https://doi.org/10.1007/s00431-018-3216-9

Brenner, M., O'Shea, M., McHugh, R., Clancy, A., Larkin, P., Luzi, D., Pecroaro, F., Olaso, E. M., Lignou, S., Alma, M., Satherlkey, R. M., Tamburis, O., Warters, A., Wolfe, I., Hillard, C., Berry, J., Alexander, D., Rigby, M., & Blair M. (2018). Principles for provision of integrated complex care for children across the acute–community interface in Europe. *The Lancet, 2*(11), 832–838. https://doi.org/10.1016/S2352-4642(18)30270-0

Brock, K., Wolfe, J., & Ullrich, C. (2018). From the child's word to clinical interventions: Novel, new and innovative approaches to symptoms in pediatric palliative care. *Children, 5*(4), 45. https://doi.org/10.3390/children5040045

Brown, M., Sunnquist, M., & Schneider, L. (2021). A framework for navigating requests for nondisclosure in pediatric palliative care. *Clinical Practice in Pediatric Psychology, 9*(3), 296–307. https://doi.org/10.1037/cpp0000415

Carnevale, F. A., Rehm, R. S., Kirk, S., & McKeever, P. (2008). What we know (and do not know) about raising children with complex continuing care needs. *Journal of Child Health Care, 12*(1), 4–6. https://doi.org/10.1177/1367493508088552

Carter, B., Bray, L., Sanders, C., van Miert, C., Hunt, A., & Moore, A. (2016). "Knowing the places of care": How nurses facilitate transition of children with complex health care needs from hospital to home. *Comprehensive Child and Adolescent Nursing, 39*(2), 139–153. https://doi.org/10.3109/01460862.2015.1134721

Cohen, E., Kuo, D. Z., Agrawal, R., Berry, J. G., Bhagat, S. K. M., Simon, T. D., & Srivastava, R. (2011). Children with medical complexity: An emerging population for clinical and research initiatives. *Pediatrics, 127*(3), 529–538. https://doi.org/10.1542/peds.2010-0910

Comtesse, H., Vogel, A., Kerstng, A., Rief, W., Steil, R., & Rosner, R. (2020). When does grief become pathological? Evaluation of the ICD-11 diagnostic proposal for prolonged grief in a treatment-seeking sample. *European Journal of PsychoTraumatology, 11*(1), 1694348. https://doi.org/10.1080/20008198.2019.1694348

Courtney, E., Kiernan, G., Guerin, S., Ryan, K., & McQuillan, R. (2018). Mothers' perspectives of the experience and impact of caring for their child with a life-limiting neurodevelopmental illness. *Child Care Health and Development, 44*(5), 704–710. https://doi.org/10.1111/cch.12580

De Clercq, E., Rost, M., Pacurari, N., Elger, B., & Wangmo, T. (2017). Aligning guidelines and medical practice: Literature review on pediatric palliative care guidelines. *Palliative and Supportive Care, 15*(4), 474–489. https://doi.org/10.1017/S1478951516000882

Department of Health. (2004). *National service framework for children, young people and maternity services: Disabled children and young people and those with complex health needs*. https://assets.publishing.service.gov.uk/government/uploads/system/uploads/attachment_data/file/199955/National_Service_Framework_for_Children_Young_People_and_Maternity_Services_-_Disabled_Children_and_Young_People_and_those_with_Complex_Health_Needs.pdf

Department of Health. (2008). *Better care, better lives: Improving outcomes and experiences for children, young people and their families living with life-limiting and life-threatening conditions*. http://www.dh.gov.uk/en/Publicationsandstatistics/Publications/PublicationsPolicyAndGuidance/DH_083106

Department of Health and Children. (2010). *Palliative care for children with life-limiting conditions in Ireland—A national policy*. https://www.gov.ie/en/publication/ed697f-palliative-care-for-children-with-life-limiting-conditions-in-irelan/#

Doyle, C. (2020). The importance of supportive relationships with general practitioners, hospitals and pharmacists for mothers who 'give medicines' to children with severe and profound intellectual disabilities. *Journal of Intellectual Disabilities*. Advance online publication. https://doi.org/10.1177/1744629520951003

Doyle, C. (2021). 'Just knowing' and the challenges of giving medicines to children with severe and profound intellectual disabilities: A hermeneutic inquiry. *British Journal of Learning Disabilities, 49*(1), 3–12. https://doi.org/10.1111/bld.12354

Eilertsen, M. E., Gunner-Steineck, A., Nyberg, T., & Kreicbergs, U. (2013). Impact of social support on bereaved siblings' anxiety: A nationwide follow-up. *Journal of Pediatric Oncology Nursing, 30*(6), 301–310. https://doi.org/10.1177/1043454213513838

Feudtner, C. (2007). Collaborative communication in pediatric palliative care: A foundation for problem-solving and decision-making. *Pediatric Clinics of North America, 54*(5), 583–607. https://doi.org/10.1016/j.pcl.2007.07.008

Fraser, L. K., Gibson-Smith, D., Jarvis, S., Norman, P., & Parslow, R. C. (2021). Estimating the current and future prevalence of life-limiting conditions in children in England. *Palliative Medicine, 35*(9), 1641–1651. https://doi.org/10.1177/0269216320975308

Fraser, L. K., Jarvis, S. W., Moran, N. E., Aldridge, J., Parslow, R. C., & Beresford, B. A. (2015). *Children in Scotland requiring palliative care: Identifying numbers and needs (The ChiSP Study)*. University of York. https://www.york.ac.uk/media/spru/projectfiles/ProjectOutput_ChispReport.pdf

Fraser, L., Miller, M., Aldridge, J., McKinney, P. A., & Parslow, R. C. (2013). *Prevalence of life-limiting and life-threatening conditions in young adults in England 2000–2010*. Report for Together for Short Lives. https://www.togetherforshortlives.org.uk/wp-content/uploads/2018/01/ExRes-Prevalence-of-Life-limiting-and-life-threatening-briefing.pdf

Friedrichsdorf, S. J., Postier, A., Dreyfus, J., Osenga, K., Sencer, S., & Wolfe, J. (2015). Improved quality of life at end of life related to home-based palliative care in children with cancer. *Journal of Palliative Medicine, 18*(2), 143–150. https://doi.org/10.1089/jpm.2014.0285

Furingsten, L., Sjögren, R., & Forsner, M. (2015). Ethical challenges when caring for dying children. *Nursing Ethics, 22*(2), 176–187. https://doi.org/10.1177/0969733014533234

Goldstein, S., & Reynolds, C. (Eds.). (2011). *Handbook of neurodevelopmental and genetic disorders in children* (2nd ed.). Guildford Press.

Grinyer, A. (2012). *Palliative and end of life care for children and young people: Home, hospice, hospital.* Wiley-Blackwell.

Guerin, S., Kiernan, G., Courtney, E., McQuillan, R., & Ryan, K. (2020). Integration of palliative care in services for children with life-limiting neurodevelopmental disabilities and their families: A Delphi study. *BMC Health Services Research, 20*(927). https://doi.org/10.1186/s12913-020-05754-w

Hain, R., & Jassal, S. S. (Eds.). (2010). *Paediatric palliative medicine*. Oxford University Press.

Hain, R., Goldman, A., Rapoport, A., & Meiring, M. (Eds.). (2021). *Oxford textbook of palliative care for children*. Oxford University Press.

Hannan J., & Gibson, F. (2005). Advanced cancer in children: How parents decide on final place of care for their dying child. *International Journal of Palliative Nursing, 11*(6), 284–291. https://doi.org/10.12968/ijpn.2005.11.6.18296

Harrop, E., & Edwards, C. (2013). How and when to refer a child for specialist paediatric palliative care. *Archives of Disease in Childhood, Education and Practice, 98*(6), 202–208. https://doi.org/10.1136/archdischild-2012-303325

Health Service Executive & Faculty of Paediatrics, RCPI (2016). *A national model of care for paediatric healthcare services in Ireland. Chapter 39: Paediatric Palliative Care.* Health Service Executive & Faculty of Paediatrics, RCPI. https://www.hse.ie/eng/about/who/cspd/ncps/paediatrics-neonatology/moc/chapters/

Health Service Executive. (2020). *Clinical governance and operational arrangements for supporting a model of care for children with life limiting conditions towards the end of life in the community in Ireland: Final report and recommendations.* https://www.hse.ie/eng/services/publications/children/operational-and-governance-framework-for-children-with-life-limiting-conditions.pdf

Heckford, E., & Beringer, A. J. (2014). Advance care planning: Challenges and approaches for pediatricians. *Journal of Palliative Medicine, 17*(9), 1049–1053. https://doi.org/10.1089/jpm.2013.0374

Hudson, J. P. (2014). *A practical guide to congenital developmental disorders and learning difficulties*. Routledge.

Hourigan, S., Fanagan, S., & Kelly, C. (2018). *Health Research Board Statistics Series 37: Annual report of the National Intellectual Disability Database Committee 2017: Main findings.* https://www.hrb.ie/fileadmin/2._Plugin_related_files/Publications/2018_pubs/Disability/NIDD/NIDD_Annual_Report_2017.pdf

Inglin, S., Hornung, R., & Bergstraesser, E. (2011). Palliative care for children and adolescents in Switzerland: A needs analysis across three diagnostic groups. *European Journal of Pediatrics, 170*(8), 1031–1038. https://doi.org/10.1007/s00431-011-1398-5

Jassal, S. S., & Hain, R. (Eds.). (2016). *Paediatric palliative medicine* (2nd ed.). Oxford University Press.

Johns, L. (2016). Bereavement care in rural and remote communities of Australia. *Austral-Asian Journal of Cancer, 14*(1), 1–3. http://hdl.handle.net/10072/99943

Keegan, O. (2013). Editorial. *Bereavement Care, 32*(3), 99. https://doi.org/10.1080/02682621.2013.851835

Kuo, D. Z., Cohen, E., Agrawal, R., Berry, J. G., & Casey, P. H. (2011). A national profile of caregiver challenges among more medically complex children with special health care needs. *Archives of Pediatric Adolescent Medicine, 165*(11), 1020–1026. https://doi.org/10.1001/archpediatrics.2011.172

Lafond, D., Patterson Kelly, K., Hinds, P., Sill, A., & Michael, M. (2015). Establishing feasibility of early palliative care consultation in pediatric hematopoietic stem cell transplantation. *Journal of Pediatric Oncology Nursing, 32*(5), 265–277. https://doi.org/10.1177/1043454214563411

Liberman, D. B., Song, E., Radbill, L. M., Pham, P. K., & Derrington, S. F. (2016). Early introduction of palliative care and advanced care planning for children with complex chronic medical conditions: A pilot study. *Child Care Health and Development, 42*(3), 439–449. https://doi.org/10.1111/cch.12332

Lindahl, B., & Lindblad, B. (2011). Family members' experiences of everyday life when a child is dependent on a ventilator: A metasynthesis study. *Journal of Family Nursing, 17*(2), 241–269. https://doi.org/10.1177/1074840711405392

Mack, J., & Wolfe, J. (2006). Early integration of pediatric palliative care: For some children, palliative care starts at diagnosis. *Current Opinion in Paediatrics, 18*(1), 10–14. https://doi.org/10.1097/01.mop.0000193266.86129.47

Marsac, M. L., Kindler, C., Weiss, D., & Ragsdale, L. (2018). Let's talk about it: Supporting family communication during end-of-life care of pediatric patients. *Journal of Palliative Medicine, 21*(6), 862–878. https://doi.org/10.1089/jpm.2017.0307

McCann, D., Bull, R., & Winzenberg, T. (2015). Sleep deprivation in parents caring for children with complex needs at home: A mixed methods systematic review. *Journal of Family Nursing, 21*(1), 86–118. https://doi.org/10.1177/1074840714562026

McCrate Protus, B., O'Neill-Hunt, M., Penfield Winters, J., & Parker, D. (2014). *Pediatric palliative care consultant: Guidelines for effective management of symptoms*. HospiScript.

McMahon, D., Twomey, M., O'Reilly, M., & Devins, M. (2018). Referrals to a perinatal specialist palliative care consult service in Ireland, 2012–2015. *Archives of Disease in Childhood, Fetal Neonatal Education, 103*(6), F573–F576.

McNeil, M. J., Namisango, E., Hunt, J., Powell, R. A., & Baker, J. N. (2020) Grief and bereavement in parents after the death of a child in low- and middle-income countries. *Children, 7*(5), 39. https://dx.doi.org/10.3390/children7050039

Mendes, M. (2013). Parents' descriptions of ideal home nursing care for their technology dependent children. *Pediatric Nursing, 39*(2), 91–96. PMID: 23705300.

Miller, S. C., Lima, J. C., & Mitchell, S. L. (2012). Influence of hospice on nursing home residents with advanced dementia who received Medicare-skilled nursing facility care near the end of life. *Journal of American Geriatric Society, 60*(11), 2035–2041. https://dx.doi.org/10.1111/j.1532-5415.2012.04204.x

Mitchell, S., Spry, J. L., Hill, E., Coad, J., Dale, J., & Plunkett, A. (2019). Parental experiences of end of life care decision-making for children with life-limiting conditions in the paediatric intensive care unit: A qualitative interview study. *British Medical Journal Open, 9*(5), e028548. https://doi.org/10.1136/bmjopen-2018-028548

Myrhaug, H. T., Brurberg, K. G., Hov, L., & Markestad, T. (2019). Survival and impairment of extremely premature Infants: A meta-analysis. *Pediatrics, 143*(2), e20180933. https://doi.org/10.1542/peds.2018-0933

National Institute for Health and Care Excellence (NICE). (2019). *End of life care for infants, children and young people with life-limiting conditions planning and management.* https://www.nice.org.uk/guidance/ng61

Office for National Statistics. (2019). *Estimates of the population for the UK, England and Wales, Scotland and Northern Ireland.* https://www.ons.gov.uk/peoplepopulationandcommunity/populationandmigration/populationestimates

Olsen, R., & Maslin-Prothero, P. (2001). Dilemmas in the provision of own-home respite support for parents of young children with complex health care needs: Evidence from an evaluation. *Journal of Advanced Nursing, 34*(5), 603–610. https://doi.org/10.1046/j.1365-2648.2001.01789.x

Papadatou, D., Kalliani, V., Karakosta, E., Liakopoulou, P., & Bluebond-Langner, M. (2021). Home or hospital as the place of end-of-life care and death: A grounded theory study of parents' decision making. *Palliative Medicine, 35*(1), 219–230. https://doi.org/10.1177/0269216320967547

Quinn, C. (in press). Optimising: How parents positively enable the care of their dying child—A grounded theory. *International Journal of Palliative Care.*

Reilly, D. E., Hastings, R. P., Vaughan, F. L., Huws, J. C., & Taylor, S. J. (2008). Parental bereavement and the loss of a child with intellectual disabilities: A review of the literature. *Intellectual and Developmental Disabilities, 46*(1), 27–43. https://doi.org/10.1352/0047-6765(2008)46[27:pbatlo]2.0.co;2

Schwantes, S., & O'Brien, H. W. (2014). Pediatric palliative care for children with complex chronic medical conditions. *Pediatric Clinics of North America, 61*(4), 797–821. https://doi.org/10.1016/j.pcl.2014.04.011

Scott, L. D. (2000.) Caregiving and care receiving among a technologically dependent heart failure population. *Advances in Nursing Science, 23*(2), 82–97. https://psycnet.apa.org/doi/10.1097/00012272-200012000-00008

Sharkey, S., Lloyd, C., Tomlinson, R., Thomas, E., Martin, A., Logan, S., & Morris, C. (2016). Communicating with disabled children when inpatients: Barriers and facilitators identified by parents and professionals in a qualitative study. *Health Expectations, 19*(3), 738–750. https://dx.doi.org/10.1111/hex.12254

Simon, T. D., Berry, J., Feudtner, C., Stone, B. L., Sheng, X., Bratton, S. L., Dean, J. M., & Srivastava, R. (2010). Children with complex chronic conditions in inpatient hospital settings in the United States. *Pediatrics, 126*(4), 647–655. https://doi.org/10.1542/peds.2009-3266

Song, G., Kwon, S. Y., Chang, Y. J., Kim, M. S., Jeong, S. H., Hahn, S. M., Han, K. T., Park, S. J., & Choi, J. Y. (2021) Paediatric palliative screening scale as a useful tool for clinicians' assessment of palliative care needs of pediatric patients: A retrospective cohort study. *BMC Palliative Care, 20*(73). https://doi.org/10.1186/s12904-021-00765-8

Tatsuno, J., Yamase, H., & Yamase, H. (2012). Grief reaction model of families who experience acute bereavement in Japan. *Nursing and Health Sciences, 14*(2), 257–264. https://doi.org/10.1111/j.1442-2018.2012.00688.x

Together for Short Lives. (2018). *A guide to children's palliative care: Supporting babies, children and young people with life-limiting and life-threatening conditions and their families* (4th ed.). https://www.togetherforshortlives.org.uk/wp-content/uploads/2018/03/TfSL-A-Guide-to-Children's-Palliative-Care-Fourth-Edition-5.pdf

Together for Short Lives. (2019). *Caring for your child at end of life: A guide for parents and carers* (1st ed.). Together for Short Lives. https://www.togetherforshortlives.org.uk/wp-content/uploads/2019/11/TfSL-Caring-for-a-child-at-end-of-life-Parents.pdf

Twamley, K., Craig, F., Kelly, P., Hollowell, D. R., Mendoza, P., & Bluebond-Langner, M. (2014). Underlying barriers to referral to paediatric palliative care services: Knowledge and attitudes of health care professionals in a paediatric tertiary care centre in the United Kingdom. *Journal of Child Health Care, 18*(1), 19–30. https://doi.org/10.1177/1367493512468363

Ulrich, C. M., Mooney-Doyle, K., & Gardy, C. (2018). Communicating with pediatric families at end-of-life is not a fantasy. *American Journal of Bioethics, 18*(1), 14–16. https://dx.doi.org/10.1080/15265161.2017.1401175

Vollenbroich, R., Borasio, G. D., Duroux, A., Grasser, M., Brandstatter, M., & Fuhrer, M. (2016). Listening to parents: The role of symptom perception in pediatric palliative home care. *Palliative and Supportive Care, 14*(1), 13–19. https://doi.org/10.1017/s1478951515000462

Walter, J. K., DeCamp, L. R., Warrier, K. S., Murphy, T. P., & Keefer, P. M. (2013). Care of the complex chronically ill child by generalist pediatricians: Lessons learned from pediatric palliative care. *Hospital Pediatrics, 3*(2), 129–138. https://doi.org/10.1542/hpeds.2012-0047

Weafer, J. (2014). *Irish attitudes to death, dying and bereavement 2004–2014.* Report for the Irish Hospice Foundation. http://edepositireland.ie/bitstream/handle/2262/83090/Weafer-J-2014-Irish-attidudes-to-death-dying-bereavement-2004-2014.pdf?sequence=1&isAllowed=y

Weidner, N. J., Cameron, M., Lee, R. C., McBride, J., Mathias, E. J., & Byczkowski, T. L. (2011). End-of-life care for the dying child: What matters most to parents. *Journal of Palliative Care, 27*(4), 279–286. PMID: 22372282.

Widger, K., Davies, D., Drouin, D. J., Beaune, L., Daoust, L., Farran, P., Humbert N., Nalewajek F., Rattray M., Rugg, M., Bishop M. (2007). Pediatric patients receiving palliative care in Canada: Results of a multicenter review. *Archives of Pediatric and Adolescent Medicine, 161*(6), 597–602. https://doi.org/10.1001/archpedi.161.6.597

Winger, A., Kvarme, L. G., Løyland, B., Kristiansen, C., Helseth S., & Ravn I. H. (2020). Family experiences with palliative care for children at home: A systematic literature review. *BMC Palliative Care, 19*(165). https://doi.org/10.1186/s12904-020-00672-4

Zablotsky, B., Black, L. I., Maenner, M. J., Schieve, L. A., Danielson, M. L., Bitsko, R .H., Blumberg, S. J., Kogan, M. D., & Boyle, C. A. (2019). Prevalence and trends of developmental disabilities among children in the United States: 2009–2017. *Pediatrics, 144*(4): e20190811. https://doi.org/10.1542/peds.2019-0811

Zadeh, S., Pao, M., & Wiener, L. (2015). Opening end-of-life discussions: How to introduce Voicing My CHOiCES™, an advance care planning guide for adolescents and young adults. *Palliative Support Care, 13*(3), 591–599. https://doi.org/10.1017/s1478951514000054

Zisook, S., & Shear, K. (2009). Grief and bereavement: What psychiatrists need to know. *World Psychiatry, 8*(2), 67. https://dx.doi.org/10.1002/j.2051-5545.2009.tb00217.x

13

End-of-Life Issues and Support Needs of People with Profound Intellectual and Multiple Disability

Hille Voss, Joanne Watson, and Melissa J. Bloomer

H. Voss (✉)
Netherlands Institute for Health Services Research (Nivel), Utrecht, The Netherlands
e-mail: h.voss@nivel.nl

J. Watson
School of Health & Social Development, Deakin University, Melbourne, VIC, Australia
e-mail: joanne.watson@deakin.edu.au

M. J. Bloomer
School of Nursing and Midwifery & Institute for Health Transformation, Deakin University, Melbourne, VIC, Australia
e-mail: m.bloomer@deakin.edu.au

School of Nursing and Midwifery, Griffith University, Brisbane, QLD, Australia

R. J. Stancliffe et al. (eds.), *End of Life and People with Intellectual and Developmental Disability*, https://doi.org/10.1007/978-3-030-98697-1_13

People with profound intellectual and multiple disability (PIMD) have significant cognitive difficulties and generally require support in most aspects of life. Most people with PIMD communicate informally and rely on others to articulate their will and preferences, making the provision of end-of-life care for people with PIMD complex and challenging. This chapter is focussed on acknowledging, interpreting and acting on the expressions of will and preference of people with PIMD. The importance of holistic and multidisciplinary end-of-life care combined with a coherent support network is highlighted. Using real-world case examples, accompanied by reflection questions, the importance of acknowledging and responding to informal communication is emphasised. Resources are provided at the end of the chapter to guide practice.

People with Profound Intellectual and Multiple Disability (PIMD)

PIMD is inconsistently defined. However, there is consensus amongst researchers and practitioners that a person with PIMD is likely to have significant cognitive difficulties and little or no apparent use or understanding of formal language (Vorhaus, 2016). Additionally, people with PIMD are likely to have severe motor disabilities and, in many cases, a sensory disability such as deafness, visual impairment or deaf-blindness with visual disabilities most prominent (Nakken & Vlaskamp, 2007). Health challenges, including epilepsy, gastrointestinal disorders, respiratory disorders and dysphagia are also common at a relatively young age (Rousseau et al., 2015; Van Timmeren et al., 2016). Due to their multiple and complex support needs, people with PIMD are dependent on others for support in almost all aspects of their life (Kamstra et al., 2017). Despite a lack of official statistics regarding the prevalence of people with PIMD, a review of the literature suggests that the estimated prevalence is between 0.13% and 0.25% of the population (Arvio & Sillanpää, 2003; Sarimski, 2016). People with PIMD often live in community settings such as residential care facilities or at home with family carers (Vlaskamp et al., 2007).

End-of-Life Care Needs of People with Profound Intellectual and Multiple Disability

Every person has the right to access timely, acceptable and affordable health care of appropriate quality (World Health Organization, 2015). This right is enshrined in Article 25 (Health) of the United Nations Convention on the Rights of Persons with Disabilities (UNCRPD) (United Nations, 2006), which mandates signatory nations to ensure that "persons with disabilities have the right to the enjoyment of the highest attainable standard of health without discrimination on the basis of disability" (p. 18). Although people with intellectual disability, especially those with PIMD, experience higher health care needs than the rest of the population, they are less likely to have access to adequate health care throughout their lives (Hosking et al., 2016).

Inequities in health care continue at the end of life (Bloomer & Walshe, 2021), at which time people with intellectual disability are less likely to have access to appropriate end-of-life care or specialist palliative care than the general population (Adam et al., 2020; see also Chapter 6). Access to the right care at the end of life is particularly challenging for people with PIMD, with referrals to specialist palliative care often occurring late in the dying trajectory (Ryan et al., 2010; Stein, 2008). Several reasons have been proposed for the inequitable access and delayed referral to specialist palliative care.

An Irish study identified that few staff from specialist palliative care services had experienced caring for a person with intellectual disability (Ryan et al., 2010), suggesting greater challenges in accessing specialist palliative care. Multiple factors are thought to contribute to this, including determining which aspects of care should take priority, who will lead care provision and how to ensure care aligns with the person's expressed preferences. There is evidence to suggest that some health professionals are reluctant to provide and manage end-of-life care for people with intellectual disability due to the potential for multiple and complex medical issues (Wark et al., 2017). Similarly, a study of staff from specialist palliative care and intellectual disability services demonstrated that both cohorts lacked confidence in their ability to provide care at the end of life (Ryan et al., 2010). Nonetheless, just as an individual's preferred place of care and death is considered for other population groups (Billingham & Billingham, 2013; Burge et al., 2015; Raijmakers et al., 2018), so too should people with PIMD and their immediate supporters be afforded the same options. But the reality is that supporting people with PIMD about care preferences at the end of life when time is often critical may not be so easy.

To address this, multiple alternative models of care in which collaboration is at the core have been proposed. Designed to promote partnership practice between specialist palliative care and intellectual disability services, McLaughlin and colleagues (2014) proposed one such model that attempts to overcome inequities for people with disability. Their model brings together personnel from intellectual disability services, specialist palliative care services and residential care facilities providing care for people with disabilities, in the provision of person-centred care at the end of life that aligns with the person's expressed wishes (McLaughlin et al., 2014). The partnership model meant that continuity of care was assured, and individuals were able to die in their place of care and were cared for by people familiar to them (McLaughlin et al., 2014). Even though changes would be necessary according to local legislation, systems and policy, a collaborative approach that prioritises partnerships in care seems to be the best way forward in end-of-life care provision, tailored according to individual needs (McLaughlin et al., 2014; see also Chapters 7 and 8).

With limited empirical evidence in this field, guidance and policy for end-of-life care and decision-making are largely based on theoretical and anecdotal evidence from high-income countries (Adam et al., 2020). A recent systematic review of the care needs of adults with intellectual disability identified that an individual's needs towards the end of life extend beyond the physical to include psychosocial, spiritual, communication and information needs—a finding which mirrors the needs of the general population (Adam et al., 2020). Yet given communication for people with PIMD is more complex, assessing, understanding and addressing the needs of people with intellectual disability are more challenging.

Communicating End-of-Life Care Needs

Understanding formal communication such as sign, speech, written text, photos or pictures is challenging for people with PIMD (Goldbart et al., 2014; Johnson et al., 2012; Ware, 2004). Most people with PIMD communicate informally, using atypical and idiosyncratic behaviours such as vocalisations, facial expression, gesticulation, shifts in respiration, eye gaze and touch (Johnson et al., 2012). People with PIMD often communicate without the intention of conveying a message to another person (Vorhaus, 2016). This means that each person deploys distinctive and unique communication methods, which need to be interpreted contextually.

An analysis conducted by Bloomer and Walshe (2021) highlights the important role that communication plays in the provision of good quality palliative care for people with intellectual disability, particularly people with PIMD. Because of the informal nature of their communication, end-of-life care needs of people with PIMD are often not acknowledged, misunderstood or ignored. For example, identifying and assessing pain for people with PIMD can be extremely difficult due to their inability to communicate verbally. People with PIMD may express pain using a variety of means, ranging from challenging behaviour to social withdrawal, making it difficult to assign meaning to their communicative attempts and therefore might lead to inadequate pain management (Zwakhalen et al., 2004). Moreover, people with PIMD are heterogeneous in terms of their communicative behaviours, creating challenges for health professionals in assessing end-of-life care while tailoring to each individual's preferred communicative behaviour (Vlaskamp & Van Der Putten, 2009).

Decision-Making and Involvement at the End of Life

Because of their dependency on others in most aspects of life, people who know the individual well, including parents, other family carers or professional support staff play a significant and often lifelong role in the lives of people with PIMD (Kamstra, 2017). Therefore, they also have a very important role to play in signalling and assessing end-of-life care needs (Voss et al., 2020). An interpretative synthesis of the literature demonstrated that parents of children with PIMD possess unique, experiential knowledge of their child's communication, wellbeing and expressions of pain. This knowledge is built on years of experience interacting intensively with their child and is crucial to providing support and care that enhances the quality of life of their child (Kruithof et al., 2020).

Perception of quality of life is often a key consideration in end-of-life decision-making for those caring for people with PIMD (Zaal-Schuller et al., 2018). Consensus about the assessment of the quality of life of a person with PIMD is crucial to make difficult, potentially life-shortening decisions regarding end-of-life care (Wagemans et al., 2010). The experiential knowledge of parents and other close family carers can be of benefit when assessing the quality of life of people with PIMD (Kruithof et al., 2020).

Although parents differ in their personal preferences as to how they want to be involved, most parents wish to actively share in the end-of-life decision-making process (Voss et al., 2020; Zaal-Schuller et al., 2016). The degree of involvement of parents may also depend on legal traditions. In the United Stated and Canada, parents are considered to be the sole decision-makers for their child with PIMD (Zaal-Schuller et al., 2016). Controversially, in Europe, the Middle East and South America, physicians bear the final responsibility for end-of-life decisions. Also, in many cases, people with PIMD do not have family carers who want or can be involved or, due to increased life expectancy, they may outlive their parents. As a consequence, some people with PIMD may face end-of-life decision-making with little support from people who know them well (Kamstra et al., 2017).

Seeking opportunities to retrieve information from the person with PIMD, and to understand what is important, what contributes to their quality of life, and their preferences for care in the event of acute or unexpected illness, is important. This is best achieved by support staff and carers who know the person well, can reflect on what is important to them and accurately record their wishes and preferences regarding treatment and care in medical or personal files, readily accessible to those involved in a person's care (Rietjens et al., 2017; Voss et al., 2021). Preferences for end-of-life care should also be documented. This is especially important in the event of an acute health crisis, as Helena's story shows.

Helena's* Story

Helena is a 22-year-old woman with PIMD. Helena lives at home with her mother and 18-year-old sister. Due to her disabilities, Helena receives support from a team of disability support workers for 18 hours per day, who attend to her everyday needs and accompany her to a disability day service for social interaction and hydrotherapy. As Helena is being settled in bed for the evening, her support worker notices that Helena is displaying signs of pain through her behaviour and vocalisations, and has a slight fever, which is treated with an analgesic. Upon return the next morning, Helena has altered consciousness and a high fever. An ambulance is called and Helena is taken to the local emergency room accompanied by her younger sister and the support worker. Her mother is away with work for a few days and the hospital staff have not yet been able to contact her. Preliminary investigations reveal that Helena has sepsis, although the cause is not clear.

Given that Helena's mother has not yet responded to calls to her phone, Helena's 18-year-old sister is approached about limitations of medical treatment; but her sister states she is not prepared or willing to make decisions. The treating doctor also approaches the support worker, who has only worked with Helena for one month, to discuss limitations of medical treatment, based on Helena's multiple health issues associated with PIMD and assumptions about her quality of life. Neither Helena's support worker nor her sister are aware of any documentation (e.g. advance care plan) that may provide some insight into her will and preference relating to her medical care. Despite the seriousness of the situation, the doctor suggests Helena is provided with intravenous hydration, antibiotics to treat her infection and analgesia to keep her comfortable, but that no further intervention is provided. In the absence of understanding Helena's will and preference, the treating doctor has made decisions about Helena's treatment, without providing alternative options.

*Assumed names are used in this story.

Reflection Questions

1. How might the support worker challenge the doctor's assumptions and the plan of care offered?
2. What tools would have been helpful to assist Helena's sister and support worker in deciding about Helena's medical care?

Assessment of Quality of Life in the Context of End-of-Life Care

The judgement an individual makes about their own quality of life is highly individual and therefore subjective. As evidenced in the literature, family carers' own expectations and emotional involvement can colour interpretations of their loved one's communication (Nieuwenhuijse et al., 2020). Simoes and Santos (2016) found inconsistencies between the perceptions of quality of life reported by a person with intellectual disability themselves compared to what support staff or family carers determined about that person's quality of life. For example, poor agreement was observed between adults with intellectual disability and their family members in relation to physical wellbeing. In general, the self-reported quality of life scores of people with intellectual disability were higher than those reported by support staff and family.

When determining quality of life for people with PIMD, attention should be paid to the potentially biased opinions and assumptions of proxies, such as parents or other family carers (Nieuwenhuijse et al., 2020; Wosinski & Newman, 2019). Especially in relation to possibly life-shortening end-of-life decisions, proxy decision-making could conceivably lead to discriminatory or even unethical practice when decisions reflect the hopes, fears and desires of the proxy, rather than those of the person with PIMD (Johnson et al., 2012; Wagemans et al., 2010).

A study focussing on end-of-life decision-making by physicians of people with intellectual disability highlighted that the evaluation and opinion of family about their loved one's quality of life carried great weight. Even if the opinions of family carers differed from their own judgement, the family's wishes influenced physician decisions, sometimes resulting in a professional ethical dilemma (Wagemans et al., 2013). Regardless of whose opinion dominates, perception of a person's quality of life is a key factor in decision-making at the end of life. As is the case for all human beings, when evaluating the quality of life of a person with PIMD, it is important to recognise that there is no *objective* truth. For end-of-life care decisions, involving different stakeholders and discussing ethical issues in an honest and open way are important to tailor the decision-making process to the individual and his or her unique circumstances (Voss et al., 2019).

Making a decision for someone knowing that such decisions may have far-reaching consequences can be accompanied by strong emotions, as in the story of Ray. Health professionals or support staff should consider the impact of the decision-making process, pay attention to feelings of those involved in a person's care and provide emotional support if needed (Voss et al., 2020, 2021).

Ray's* Story

Ray is an 8-year-old boy who was operated on when he was a baby for a benign cardiac tumour. During the operation, serious complications occurred, and Ray was resuscitated for a long time. As a result, Ray acquired profound and multiple disabilities. Ray lives in a group home, managed by a disability service provider, and receives 24 hour care. Due to Ray's challenges with eating and swallowing, a percutaneous endoscopic gastrostomy (PEG-tube) provides artificial nutrition and hydration.

Ray's parents are actively involved in his care and medical decision-making. In conversations with Ray's physicians, his parents expressed that they do not want Ray to experience fear or pain and discussed that providing care aimed at comfort was in Ray's best interest. When Ray was six years old, his parents and the physician discussed and signed a non-resuscitation declaration for Ray and noted that future hospital admissions should be avoided as much as possible.

Ray appears to be very uncomfortable as a result of his spasms and regularly vomits food that he receives through the PEG-tube. Ray experiences daily epileptic seizures and cries continuously. Only in the water does Ray appear relaxed. The specialist physician from the hospital indicated that future treatment possibilities for a respiratory infection are limited. Ray's parents, physician and support workers discuss Ray's quality of life and questions are raised about Ray's PEG-tube. Ray's parents feel there is an important decision to make about continuing or removing the tube feeding.

*Assumed names are used in this story.

Reflection Questions

1. What is important to consider when making a decision about Ray's PEG-tube?
2. Who should be involved in decision-making about Ray's PEG-tube? In what way should they be involved?
3. How can Ray's parents be supported in the decision-making process?

Preferences About End-of-Life Care

Because people with PIMD are highly dependent on others and are in need for lifelong care and support from family carers and support staff, they are commonly perceived as passive recipients of care rather than as citizens or agents with capacity to contribute to their own care choices (Vorhaus, 2016). It is often assumed, due to a person with PIMD's limited communicative and cognitive skills, that they are unable to live self-determined lives, defined as acting volitionally and making things happen in one's life (Wehmeyer & Garner, 2003). Although being self-determined has been linked to an increased quality of life, research has shown that people with intellectual disability are not routinely provided opportunities for choice-making and the expression of preferences (Stancliffe, 2001). Evidence-based practices supporting the decision-making of people with PIMD, including end-of-life decision-making, are rare (Kirkendall et al., 2017; Noorlandt et al., 2020; Watson, 2016).

Consequently, the preferences of people with PIMD around end-of-life treatment, their preferred location for end-of-life care and other wishes relating to death and dying are often not known (Voss et al., 2019). A Dutch survey study on the considerations and beliefs of physicians and support staff on decision-making about medical interventions at the end of life reported that people with PIMD had their end-of-life preferences taken into account only 2.9% of the time (Bekkema, de Veer, Wagemans, et al., 2014). More recent evidence demonstrates that even specialist palliative care clinicians, who are skilled in talking about prognosis, dying and death, experience difficulties in talking with people with intellectual disability (Foo et al., 2021). Studies on advance care planning showed that wishes or preferences regarding future care of people with intellectual disability, especially of people with PIMD, were often not recorded or discussed, with decisions instead made by family carers or health professionals (Voss et al., 2019, 2020). This indicates that there is a lack of acceptance and understanding of the unique needs and strengths of people with PIMD, particularly in communication about end-of-life care preferences (Johnson et al., 2012; Watson et al., 2017).

Autonomy in Decision-Making

The UNCRPD was the first international human rights treaty to specifically recognise that people with disabilities have a right to autonomy (United Nations, 2006). Article 12 of the UNCRPD recognises that "persons with disabilities enjoy legal capacity on an equal basis with others" (p. 9); in other words, an individual cannot lose his/her legal capacity to act simply because of a disability (see also Chapter 5). Also, Article 12 stresses the importance of supporting decision-making stating that signatory nations "shall take appropriate measures to provide access by persons with disabilities to the support they may require in exercising their legal capacity" (p. 9). Article 25 requires "health professionals to provide care of the same quality to people with disabilities as to others, including on the basis of free and informed consent" (p. 16).

In addition, the *Caring at the End of Life* position statement of the American Association on Intellectual and Developmental Disabilities

(AAIDD) highlights respect for autonomy as one of four major principles and notes that support staff should always attempt to discover the self-determined wishes and needs of the person with intellectual disability and honour those wishes (American Association on Intellectual & Developmental Disabilities, 2020). With support from people who know them well, people with PIMD can and do express their likes, dislikes and desires (Gjermestad, 2017; Watson, 2016). Through an open, active and reflective attitude, individuals with PIMD can be supported to form and express preferences in relation to many life decisions, including those at the end of life (Bekkema, de Veer, Hertogh, et al., 2014). Within this context, autonomy is viewed as a relational construct, best realised in collaboration with supporters in the person's life who know him or her well (Bekkema, de Veer, Hertogh, et al., 2014; Watson et al., 2017).

Despite the importance of autonomy in end-of-life care, guardianship laws and practice still dominate (United Nations, 2007), arguably representing a risk-averse position about self-decision-making. A person with PIMD's legal capacity for decision-making is often determined on the basis of cognitive assessments. These assessments define decision-making capability as dependent on the existence of a set of cognitive abilities such as problem solving, memory function, rationality and language (Kirkendall et al., 2017). Due to the nature of these assessments, people with PIMD are likely to be assessed as having no or very limited decision-making capacity (Johnson et al., 2012; Watson et al., 2019). Therefore, many people with PIMD are restricted in their legal capacity and the default position is typically the appointment of a legal guardian who makes substitute decisions, including end-of-life decisions, for them (European Association of Service Providers for Persons with Disabilities, 2019).

Supported Decision-Making

To address inequities in self-determination and autonomy, supported decision-making, instead of substitute decision-making, is increasingly recommended as best practice (Arstein-Kerslake, 2017; United Nations,

2007). Supported decision-making for people with PIMD can be defined as a process of enhancing the decision-making capacity of an individual with PIMD through collaborative support from a group of people in that person's life who know them well (Watson, 2016). Involving the person's circle of support, comprised of key people in the person's life who know their history, preferences and personal characteristics or are willing to explore them, is of crucial importance for a good quality of life, planning for end-of-life care and death (Hostyn & Maes, 2009; Voss et al., 2020).

Supported decision-making is a person-centred approach. It shifts the focus away from the person's disability and emphasises their individual competence by recognising and acknowledging the person and his/her unique circumstances (Browning et al., 2014). The individual is the decision maker and the person supporting the individual should enable that person to exercise their legal capacity to the greatest extent possible, according to the latter's wishes or best interests (United Nations, 2007, p. 89). A component of this support is ensuring information is presented in a way that maximises the person's ability to understand (Tuffrey-Wijne et al., 2007).

Supported decision-making can take many forms. For people with PIMD, supported decision-making is best accompanied by close observation of communication within social interactions to determine the reciprocal influence of both communication partners (Watson et al., 2017). Supported decision-making models and frameworks primarily focussed on decision-making support for people with mild, rather than more severe, intellectual disability are most prevalent. Watson's (2016) supported decision-making approach for people with PIMD is an exception. Within this model, supported decision-making is characterised in terms of the existence of two distinct, but interdependent roles played by (a) the person with a disability (supported) and (b) the circle of support in the decision-making process. Supporter responsiveness is the key component that can change, making the enablement of responsiveness a crucial strategy for supporting decision-making relating to end-of-life care for people with PIMD (Watson et al., 2017). Watson and colleagues (2017) have characterised this responsiveness as comprising three phases: (a) acknowledgement of a person's expression of will and preference, (b) collaboratively interpreting that expression and (c) acting on that preference.

Taking the Will and Preferences of People with PIMD into Account: A Narrative Approach

To understand the end-of-life will and preferences of a person with PIMD, it is of utmost importance to know a person's history and personal story (Johnson et al., 2012; Watson et al., 2019). The common understanding of narratives and storytelling is that they are told or communicated verbally. However, people with PIMD may have limited opportunities for such storytelling (Vorhaus, 2016). Abandoning this traditional view of narratives, stories of persons with PIMD are told through their interaction with others, including sharing stories, photos and videos and acknowledging non-verbal communication, such as gestures, sounds and eye movement (Watson et al., 2017). Such a narrative approach is seen as an important way of transferring knowledge, which can be helpful in assessing end-of-life care needs and decision-making (Watson et al., 2019).

It is important for staff who provide support for people with PIMD to understand that knowledge regarding a person's history and life story is not just acquired through relationship longevity (Watson et al., 2017). Studies showed that when family carers share life stories and images with support workers, these caregivers may acquire a deeper understanding of the person with PIMD. The study of Kruithof et al. (2020) on parental knowledge showed that the in-depth knowledge of their child with PIMD can be transferred from parents to staff "by example" (p. 1146). By looking at the interaction between parent and child, parents can "act out" (p. 1147) their embodied knowledge and support staff can learn how best to interact with the child (Kruithof et al., 2020). In this way, the views and preferences of people with PIMD can be acknowledged and used to provide end-of-life care in line with the will and preferences of the person (Watson et al., 2017, 2019).

Simone's story shows how her supporters gather and share information about Simone's past experiences, collaboratively interpret expressions, and act upon her will and preferences in decision-making at the end of life.

Simone's* Story Simone was a 40-year-old woman with PIMD who lived in a group home managed by a large disability service provider. She communicated using a range of informal communication methods. She smiled and laughed a lot. Due to her complex communication and support needs, Simone was heavily reliant on those who knew her well, her mother, father and paid support staff, to have her communication acknowledged, interpreted and acted upon.

Simone developed aspiration pneumonia, a complication often experienced by people with PIMD due to challenges with eating and swallowing (dysphagia). As a consequence, Simone was offered a tracheotomy, an intervention that her treating medical team believed may give her a few more weeks or months to live. Simone's circle of support used a supported decision-making framework developed by Watson and Joseph (2015) to assist her in making this difficult decision about her end-of-life care.

Simone's supporters gathered information from her medical team about her options and spent time collaboratively weighing these up. They drew heavily from their relationships with Simone as well as her past experiences of tracheotomy. One of Simone's support workers recalled Simone's reaction to a tracheotomy three years earlier. "I remember… she hated it. It was horrible. She was so distressed and seemed so scared". Simone's mother replied, "Don't you worry love, I remember. I videoed her". Simone's circle of support watched the video Simone's mother had taken together and reviewed diary notes recorded at that time. These memory tools reminded them of how Simone's behaviour had changed significantly at the time of her last tracheotomy. Her circle of support agreed that Simone's behaviour during this time was due to the distress she experienced with this alternative method of breathing and, as articulated by her father, provided evidence that "if she had a choice she wouldn't want that bloody thing again". Guided by the preferences Simone had communicated to them through the documentation of her past experiences of having a tracheotomy, the group collectively made the difficult decision that a tracheotomy would not be carried out. Simone died a number of days later with her parents and favourite support worker by her side.

*Assumed names are used in this story.

Reflection Questions

1. How were Simone's will and preference taken into account at the end of her life?
2. What is the role of people who know someone well in decision-making at the end of life?
3. How can a person's historical experiences guide those supporting them to make decisions about their end-of-life care?

Placing End-of-Life Preferences of the Person at the Centre

Providing end-of-life care is holistic and multidisciplinary. The contribution of a diverse range of professionals and regular interdisciplinary meetings is important to enhance the quality of end-of-life care (Tuffrey-Wijne & Davidson, 2018). Irrespective of the care location/setting, end-of-life care needs should extend beyond physical care needs including pain assessment and management to a focus on psychosocial, spiritual, social and communication needs (Adam et al., 2020).

Supporting psychosocial and spiritual wellbeing at the end of life for people with intellectual disability is seen as an important component of person-centred care (Tuffrey-Wijne et al., 2009). Despite this recognition, little is known about the religious, psychosocial and spiritual needs of people with PIMD (Loynes & O'Hara, 2015). Although conversations about complex issues like existential distress and the meaning of life (and death) may be challenging for people with PIMD, it is important to find appropriate ways to assess and meet their spiritual and psychosocial needs. Pastoral carers may be able to assist in supporting spiritual health and wellbeing, and in dealing with the impending death and farewell process (Demmons, 2008; Voss et al., 2019).

For those with PIMD, the importance and presence of a strong and coherent support network made up of people who know them well, such as family and support workers, was identified as key at the end of life (Adam et al., 2020). This network of support is essential for the process of supported decision-making in end-of-life care for people with PIMD, as they play an important role in collaboratively acknowledging, interpreting and acting on a person's expression of will and preference (Watson et al., 2017). The imminence of the end of life often leads to an intensified caring relationship between the person with PIMD, family and support workers. Honest, proactive and repeated discussion on what constitutes good care is important to jointly provide the best possible warm, comforting care for the person with PIMD at the end of life, in line with their wishes and needs (Bekkema, de Veer, et al., 2015).

Resources

1. *Bereavement and Loss Learning Resource Pack and Training*. For those supporting bereaved people with PIMD. Pamis Scotland. http://pamis.org.uk/resources/bereavement-and-loss/
2. *Decision-making about the best place of palliative care for people with intellectual disabilities*: A guide for care staff and healthcare professionals providing palliative care for people with intellectual disabilities (Bekkema, Tuffrey-Wijne, et al., 2015) https://www.nivel.nl/sites/default/files/bestanden/Handreiking-decisionmaking-palliative-care-disabilities.pdf
3. *Listening to Those Rarely Heard—A Guide for Supporters*. A training package to guide supporters of people with severe to profound disability through the supported decision-making process (Watson & Joseph, 2015). https://shop.scopeaust.org.au/shop/listening-rarely-heard-guide-supporters/
4. *Listening to Those Rarely Heard*. A video that is part of a training package for people who know someone with a severe to profound

intellectual disability. Scope Victoria Australia. https://vimeo.com/21176882
5. *Representing the Rights of Persons with Disabilities.* TED Talk by Joanne Watson (2016). https://www.youtube.com/watch?v=qMerG7CULJE

References

Adam, E., Sleeman, K. E., Brearley, S., Hunt, K., & Tuffrey-Wijne, I. (2020). The palliative care needs of adults with intellectual disabilities and their access to palliative care services: A systematic review. *Palliative Medicine, 34*(8), 1006–1018. https://doi.org/10.1177/0269216320932774

American Association on Intellectual and Developmental Disabilities. (2020). *Position statement: Caring at the end of life*. AAIDD. https://www.aaidd.org/news-policy/policy/position-statements/caring-at-the-end-of-life

Arstein-Kerslake, A. (2017). Restoring voice to people with cognitive abilities: Realizing the right to equal recognition before the law. *Cambridge University Press*. https://doi.org/10.1017/9781316493526

Arvio, M., & Sillanpää, M. (2003). Prevalence, aetiology and comorbidity of severe and profound intellectual disability in Finland. *Journal of Intellectual Disability Research, 47*(2), 108–112. https://doi.org/10.1046/j.1365-2788.2003.00447.x

Bekkema, N., de Veer, A. J. E., Hertogh, C. M. P. M., & Francke, A. L. (2014). Respecting autonomy in the end-of-life care of people with intellectual disabilities: A qualitative multiple-case study. *Journal of Intellectual Disability Research, 58*(4), 368–380. https://doi.org/10.1111/jir.12023

Bekkema, N., de Veer, A. J. E., Wagemans, A. M. A., Hertogh, C. M. P. M., & Francke, A. L. (2014). Decision making about medical interventions in the end-of-life care of people with intellectual disabilities: A national survey of the considerations and beliefs of GPs, ID physicians and care staff. *Patient Education and Counseling, 96*(2), 204–209. https://doi.org/10.1016/j.pec.2014.05.014

Bekkema, N., de Veer, A. J. E., Hertogh, C. M. P. M., & Francke, A. L. (2015). 'From activating towards caring': Shifts in care approaches at the end of life of people with intellectual disabilities; a qualitative study of the perspectives of relatives, care-staff and physicians. *BMC Palliative Care, 14*(1), 1–10. https://doi.org/10.1186/s12904-015-0030-2

Bekkema, N., Tuffrey-Wijne, I., Wagemans, A. M. A., Hertogh, C. M. P. M., Francke, A. L., & de Veer, A. J. E. (2015). *Decision-making about the best place of palliative care for people with intellectual disabilities: A guide for care staff and healthcare professionals providing palliative care for people with intellectual disabilities.* http://www.nivel.nl/sites/default/files/bestanden/Handreiking-decisionmaking-palliative-care-disabilities.pdf

Billingham, M. J., & Billingham, S.-J. (2013). Congruence between preferred and actual place of death according to the presence of malignant or non-malignant disease: A systematic review and meta-analysis. *BMJ Supportive and Palliative Care, 3*(2), 144–154. https://doi.org/10.1136/bmjspcare-2012-000292

Bloomer, M. J., & Walshe, C. (2021). What mediates end-of-life care choices? *The Lancet Public Health, 6*(3), e139–e140. https://doi.org/10.1016/S2468-2667(21)00005-0

Bloomer, M. J., & Watson, J. Improving palliative care for people with intellectual disabilities: Communication is key. *Palliative Medicine,* doi:https://doi.org/10.1177/02692163211046445

Browning, M., Bigby, C., & Douglas, J. (2014). Supported decision making: Understanding how its conceptual link to legal capacity is influencing the development of practice. *Research and Practice in Intellectual and Developmental Disabilities, 1*(1), 34–45. https://doi.org/10.1080/23297018.2014.902726

Burge, F., Lawson, B., Johnston, G., Asada, Y., McIntyre, P. F., & Flowerdew, G. (2015). Preferred and actual location of death: What factors enable a preferred home death? *Journal of Palliative Medicine, 18*(12), 1054–1059. https://doi.org/10.1089/jpm.2015.0177

Demmons, T. (2008). Embodied encounter through imagination and the arts: Toward a (Barthian) theology and praxis of pastoral care and counseling for persons with intellectual disabilities. *Journal of Religion, Disability & Health, 12*(4), 365–375. https://doi.org/10.1080/15228960802515642

European Association of Service Providers for Persons with Disabilities. (2019). *Legal capacity*. EASPD. https://easpd.eu/key-areas-of-work/legal-capacity/

Foo, B., Wiese, M., Curryer, B., Stancliffe, R. J., Wilson, N. J., & Clayton, J. M. (2021). Specialist palliative care staff's varying experiences of talking with people with intellectual disability about their dying and death: A thematic analysis of in-depth interviews. *Palliative Medicine, 35*(4), 738–749. https://doi.org/10.1177/0269216321998207

Gjermestad, A. (2017). Narrative competence in caring encounters with persons with profound intellectual and multiple disabilities. *International Practice Development Journal, 7* (Suppl). https://doi.org/10.19043/ipdj.7SP.007

Goldbart, J., Chadwick, D., & Buell, S. (2014). Speech and language therapists' approaches to communication intervention with children and adults with profound and multiple learning disability. *International Journal of Language & Communication Disorders, 49*(6), 687–701. https://doi.org/10.1111/1460-6984.12098

Hosking, F. J., Carey, I. M., Shah, S. M., Harris, T., DeWilde, S., Beighton, C., & Cook, D. G. (2016). Mortality among adults with intellectual disability in England: Comparisons with the general population. *American Journal of Public Health, 106*(8), 1483–1490. https://doi.org/10.2105/AJPH.2016.303240

Hostyn, I., & Maes, B. (2009). Interaction between persons with profound intellectual and multiple disabilities and their partners: A literature review. *Journal of Intellectual and Developmental Disability, 34*(4), 296–312. https://doi.org/10.3109/13668250903285648

Johnson, H., Watson, J., Iacono, T., Bloomberg, K., & West, D. (2012). Assessing communication in people with severe-profound disabilities: Co-constructing competence. *Journal of Clinical Practice in Speech-Language Pathology, 14*(2), 64–68. https://doi.org/10.1046/j.1365-2648.2003.02884.x

Kamstra, A. (2017). *Who cares. Research into maintaining, strengthening, and expanding the informal social networks of people with profound intellectual and multiple disabilities* (PhD thesis). University of Groningen. https://www.rug.nl/research/portal/en/publications/who-cares(b77b6026-9b61-40af-b7ae-0384c7728d69).html.

Kamstra, A., van der Putten, A. A. J., & Vlaskamp, C. (2017). Efforts to increase social contact in persons with profound intellectual and multiple disabilities: Analysing individual support plans in the Netherlands. *Journal of Intellectual Disabilities, 21*(2), 158–174. https://doi.org/10.1177/1744629516653037

Kirkendall, A., Linton, K., & Farris, S. (2017). Intellectual disabilities and decision making at end of life: A literature review. *Journal of Applied Research in Intellectual Disabilities, 30*(6), 982–994. https://doi.org/10.1111/jar.12270

Kruithof, K., Willems, D., van Etten-Jamaludin, F., & Olsman, E. (2020). Parents' knowledge of their child with profound intellectual and multiple disabilities: An interpretative synthesis. *Journal of Applied Research in Intellectual Disabilities, 33*(63), 1141–1150. https://doi.org/10.1111/jar.12740

Loynes, B., & O'Hara, J. (2015). How can mental health clinicians, working in intellectual disability services, meet the spiritual needs of their service users? *Advances in Mental Health and Intellectual Disabilities, 9*(1), 9–18. https://doi.org/10.1108/AMHID-10-2014-0035

McLaughlin, D., Barr, O., McIlfatrick, S., & McConkey, R. (2014). Developing a best practice model for partnership practice between specialist palliative care and intellectual disability services: A mixed methods study. *Palliative Medicine, 28*(10), 1213–1221. https://doi.org/10.1177/0269216314550373

Nakken, H., & Vlaskamp, C. (2007). A need for a taxonomy for profound intellectual and multiple disabilities. *Journal of Policy and Practice in Intellectual Disabilities, 4*(2), 83–87. https://doi.org/10.1111/j.1741-1130.2007.00104.x

Nieuwenhuijse, A. M., Willems, D. L., & Olsman, E. (2020). Physicians' perceptions on quality of life of persons with profound intellectual and multiple disabilities: A qualitative study. *Journal of Intellectual & Developmental Disability, 45*(2), 176–183. https://doi.org/10.3109/13668250.2019.1580117

Noorlandt, H. W., Echteld, M. A., Tuffrey-Wijne, I., Festen, D. A. M., Vrijmoeth, C., van der Heide, A., & Korfage, I. J. (2020). Shared decision-making with people with intellectual disabilities in the last phase of life: A scoping review. *Journal of Intellectual Disability Research, 64*(11), 881–894. https://doi.org/10.1111/jir.12774

Raijmakers, N. J. H., de Veer, A. J. E., Zwaan, R., Hofstede, J. M., & Francke, A. L. (2018). Which patients die in their preferred place? A secondary analysis of questionnaire data from bereaved relatives. *Palliative Medicine, 32*(2), 347–356. https://doi.org/10.1177/0269216317710383

Rietjens, J. A., Sudore, R. L., Connolly, M., van Delden, J. J., Drickamer, M. A., Droger, M., van der Heide, A., Heyland, D. K., Houttekier, D., & Janssen, D. J. (2017). Definition and recommendations for advance care planning: An international consensus supported by the European Association for Palliative Care. *The Lancet Oncology, 18*(9), e543–e551. https://doi.org/10.1016/S1470-2045(17)30582-X

Rousseau, M., Mathieu, S., Brisse, C., Motawaj, M., Grimont, E., Auquier, P., & Billette de Villemeur, T. (2015). Aetiologies, comorbidities and causes of death in a population of 133 patients with polyhandicaps cared for at specialist rehabilitation centres. *Brain Injury, 29*(7–8), 837–842. https://doi.org/10.3109/02699052.2015.1004757

Ryan, K., McEvoy, J., Guerin, S., & Dodd, P. (2010). An exploration of the experience, confidence and attitudes of staff to the provision of palliative care to people with intellectual disabilities. *Palliative Medicine, 24*(6), 566–572. https://doi.org/10.1177/0269216310371413

Sarimski, K. (2016). *Soziale Teilhabe von Kindern mit komplexer Behinderung in der Kita* [Social Participation of Children with Complex Disabilities in Day Care]. Reinhardt.

Simoes, C., & Santos, S. (2016). The quality of life perceptions of people with intellectual disability and their proxies. *Journal of Intellectual & Developmental Disability, 41*(4), 311–323. https://doi.org/10.3109/13668250.2016.1197385

Stancliffe, R. J. (2001). Living with support in the community: Predictors of choice and self-determination. *Mental Retardation and Developmental Disabilities Research Reviews, 7*(2), 91–98. https://doi.org/10.1002/mrdd.1013

Stein, G. L. (2008). Providing palliative care to people with intellectual disabilities: Services, staff knowledge, and challenges. *Journal of Palliative Medicine, 11*(9), 1241–1248. https://doi.org/10.1089/jpm.2008.0130

Tuffrey-Wijne, I., Bernal, J., Hubert, J., Butler, G., & Hollins, S. (2009). People with learning disabilities who have cancer: An ethnographic study. *British Journal of General Practice, 59*(564), 503–509. https://doi.org/10.3399/bjgp09X453413

Tuffrey-Wijne, I., & Davidson, J. (2018). Excellence in palliative and end-of-life care provision for people with intellectual disability. *International Journal of Palliative Nursing, 24*(12), 598–610. https://doi.org/10.12968/ijpn.2018.24.12.598

Tuffrey-Wijne, I., Hogg, J., & Curfs, L. (2007). End-of-life and palliative care for people with intellectual disabilities who have cancer or other life-limiting illness: A review of the literature and available resources. *Journal of Applied Research in Intellectual Disabilities, 20*(4), 331–344. https://doi.org/10.1111/j.1468-3148.2006.00350.x

United Nations. (2006). *Convention on the rights of persons with disabilities*. UNCRPD. https://www.un.org/development/desa/disabilities/convention-on-the-rights-of-persons-with-disabilities.html#Fulltext

United Nations. (2007). *From exclusion to equality: Realizing the rights of persons with disabilities.* https://www.un.org/disabilities/documents/toolaction/ipuhb.pdf

Van Timmeren, E., Van der Putten, A., & van Schrojenstein Lantman-de Valk, H., Van der Schans, C., & Waninge, A. (2016). Prevalence of reported physical health problems in people with severe or profound intellectual and motor disabilities: A cross-sectional study of medical records and care plans. *Journal of Intellectual Disability Research, 60*(11), 1109–1118. https://doi.org/10.1111/jir.12298

Vlaskamp, C., Hiemstra, S. J., Wiersma, L. A., & Zijlstra, B. J. (2007). Extent, duration, and content of day services' activities for persons with profound intellectual and multiple disabilities. *Journal of Policy and Practice in Intellectual Disabilities, 4*(2), 152–159. https://doi.org/10.1111/j.1741-1130.2007.00112.x

Vlaskamp, C., & Van Der Putten, A. (2009). Focus on interaction: The use of an individualized support program for persons with profound intellectual and multiple disabilities. *Research in Developmental Disabilities, 30*(5), 873–883. https://doi.org/10.1016/j.ridd.2008.12.005

Vorhaus, J. (2016). *Giving voice to profound disability: Dignity, dependence and human capabilities*. Routledge.

Voss, H., Vogel, A., Wagemans, A., Francke, A., Metsemakers, J., Courtens, A., & de Veer, A. (2019). Advance care planning in the palliative phase of people with intellectual disabilities: Analysis of medical files and interviews. *Journal of Intellectual Disability Research, 63*(10), 1262–1272. https://doi.org/10.1111/jir.12664

Voss, H., Vogel, A., Wagemans, A. M. A., Francke, A. L., Metsemakers, J. F. M., Courtens, A. M., & de Veer, A. J. E. (2020). What is important for advance care planning in the palliative phase of people with intellectual disabilities? A multi-perspective interview study. *Journal of Applied Research in Intellectual Disabilities, 33*(2), 160–171. https://doi.org/10.1111/jar.12653

Voss, H., Vogel, A. G., Wagemans, A. M., Francke, A. L., Metsemakers, J. F., Courtens, A. M., & de Veer, A. J. (2021). Development, implementation, and evaluation of an Advance Care Planning program for professionals in palliative care of people with intellectual disability. *Intellectual and Developmental Disabilities, 59*(1), 39–54. https://doi.org/10.1352/1934-9556-59.1.39

Wagemans, A., & van Schrojenstein Lantman-de Valk, H., Proot, I., Metsemakers, J., Tuffrey-Wijne, I., & Curfs, L. (2013). The factors affecting end-of-life decision-making by physicians of patients with intellectual disabilities in the Netherlands: A qualitative study. *Journal of Intellectual Disability*

Research, 57(4), 380–389. https://doi.org/10.1111/j.1365-2788.2012.01550.x

Wagemans, A., Van Schrojenstein Lantman-de-Valk, H., Tuffrey-Wijne, I., Widdershoven, G., & Curfs, L. (2010). End-of-life decisions: An important theme in the care for people with intellectual disabilities. *Journal of Intellectual Disability Research, 54*(6), 516–524. https://doi.org/10.1111/j.1365-2788.2010.01276.x

Ware, J. (2004). Ascertaining the views of people with profound and multiple learning disabilities. *British Journal of Learning Disabilities, 32*(4), 175–179. https://doi.org/10.1111/j.1468-3156.2004.00316.x

Wark, S., Hussain, R., Müller, A., Ryan, P., & Parmenter, T. (2017). Challenges in providing end-of-life care for people with intellectual disability: Health services access. *Journal of Applied Research in Intellectual Disabilities, 30*(6), 1151–1159. https://doi.org/10.1111/jar.12408

Watson, J. (2016). *The right to supported decision-making for people rarely heard* [PhD thesis, Deakin University]. Melbourne. http://dro.deakin.edu.au/eserv/DU:30083812/watson-theright-2016A.pdf

Watson, J., & Joseph, R. (2015). *People with severe to profound intellectual disabilities leading lives they prefer through supported decision-making: A guide for supporters*. The Scope Shop. https://shop.scopeaust.org.au/shop/listening-rarely-heard-guide-supporters/

Watson, J., Voss, H., & Bloomer, M. J. (2019). Placing the preferences of people with profound intellectual and multiple disabilities at the center of end-of-life decision making through storytelling. *Research and Practice for Persons with Severe Disabilities, 44*(4), 267–279. https://doi.org/10.1177/1540796919879701

Watson, J., Wilson, E., & Hagiliassis, N. (2017). Supporting end of life decision making: Case studies of relational closeness in supported decision making for people with severe or profound intellectual disability. *Journal of Applied Research in Intellectual Disabilities, 30*(6), 1022–1034. https://doi.org/10.1111/jar.12393

Wehmeyer, M. L., & Garner, N. W. (2003). The impact of personal characteristics of people with intellectual and developmental disability on self-determination and autonomous functioning. *Journal of Applied Research in Intellectual Disabilities, 16*(4), 255–265. https://doi.org/10.1046/j.1468-3148.2003.00161.x

World Health Organization. (2015). *WHO Global disability action plan 2014–2021. Better health for all people with disability*. WHO. https://www.who.int/health-topics/disability#tab=tab_1

Wosinski, B., & Newman, C. J. (2019). Physicians' attitudes when faced with life-threatening events in children with severe neurological disabilities. *Developmental Neurorehabilitation, 22*(1), 61–66. https://doi.org/10.1080/17518423.2018.1461946

Zaal-Schuller, I. H., De Vos, M., Ewals, F. V., Van Goudoever, J., & Willems, D. L. (2016). End-of-life decision-making for children with severe developmental disabilities: The parental perspective. *Research in Developmental Disabilities, 49*, 235–246. https://doi.org/10.1016/j.ridd.2015.12.006

Zaal-Schuller, I. H., Willems, D. L., Ewals, F. V., van Goudoever, J., & de Vos, M. (2018). Considering quality of life in end-of-life decisions for severely disabled children. *Research in Developmental Disabilities, 73*, 67–75. https://doi.org/10.1016/j.ridd.2017.12.015

Zwakhalen, S. M., Van Dongen, K. A., Hamers, J. P., & Abu-Saad, H. H. (2004). Pain assessment in intellectually disabled people: Non-verbal indicators. *Journal of Advanced Nursing, 45*(3), 236–245. https://doi.org/10.1046/j.1365-2648.2003.02884.x

14

The Process of Dying

Karen Ryan

Peter's* story

Peter is a 36-year-old man with a severe intellectual disability secondary to cerebral palsy who lives in a community group home. He is non-verbal, has spastic quadriparesis and is fully dependent. He has experienced repeated aspiration pneumonias in the last two years and has become progressively more debilitated. Peter has been sent into hospital each time that he developed an infection, but he has become increasingly more distressed when in hospital. His carers feel that he is frightened at leaving his familiar home environment and that this has triggered episodes of challenging behaviour. They do not think that the hospital

K. Ryan (✉)
Consultant in Palliative Medicine, St Francis Hospice and Mater Misericordiae University Hospital, Dublin, Ireland
e-mail: karen.ryan@ucd.ie

R. J. Stancliffe et al. (eds.), *End of Life and People with Intellectual and Developmental Disability*, https://doi.org/10.1007/978-3-030-98697-1_14

staff know how to manage Peter's distress. They are concerned that he is weakening despite the treatment that he has received and has begun to ask themselves whether he is benefitting from the hospitalisations or not. His carers are conflicted—on the one hand, they are reluctant to speak about this to the hospital staff because they are afraid that they might deprive Peter of treatment but on the other, they do not want him to spend his final days afraid and distressed.

*Assumed names are used in this story.

A Good Death

As every person's lived experience is unique, so too is their dying. What is meant by a good death is also something that is deeply individual. This means that quality end-of-life care is a dynamic process that is negotiated and re-negotiated among those who are dying, their families and care-givers. Although there is no one definition of a good death, the attributes that are most commonly associated with this concept in the general population are a sense of life completion, being free from suffering, being treated with dignity and respect and being surrounded by family (Meier et al., 2016). Those attributes have also been found to be important to people with intellectual disability, with particular emphasis being placed on maintaining presence and relationships, communicating and honouring wishes and meeting spiritual needs (Bekkema et al., 2015; Cithambarm et al., 2021).

Recognising Dying

A first step in providing good end-of-life care is recognising when a person is reaching the terminal stage of illness. Of course, not all deaths may be anticipated—in the general population, it is estimated that sudden, unexpected death may account for approximately 10% of deaths classified as natural (Ito et al., 2021; Manolis et al., 2019). People with intellectual disability have been observed to be at higher risk of unexpected death compared to the general population and some estimate that the level of sudden death is almost twice as high as the wider population (Hunt et al., 2020).

Imminently or actively dying is a term used to describe individuals who are rapidly approaching end of life. It is most often understood to mean that a person has a likely prognosis of hours or days. Making a diagnosis of dying has been described as 'part art, part science' (Kennedy et al., 2014, p. 263) and any of a range of physical, social, spiritual and psychological indicators may suggest that a person is imminently dying (Table 14.1). It is important that a competent professional is involved when making decisions about the care of a dying person, because treatable, acute exacerbations of some illnesses can sometimes present similar symptoms. When there is uncertainty or disagreement between those involved in care, additional advice should be sought from healthcare professionals with experience of providing care to people with serious illness (see also Ethical Dilemmas below).

Table 14.1 Body function by possible signs of imminent death

Body function	Possible sign of imminent death
Gastrointestinal	Decreased interest in food or fluids Swallowing difficulties Weight loss
Neurological and musculoskeletal	Frailty and weakness Profound fatigue Delirium Reduced level of consciousness
Respiratory	Altered breathing pattern Noisy airway secretions Weak or absent cough
Circulatory	Altered pulse and cool peripheries Mottled or discoloured skin Cyanosis
Urinary	Decreased or absent urinary output
Psychosocial and spiritual	Withdrawal and reduced social interaction Personality and mood changes ranging from restlessness, anxiety, fear to peaceful awareness of impending death

Responding to Signs That a Person May Be Dying

The recognition that a person may be beginning to die should trigger a number of actions. Firstly, consideration should be given as to whether the deterioration is expected or unexpected. If the deterioration is unexpected, assessment should be carried out by a healthcare professional who is competent to judge whether the change has a realistic prospect of reversibility or not. If the professional judges that the change in condition is potentially reversible, prompt action should be taken to attempt this, provided that is in accordance with the person's wishes (if the person has capacity to be involved in making that decision; see also Chapters 4, 5, and 7). If the person lacks capacity, decisions should be made according to the local, relevant legislative framework. This will likely involve consultation with a proxy or substitute decision maker and/or consideration of best interests. Prior wishes or decisions about what to do in the event of a deterioration may have previously been documented, and this also should inform decision-making.

Providing a decision is made that the person appears to be dying and in situations where the goals of care are focused on comfort, the next actions are to:

- Talk to the person and/or family.
- Agree on an individual plan of care, which includes consideration of place of care, food and drink, symptom control and psychological, social and spiritual support.
- Ensure needed medications are available for symptom control and that only medications consistent with the goals of care are continued.
- Ensure that all supports and resources needed to deliver the plan of care are in place.
- Advise other health and social care professionals.
- Prepare family, friends and any disability staff who may be involved.

Communicating with the Person with Intellectual Disability about Serious Illness and Death

Open and honest communication with the person who is dying, and those identified as important to them, is regarded by Western healthcare systems as fundamental to the quality of end-of-life care. Although individuals from the general population may vary about the extent to which they want to know about the facts of their illness and to be involved in decision-making, it is generally considered essential to provide adults with at least the opportunity to talk. For any person, communication opportunities should be proactive but sensitive; they should be respectful in pace and tone and take account of what the dying person wants and is able to discuss at any particular point in time.

Death remains a taboo subject for many members of society but studies have shown that people with intellectual disability are afforded fewer opportunities to engage in conversations than members of the general population (Kirkendall et al., 2017). Staff have been widely reported as lacking skills and experience to talk about serious illness and death, and to be fearful of initiating conversations about these topics. Commonly, the reasons given for avoiding speaking of serious illness range from a belief that the person with intellectual disability won't understand, to a concern that either the truth would be too upsetting, to a fear of acting against the wishes of others who do not want the person with intellectual disability to be told (Tuffrey-Wijne et al., 2010). As a result, people with intellectual disability are frequently not involved in decisions relating to their end-of-life care (Wagemans et al., 2015). The failure to provide people with intellectual disability with the opportunity to speak of their wishes, worries or concerns may result in missed opportunities and heightened anxiety with resultant negative impact on quality of life (see also Chapters 4 and 15). While training has an important role in enabling staff to engage in conversations with people with intellectual disability, it has been noted that education alone is not sufficient and that staff also need considerable practical and emotional support that should include reflective practice, supervision and team discussion (Tuffrey-Wijne & Rose, 2017).

Decisions around open communication should be made in the same way as those that concern the general population; the decision-making process should be grounded on ethical principles while considering individual circumstances. Specifically, as part of the individualised assessment, carers should make assessments of the person's levels of emotional and conceptual understanding. While there is a relative lack of research in the area, it is thought that the way in which loss and grief are experienced is dependent on the intellectual and socio-emotional age of the person (Klass et al., 2014). Table 14.2 is based on the work of Meusen-van de Kerkhof et al., and it summarises the way in which people with intellectual disability are thought to experience the death of someone they have built a relationship with (Meeusen-van de Kerkhof et al., 2006; Ryan & Dodd, 2008). It also provides practical suggestions for strategies to provide support to the person with an intellectual disability.

If a decision is made that it is appropriate to engage in communication about dying and death, then the communication needs and expectations of the person with intellectual disability should be established, considering whether they would like a person important to them to be present, how much information they would like to have about their condition, and their current level of insight or understanding that they may be nearing death. Cultural, religious, social or spiritual needs or preferences should be taken into account.

Many people with intellectual disability have communication impairments and interactions often need to be undertaken on a gradual or staged basis and time spent building a relationship of trust (Ryan et al., 2016). Longer periods of time should be set aside for the process of communication, and conversation should be in the form of simple language and short sentences, appropriate to the adult's developmental level. Professionals should remain sensitive to non-verbal cues and may consider the use of a variety of approaches to the process of communication, such as through art.

The usual models for breaking bad news rely on individuals having the ability to recognise a 'warning shot' (Tuffrey-Wijne, 2012, p. 6), and this subtlety may not be recognised by people with intellectual disability.

Table 14.2 Level of intellectual disability (ID) by how person thought to experience death by possible supports

Level of ID	How person thought to experience death	Possible supports
Mild	Clear realisation of death	Engage in open communication where desired; engage with spiritual or religious supports where desired
	Logical thought coupled to specific events	Offer closeness: share experience and offer participation in rituals or behaviours associated with loss and grief
	Grieving process comparable to general population	Engage in reminiscence, life story work; explore the feelings that lie behind reactions
Moderate	Limited realisation of death	Offer the opportunity to express grief; give logical explanations to questions, make connections visible where appropriate and desired; engage with spiritual or religious supports where desired
	Searching for logical explanations of illness and death; growing comprehension of the irreversibility of death	Offer closeness: Be there for the other person; make use of stories and photographs to support reminiscence; maintain supportive environment
	Basic capability of putting oneself in another person's place; projection of own feelings onto other person; feelings of guilt and fear as reactions to other people's grief	
	Grieving reactions may be delayed	

(continued)

Table 14.2 (continued)

Level of ID	How person thought to experience death	Possible supports
Severe	Limited realisation of death, death seen as temporary, magical thinking, realisation of death as irreversible starts to grow as deceased person is increasingly missed	Keep daily life organised and familiar as far as is possible Offer closeness: Be there for the other person
	Beginning of linking up events such as sickness and death, understanding of death linked to concrete experiences	Make concept of death clear by answering questions (if asked) on a concrete level; adjust imaginary images of death to prevent fear; engage with spiritual or religious supports where desired
	Egocentric way of thinking Limited development of words, expresses emotions non-verbally	Watch own (model) behaviour, make own emotions known, watch own language
	Model behaviour of others has strong influence on experience of illness, loss and death	Use a variety of approaches to explore emotions e.g. play, music, art
Profound	No realisation of death, loss mainly causes a disruption of bonding, security and confidence, understanding is based on sensual (body-centred) impressions and experiences	Offer closeness: Physical contact Keep daily life organised and familiar as far as is possible
	Communication is non-verbal	Important resources: posture, facial expression, intonation of voice, making use of one's favourite senses, respectful touching
	Reactions of mourning emerge only at a later time	Allow people to concretely experience change but within the space of a secure environment; offer concrete experiences to help cope with loss and grief; engage with spiritual or religious supports where desired

Indeed, it may contribute to distress if the individual becomes frustrated by the indirect nature of the communication. People with intellectual disability may not understand the need to ask questions to get information, and they are also often likely to acquiesce to what they believe the healthcare professional wants to hear. It is important, therefore, to maximise their sense of ease and confidence to minimise this tendency.

People with intellectual disability may find it difficult to understand information, and it is helpful to recapitulate previously stated material. This also provides the opportunity to check understanding of what has been discussed. People with intellectual disability may have a limited ability to express emotions verbally and instead sometimes express emotions in atypical ways or through behavioural disturbance. Caregivers should recognise these non-verbal methods of communication and respond appropriately. A useful guide about breaking bad news to people with intellectual disability has been developed by Tuffrey-Wijne (2013) (See Resources at the end of this chapter).

Communication with Family

The people close to a person with intellectual disability have often had an important role in supporting the individual through interactions with health and social care services over the course of their life; indeed, it has been observed that a lack of effective carer involvement is associated with poorer end-of-life outcomes (Heslop et al., 2014; Tuffrey-Wijne et al., 2016). Support from family is also of value in ensuring that the person with intellectual disability receives high quality care towards the end of life and people with intellectual disability have spoken of how important it is for many of them to have familiar people present at this time (Bekkema et al., 2016). It is important to remember that family members will themselves be emotionally affected and, as far as possible, their needs for support should be met and their feelings respected, although the main focus of care should remain on the dying individual.

As with the general population, the circumstances affecting relationships between the person with intellectual disability and their family may vary, and where possible healthcare professionals should check to make sure that the person with intellectual disability agrees to the sharing of information and involvement of family. When family members are involved, they can facilitate conversations with the person with intellectual disability and support them to make choices and help them to communicate their wishes. Care should be taken, however, to ensure that health and social care professionals do not use family members as communication proxies in order to avoid direct engagement with the person with intellectual disability (Ryan et al., 2011a; Tuffrey-Wijne et al., 2007). In some jurisdictions and in certain circumstances, family members have legal authority to make healthcare decisions on behalf of the person when the individual lacks capacity to make their own choices.

Plan and Do

An individualised plan for end-of-life care should pay attention to symptom control (e.g. relief of pain and other discomforts) and the person's physical, emotional, psychosocial and spiritual needs. It should also consider issues such as food and fluid, medication management, preferred care setting and resource requirements. The care plan should be documented, shared with those involved in the person's care and readily available at any time that information is needed (see also Chapters 4 and 7).

A competent, and preferably familiar, professional should establish the capacity of the individual with intellectual disability to be involved in development of the care plan and the level of involvement that the person wants. The professional should also establish whether any advance care plan exists that is valid and applicable to the current situation. Even in situations where involvement of the person with intellectual disability is either not possible or desired, then the values and preferences of the person with intellectual disability should still play a central role in informing the development of the care plan. This plan should be reviewed on a regular basis and updated according to the evolving needs and condition of the dying person and those important to them.

As early as possible, the needed resources should be considered (e.g. equipment, care during out-of-hours periods, assistance needed for medication administration). While it is often possible to meet the needs and wishes of the dying person and/ or their family in their preferred place of care, this is not always the case. Situations can arise where care needs exceed the ability of usual carers to provide care, where care cannot be provided safely, or where the person with intellectual disability or their family wants care that healthcare professionals consider to be futile or in conflict with the best interests of the person (Bekkema et al., 2015). Prompt referral to, and input from, specialist palliative care services should be sought to help manage any such situations. Other strategies that may be of help in resolving the issue include seeking input from a more experienced colleague, holding a case conference or family meeting, involving an independent advocate or using a mediation service, if available.

Symptom Assessment

Self-report measures are the gold-standard of symptom assessment, but their use poses challenges for individuals with impairments of communication or for people who experience reduced levels of consciousness towards the end of life. In the past, it was suggested that people with severe disabilities demonstrate pain insensitivity or pain intolerance, but it is now recognised that an inability to verbally communicate the presence of pain does not mean that pain is absent. Carers should be alert to the variety of ways that people with intellectual disability may use to signal that they are distressed, such as physical or behavioural change or the use of a specific word or sound.

The term *alternative communication* is used to describe the non-verbal signs and behaviours that people use in the place of language (Glennen, 1997). While research on alternative communication has tended to focus on the expression of physical pain, the recognition of *total pain* is of particular importance in the palliative care setting. Dame Cicely Saunders, the founder of the modern hospice movement, defined the concept of total pain as the suffering that encompasses all of a person's physical, psychological, social, spiritual and practical struggles (Richmond, 2005). It can be challenging to distinguish expressions of physical pain from these other dimensions of distress, and it is important that assessment tools are used carefully to ensure that all distress expressions are not treated with analgesics. The Disability Distress Assessment Tool (DisDAT) was developed in order to address this issue (Regnard et al., 2003; see Resources at the end of this chapter). Its use involves documenting an individual's usual patterns of communication when they are content. Distress can then be identified if the baseline changes. Once distress has been recognised, a clinical decision checklist guides carers in determining cause and management.

Symptom Management

The management of symptoms at end of life utilises both pharmacological and non-pharmacological approaches. For example, facial fans or open windows may be helpful for breathlessness, and repositioning may reduce noisy respiratory secretions. As a general principle, for people who are starting treatment but who have not previously been given medications for symptom management, it is advisable to start with the lowest effective dose and titrate as clinically indicated. Prescribed medications should always be targeted at specific symptoms, have a clinical rationale for the dose used, be regularly reviewed and adjusted as needed for effect. Medications should be given by the oral route where possible. Pre-existing or emerging symptoms that are present with any frequency require the use of regular, around the clock medications. *As required* medications are prescribed for the management of symptoms that may arise at some stage during the dying process (see Anticipatory Medications below).

A continuous subcutaneous infusion (syringe pump or syringe driver) is a reliable and minimally invasive way to deliver many necessary medications when the person can no longer manage them orally. A subcutaneous infusion should be considered whenever more than two or three doses of any *as required* medications have been given within the last

24 hours. Explanation about the use of the infusion should be provided to the dying person where possible, and/or their family. Other than in exceptional circumstances, this should be done before the infusion is started.

Prescribers should follow local prescribing guidance and specialist palliative care advice should be sought in a timely manner if the individual's symptoms do not respond to treatment or if undesirable side effects are noted. As a general principle, medications for symptom control at the end of life are managed carefully to ensure that side effects are minimised. Healthcare professionals understand how precious remaining time is to both patient and family, and how important it is for the person to be as interactive as possible. Most commonly, any drowsiness that the dying person experiences while receiving medications is caused by the natural process of dying. However, sometimes the presence of organ failure means that metabolism of medications is affected, and it is difficult to use medications without side effects occurring. Occasionally, it may not be possible to use medications at the level necessary to control symptoms without associated sedation. If it is necessary to use medications in this way, this must be discussed with the dying person where possible, and those important to them.

A comprehensive guide to symptom management is outside the scope of this chapter, however, a high-level overview of common approaches to symptom management is provided in Table 14.3.

Anticipatory Medications

If a person is in the terminal phase of illness, it is helpful if *anticipatory medications* (also known as *just in case* or *as required* medications) are available for symptom control. This ensures that the medications can be given without unnecessary delay if they are needed. Anticipatory medications comprise medications that might be required to manage predictable and distressing symptoms, or medications that can be given by a subcutaneous route if the person cannot manage necessary oral medications. Typical medications include:

Table 14.3 Symptom by common management approaches in the last days of life

Symptom	Common management approach
Pain	Assess for the likely cause and treat any underlying, easily reversible causes where appropriate and possible, e.g. urinary catheter for retention, repositioning or pressure relieving strategies for immobility Use an assessment tool such as DisDAT (see Resources at the end of this chapter) where appropriate Follow the usual principles of pain management and consider, where possible, the person's preferences for how medications should be given Ordinarily, this will involve the use of non-opioid analgesia for mild pain, and weak or strong opioids for moderate or severe pain
Breathlessness	Assess for the likely cause, and treat any underlying, easily reversible causes where appropriate and possible Opioid medication may be used to manage refractory breathlessness and the use of benzodiazepines may be helpful where anxiety is associated with breathlessness
Respiratory secretions	Assess for the likely cause and establish whether the noise has an impact on the dying person or those important to that person Treat any factors that are considered likely to be reversible and where treatment is not excessively burdensome e.g. cessation of artificial hydration, repositioning Reassure carers that secretions are unlikely to cause discomfort and may be relieved by non-drug-based interventions such as repositioning and suction Consider a trial of an antisecretory agent (e.g. glycopyrronium, hyoscine butylbromide, hyoscine hydrobromide) to treat noisy respiratory secretions if they are causing distress to the dying person Monitor at least twice daily and consider changing or stopping drugs if: • Noisy respiratory secretions continue and are still causing distress after 12 hours, or • There are unacceptable side effects, such as dry mouth, urinary retention, delirium, agitation and unwanted levels of sedation

Symptom	Common management approach
Nausea and vomiting	Assess for the likely cause, and treat any underlying, easily reversible causes where appropriate and possible If already controlled with an oral anti-emetic, use the same drug in a subcutaneous infusion Nausea and vomiting of new onset may require the introduction of new anti-emetic in a subcutaneous infusion
Anxiety	Assess for the likely cause, and use non-pharmacological strategies to address distress, where possible Anxiolytic medications, such as benzodiazepines, can be used where non-pharmacological strategies are ineffective. Benzodiazepines can be given by a variety of routes including oral, buccal, and subcutaneous (including subcutaneous infusion)
Delirium	Assess for the likely cause, and treat any underlying, easily reversible causes where appropriate and possible Use non-pharmacological strategies to alleviate agitation, where possible e.g. presence, reassurance, environmental modifications Neuroleptic and anxiolytic medications may be used in the management of refractory agitation associated with delirium

- Opioid for management of pain and/or breathlessness.
- Anti-sickness medication for management of nausea and/or vomiting.
- Anxiolytic for management of anxiety, agitation or breathlessness.
- Anti-secretory medication for management of respiratory secretions.

It may be necessary to consider an anticonvulsant medication also for those people with intellectual disability who have pre-existing seizure disorders, as seizures may occur with more frequency in the dying phase.

Spiritual and/or Religious Care

The Spiritual Care Taskforce of the European Association for Palliative Care defines spirituality as 'the dynamic dimension of human life that relates to the way persons (individual and community) experience, express and/or seek meaning, purpose and transcendence, and the way they connect to the moment, to self, to others, to nature, to the significant and/or the sacred' (Nolan et al., 2011, p. 88). Spiritual care has been observed to be an intrinsic and essential component of palliative care, yet the least developed and most neglected dimension of palliative care (Gijsberts et al., 2019). Research on spiritual care for people with intellectual disability is also lacking, although participants with intellectual disability in a recent study by Cithambarm et al. (2021) people with intellectual disabilities spoke of the importance of meeting the spiritual needs of the dying in end-of-life care. In a study by Bekkema et al. (2016), people with intellectual disabilities also spoke about the importance of the support of spiritual caregivers, such as pastors, chaplains or general counsellors. In the study, participants spoke of the value of looking back on life, talking about fears, existential issues or religion and preparing the person for the terminal phase and imminent death. It was noted that this could help to bring peace and reduce fear.

Ethical Dilemmas

The provision of end-of-life care may involve decisions that are clinically complex and emotionally distressing. Some decisions may involve ethical dilemmas and legal uncertainties that further complicate the decision-making process. The input of experienced colleagues or the specialist palliative care team should be sought to help manage such issues, and practice should be in accordance with local policies, professional codes of conduct and national legislation. A detailed consideration of the range of ethical dilemmas that may arise at the end of life is beyond the scope of this chapter, but the issue of food and fluids is considered below, given the frequency with which this dilemma arises.

Food and Fluids

Food and drink are basic needs of human beings. All patients are entitled to appropriate food and drink and to assistance from carers if they need help to eat and drink. Ordinarily, in situations where a person is unable to take sufficient food and drink, healthcare professionals will consider whether hydration and/or nutrition should be given by other routes such as subcutaneous, intravenous or enteral feeding routes. This is often referred to as *clinically assisted nutrition and hydration* or CANH. A person who is dying will not obtain the same benefits from these interventions as a person who has a treatable condition but clear-cut evidence about benefits, burdens and risks is lacking (National Institute for Health and Care Excellence [NICE], 2015). Proposed benefits of CANH include preventing thirst, delirium, hypercalcemia and opioid toxicity through increasing renal perfusion. Proposed burdens or risks include precipitating or worsening of respiratory secretions, peripheral and pulmonary oedema, and contributing to gastrointestinal overload leading to nausea, vomiting, bloating and cramps (Carter, 2020).

Dying people should be supported to drink and take small amounts of oral intake, if they wish to and are able. As part of the natural process of dying, most people's sense of hunger and thirst diminish, and small amounts of intake are generally adequate to provide comfort. Carers should observe for any signs of difficulty, such as coughing or aspiration.

Swallow is often impaired at the end of life, and the risks and benefits of intake should be considered and communicated with the person, where appropriate, and those involved in care. Strategies such as positioning and consistency of food or fluid may enable to the person to manage intake more successfully. It is important to attend to frequent mouth care and to include the management of dry mouth in the care plan if needed. People important to the dying person should be supported to help with mouth care or oral intake if they would like to do so.

For those people with intellectual disability who ordinarily are dependent on CANH, the usual methods of providing intake should be continued. Consideration should be given to reducing rate or volume of intake, if there is concern that the person may be less tolerant of intake. The person should be carefully monitored and if there are signs of harm to the dying person or if the person does not want or tolerate it, then CANH should be stopped.

In situations where the issue of commencement of CANH arises, it should be remembered that in most Western jurisdictions, nutrition and hydration provided by artificial means are regarded in law as medical treatment and should be treated in the same way as other medical interventions. However, food and fluid hold psychological, social and symbolic significance (Del Rio et al., 2012) and some see CANH as part of basic nurture that should almost always be continued. For this reason, it is considered to be especially important that the views of the patient and of those close to them are listened to and considered (including their cultural and religious views). Sensitive explanation of the issues to be considered should be undertaken, including the benefits, burdens and risks of providing CANH. While the administration of clinically assisted hydration is a relatively straightforward procedure, the commencement of clinically assisted nutrition requires hospitalisation and is far more technically complex and associated risks are significantly higher.

Given the emotional significance of CANH, disagreement may arise between healthcare professionals and the person with intellectual disability and/or their family, or between team members. If there is disagreement regarding CANH, all efforts should be made to resolve the issue. Local policies, procedures and legal frameworks should guide the process to find resolution.

Communication and Care of Health and Social Care Professionals

A number of studies have shown that staff working in intellectual disability services want to help people with intellectual disability die in their preferred place of care (usually their home setting) but that they are often under-prepared to do so (Dunkley & Sales, 2014). While end-of-life care is generally regarded as an important and rewarding aspect of their responsibilities, staff may nevertheless experience stress and disenfranchised grief when providing such care (Ryan et al., 2011b). This is because staff frequently lack palliative care experience and skills, and organisations have often not developed policies or procedures governing care provision and because staff commonly experience grief that is not acknowledged or supported (Lord et al., 2017). Strategies to address these challenges include providing palliative care training to staff, fostering a more open culture around death, implementing policies and procedures and ensuring adequate staffing and resources are in place to provide care (see also Chapter 16). Staff may also benefit from time and space for reflection, validation, debriefing and feedback. Collaboration with specialist palliative care services should provide a further layer of support when required.

Preparation of Peers with Intellectual Disability

The preparation that a person with intellectual disability receives for the impending death of a peer is often minimal because of a tendency to withhold information about illness and death. Although this may be done to protect the person from anxiety, it effectively makes their grief a more acute experience, and the development of long-term mental health problems becomes more likely. These problems may be compounded by the exclusion of the person with intellectual disability from rituals of grieving, such as funerals (McRitchie et al., 2014) (see also Chapter 10).

Dealing with death and bereavement is difficult, and people with intellectual disability are particularly disadvantaged when facing this challenge because of social exclusion, disempowerment and impairments of cognition, adaptive skills and communication. It is therefore essential that carers engage in forward planning whenever a loss is anticipated to reduce the risk of complicated bereavement reactions developing. While preparation for loss and change is not an easy option, it is one that can lead to increased emotional growth, self-awareness and empowerment (Leick & Davidsen-Nielsen, 1991). Carers should therefore sensitively support these peers to maintain friendships with their peers, even towards the end of life, to ensure they are enabled to participate in all aspect of the life cycle, including death.

Managing the Uncertainty Inherent in the Dying Process

While most people experience a period of ill-health prior to death, there is significant variation in both how long and how unwell individuals are before death. Three common illness trajectories have been described for patients with progressive chronic illness (Murray et al., 2005):

- A trajectory with steady progression and usually a clear terminal phase.
- A trajectory with gradual decline, punctuated by episodes of acute deterioration and some recovery before a relatively sudden, final deterioration.
- A trajectory with prolonged gradual decline.

Recognition of these trajectories of illness may allow health and social care professionals to think ahead and to engage with individuals, families and disability staff (where relevant) about what kind of care they would like towards the end of life. However, despite knowledge of these frameworks, clinical assessment of an individual patient's prognosis remains an inexact and challenging science (Stone et al., 2021). To further complicate matters, sometimes people who appear to be dying may stabilise for a period before again weakening. Indeed, sometimes people may unexpectedly recover from the acute episode that has precipitated their deterioration. Honest communication about prognostic uncertainty with the dying person wherever possible, and/or those important to them is essential, although less experienced clinicians may find these conversations difficult. Trust is an important component of relationships between clinicians, patients and families, and it may be negatively impacted when clinicians do not discuss prognostic uncertainty in uncertain end-of-life contexts (Evans et al., 2009; Robinson et al., 2021). Advice from more experienced colleagues or involvement of palliative care specialists is often of value in supporting communication and maintaining the trust of the person who is dying and their family in such situations.

Conclusion

People with intellectual disability who are coming to the end of their lives need high quality care and treatment to support them to live as well as possible until they die, and to die with dignity. It is essential that all staff are prepared to provide such care and that they possess the necessary empathy and expertise to provide care that maximises comfort and well-being. Many people who are dying will experience some degree of physical, psychological, emotional, social or spiritual distress. Prompt recognition and response to distress should be the norm and in situations where problems are complex, collaboration with specialist palliative care teams can help to address any challenges that may arise. Although the person with intellectual disability who is dying is the main focus of care, it is important to remember that family members, peers with intellectual disability, and formal carers also are experiencing grief and loss, and their needs should also be considered and addressed.

Reflection

Mary's* story

Mary is a 61-year-old woman with a moderate intellectual disability. She has lived for the past 28 years in a group home in the community with two other ladies who have also moderate intellectual disability, and they can be described as best friends. Her parents are both deceased, and she has one sister who has limited contact and visits at Christmas and for her birthday. The group home has the support of two social care workers, and there is no awake staffing at nighttime. Both staff members have worked there for about 10 years and know Mary very well. Over the past six months, it was noticed that Mary appeared unwell and had lost a lot of weight with a change in bowel pattern. Sadly, following referral to the local hospital, she was diagnosed with metastatic gastric cancer two months ago. At time of diagnosis Mary was not fit enough for systemic treatment, and a decision was made to focus on comfort care at home instead. Over the past week Mary has deteriorated significantly, she is now spending much of the day sleeping, she has minimal oral intake only and she is breathless on minimal exertion (such as when her carers attend to her personal hygiene). She has increasing abdominal pain and appears frightened at times when she is awake. In the last two days, her urinary output has decreased. Her sister, peers and staff are very upset by this deterioration. They know that Mary previously said that she wanted to remain in her home with her friends, but they are worried as to how they will manage. Additional staffing has been provided with 24-hour nursing support. The local specialist palliative care team have been contacted and visited for the first time yesterday to provide support. Mary has been commenced on a syringe driver to control her symptoms, and this appears to be having some effect. Her favourite music, flowers from her garden and a gentle hand and head message appear to bring additional comfort. The team is also offering support to the staff, peers and family as they prepare for Mary's death. However, staff continue to feel very anxious and have never witnessed someone die.

*Assumed names are used in this story.

1. What symptoms and changes indicate that Mary is approaching her final days?
2. Do you think that it will be possible to meet Mary's wish to continue to be cared for at home?
3. What further symptoms or changes might be expected in the coming days and what plan would you put in place to manage them?
4. What three things could you do to support peers in the group home?
5. What three things could you do to support staff in the group home?

Resources

1. Tuffrey-Wijne, I. (2013). *How to break bad news to people with intellectual disabilities: A guide for carers and professionals*. London: Jessica Kingsley. Guidelines for practitioners, families and carers to ease the process of explaining bad news. http://www.breakingbadnews.org
2. *Distress and Discomfort Assessment Tool (DisDat).* A resource for understanding distress in people with communication difficulties. https://www.stoswaldsuk.org/how-we-help/we-educate/education/resources/disability-distress-assessment-tool-disdat/
3. *Trinity Centre for Ageing and Intellectual Disability (TCAID)* - How to support people with an intellectual disability with a diagnosis of COVID-19. https://www.tcd.ie/tcaid/about/webinars.php

References

Bekkema, N., de Veer, A. J. E., Hertogh, C. M. P. M., & Francke, A. L. (2016). Perspectives of people with mild intellectual disabilities on care relationships at the end of life: A group interview study. *Palliative Medicine, 30*(7), 625–633. https://doi.org/10.1177/0269216316640421

Bekkema, N., de Veer, A. J. E., Wagemans, A. M. A., Hertogh, C. M. P. M., & Francke, A. L. (2015). To move or not to move?: A national survey among professionals on beliefs and considerations about the place of end-of-life care for people with intellectual disabilities. *Journal of Intellectual Disability Research, 59*(3), 226–237. https://doi.org/10.1111/jir.12130

Carter, A. N. (2020). To what extent does clinically assisted nutrition and hydration have a role in the care of dying people? *Journal of Palliative Care, 35*(4), 209–216. https://doi.org/10.1177/0825859720907426

Cithambarm, K., Duffy, M., & Courtney, E. (2021). What constitutes good quality end-of-life care? perspectives of people with intellectual disabilities and their families. *Journal of Policy and Practice in Intellectual Disabilities, 18*(3), 207–216. https://doi.org/10.1111/jppi.12376

Del Rio, M., Shand, B., Bonati, P., Palma, A., Maldonado, A., Taboada, P., & Nervi, F. (2012). Hydration and nutrition at the end of life: A systematic review of emotional impact, perceptions, and decision-making among patients, family, and health care staff. *Psycho-Oncology, 21*(9), 913–921. https://doi.org/10.1002/pon.2099

Dunkley, S., & Sales, R. (2014). The challenges of providing palliative care for people with intellectual disabilities: A literature review. *International Journal of Palliative Nursing, 20*(6), 279-284. https://doi.org/10.12968/ijpn.2014.20.6.279

Evans, L. R., Boyd, E. A., Malvar, G., Apatira, L., Luce, J. M., Lo, B., & White, D. B. (2009). Surrogate decision-makers' perspectives on discussing prognosis in the face of uncertainty. *American Journal of Respiratory and Critical Care Medicine, 179*(1), 48–53. https://doi.org/10.1164/rccm.200806-969OC

Gijsberts, M.-J.H.E., Liefbroer, A. I., Otten, R., & Olsman, E. (2019). Spiritual care in palliative care: A systematic review of the recent European literature. *Medical Sciences, 7*(2), 25. https://doi.org/10.3390/medsci7020025

Glennen, S. (1997). Introduction to augmentative and alternative communication. In S. Glennen & D. C. DeCoste (Eds.), *The handbook of augmentative and alternative communication* (pp. 3–19). Singular Publishing Group.

Heslop, P., Blair, P. S., Fleming, P., Hoghton, M., Marriott, A., & Russ, L. (2014). The confidential inquiry into premature deaths of people with intellectual disabilities in the UK: A population-based study. *The Lancet, 383*(9920), 889–895. https://doi.org/10.1016/S0140-6736(13)62026-7

Hunt, K., Bernal, J., Worth, R., Shearn, J., Jarvis, P., Jones, E., Lowe, K., Madden, P., Barr, O., Forrester-Jones, R., Kroll, T., McCarron, M., Read, S., & Todd, S. (2020). End-of-life care in intellectual disability: A restrospective cross-sectional study. *BMJ Supportive & Palliative Care, 10*(4), 469–477. https://doi.org/10.1136/bmjspcare-2019-001985

Ito, S., Morita, T., Uneno, Y., Taniyama, T., Matsuda, Y., Kohara, H., Maeda, I., Nakayama, T., Mori, M., & the EASED Investigators. (2021). Incidence and associated factors of sudden unexpected death in advanced cancer patients: A multicenter prospective cohort study. *Cancer Medicine, 10*(14), 4939–4947. https://doi.org/10.1002/cam4.4030

Kennedy, C., Brooks-Young, P., Brunton Gray, C., Larkin, P., Connolly, M., Wilde-Larsson, B., Larsson, M., Smith, T., & Chater, S. (2014). Diagnosing dying: An integrative literature review. *BMJ Supportive & Palliative Care, 4*(3), 263–270. https://doi.org/10.1136/bmjspcare-2013-000621

Kirkendall, A., Linton, K., & Farris, S. (2017). Intellectual disabilities and decision making at end of life: A literature review. *Journal of Applied Research in Intellectual Disabilities, 30*(6), 982–994. https://doi.org/10.1111/jar.12270

Klass, D., Silverman, P. R., & Nickman, S. (2014). Continuing bonds: New understandings of grief. *Taylor & Francis*. https://doi.org/10.4324/978131 5800790

Leick, N., & Davidsen-Nielsen, M. (1991). *Healing pain: Attachment, loss, and grief therapy*. Psychology Press.

Lord, A. J., Field, S., & Smith, I. C. (2017). The experiences of staff who support people with intellectual disability on issues about death, dying and bereavement: A metasynthesis. *Journal of Applied Research in Intellectual Disabilities, 30*(6), 1007–1021. https://doi.org/10.1111/jar.12376

Manolis, T. A., Manolis, A. A., Melita, H., & Manolis, A. S. (2019). Sudden unexpected death in epilepsy: The neuro-cardio-respiratory connection. *Seizure, 64*, 65–73. https://doi.org/10.1016/j.seizure.2018.12.007

McRitchie, R., McKenzie, K., Quayle, E., Harlin, M., & Neumann, K. (2014). How adults with an intellectual disability experience bereavement and grief: A qualitative exploration. *Death Studies, 38*(3), 179–185. https://doi.org/10.1080/07481187.2012.738772

Meeusen-van de Kerkhof, R., Van Bommel, H., Van De Wouw, W., & Maaskant, M. (2006). Perceptions of death and management of grief in people with intellectual disability. *Journal of Policy and Practice in Intellectual Disabilities, 3*(2), 95–104. https://doi.org/10.1111/j.1741-1130.2006.00060.x

Meier, E. A., Gallegos, J. V., Thomas, L. P. M., Depp, C. A., Irwin, S. A., & Jeste, D. V. (2016). Defining a good death (successful dying): Literature review and a call for research and public dialogue. *The American Journal of Geriatric Psychiatry, 24*(4), 261–271. https://doi.org/10.1016/j.jagp.2016.01.135

Murray, S. A., Kendall, M., Boyd, K., & Sheikh, A. (2005). Illness trajectories and palliative care. *BMJ, 330*(7498), 1007–1011. https://doi.org/10.1136/bmj.330.7498.1007

National Institute for Health and Care Excellence (NICE). (2015). *Care of dying adults in the last days of life: NICE guidelines (NG 31)*. www.nice.org.uk/guidance/NG31

Nolan, S., Saltmarsh, P., & Leget, C. (2011). Spiritual care in palliative care: Working towards an EAPC task force. *European Journal of Palliative Care, 18*(2), 86-89. https://www.eapcnet.eu/wp-content/uploads/2021/03/EJPC-2011-18-2.pdf

Regnard, C., Mathews, D., Gibson, L., & Clarke, C. (2003). Difficulties in identifying distress and its causes in people with severe communication problems. *International Journal of Palliative Nursing*, *9*(4), 173-176. https://doi.org/10.12968/ijpn.2003.9.4.11501

Richmond, C. (2005). Dame Cicely Saunders. *British Medical Journal, 331*(7510), 238. https://www.ncbi.nlm.nih.gov/pmc/articles/PMC1179787/

Robinson, J., Pilbeam, C., Goodwin, H., Raphael, D., Waterworth, S., & Gott, M. (2021). The impact of uncertainty on bereaved family's experiences of care at the end of life: A thematic analysis of free text survey data. *BMC Palliative Care, 20*(1), 60. https://doi.org/10.1186/s12904-021-00748-9

Ryan, K., & Dodd, P. (2008). Congenital developmental disabilities. In D. Walsh (Ed.), *Palliative medicine* (pp. 1027–1033). Elsevier.

Ryan, K., Guerin, S., Dodd, P., & McEvoy, J. (2011a). Communication contexts about illness, death and dying for people with intellectual disabilities and life-limiting illness. *Palliative & Supportive Care, 9*(2), 201–208. https://doi.org/10.1017/S1478951511000137

Ryan, K., Guerin, S., Dodd, P., & McEvoy, J. (2011b). End-of-life care for people with intellectual disabilities: Paid carer perspectives. *Journal of Applied Research in Intellectual Disabilities, 24*(3), 199–207. https://doi.org/10.1111/j.1468-3148.2010.00605.x

Ryan, K., Guerin, S., & McEvoy, J. (2016). The nature and importance of quality of therapeutic relationships in the delivery of palliative care to people with intellectual disabilities. *BMJ Palliative and Supportive Care, 6*(4), 430–436. https://doi.org/10.1136/bmjspcare-2013-000619

Stone, P., Vickerstaff, V., Kalpakidou, A., Todd, C., Griffiths, J., Keeley, V., Spencer, K., Buckle, P., Finlay, D., & Omar, R. Z. (2021). Prognostic tools or clinical predictions: Which are better in palliative care? *PLoS ONE, 16*(4). https://doi.org/10.1371/journal.pone.0249763

Tuffrey-Wijne, I. (2012). A new model for breaking bad news to people with intellectual disabilities. *Palliative Medicine, 27*(1), 5–12. https://doi.org/10.1177/0269216311433476

Tuffrey-Wijne, I. (2013). *How to break bad news to people with intellectual disabilities: A guide for carers and professionals*. Jessica Kingsley. http://www.breakingbadnews.org

Tuffrey-Wijne, I., Bernal, J., & Hollins, S. (2010). Disclosure and understanding of cancer diagnosis and prognosis for people with intellectual disabilities: Findings from an ethnographic study. *European Journal of Oncology Nursing, 14*(3), 224–230. https://doi.org/10.1016/j.ejon.2010.01.021

Tuffrey-Wijne, I., McEnhill, L., Curfs, L., & Hollins, S. (2007). Palliative care provision for people with intellectual disabilities: Interviews with specialist palliative care professionals in London. *Palliative Medicine, 21*(6), 493–499. https://doi.org/10.1177/0269216307082019

Tuffrey-Wijne, I., Abraham, E., Goulding, L., Giatras, N., Edwards, C., Gillard, S., & Hollins, S. (2016). Role confusion as a barrier to effective carer involvement for people with intellectual disabilities in acute hospitals: Findings from a mixed-method study. *Journal of Advanced Nursing, 72*(11), 2907–2922. https://doi.org/10.1111/jan.13041

Tuffrey-Wijne, I., & Rose, T. (2017). Investigating the factors that affect the communication of death-related bad news to people with intellectual disabilities by staff in residential and supported living services: An interview study. *Journal of Intellectual Disability Research, 61*(8), 727–736. https://doi.org/10.1111/jir.12375

Wagemans, A., & van Schrojenstein Lantman-de Valk, H., Proot, I., Metsemakers, J., Tuffrey-Wijne, I., & Curfs, L. (2015). End-of-life decision-making for people with intellectual disability from the perspective of nurses. *Journal of Policy and Practice in Intellectual Disabilities, 12*(4), 294–302. https://doi.org/10.1111/jppi.12140

15

Use of Do-Not-Resuscitate Orders

Philip McCallion, Lisa A. Ferretti, and Mary McCarron

This chapter begins with some definitions. The specifics and wording by jurisdiction may vary, but a do-not-resuscitate (DNR) order is usually a legally binding physician's order stating that no steps will be taken to restart a patient's heart or restore breathing if the patient experiences cardiac arrest or respiratory arrest. Such an order is usually written in response to an advance directive or other declaration (including verbal)

P. McCallion (✉) · L. A. Ferretti
School of Social Work, Temple University, Philadelphia, PA, USA
e-mail: philip.mccallion@temple.edu

L. A. Ferretti
e-mail: lisa.ferretti@temple.edu

M. McCarron
Trinity Centre for Ageing and Intellectual Disability, Trinity College Dublin, Dublin, Ireland
e-mail: mccarrm@tcd.ie

R. J. Stancliffe et al. (eds.), *End of Life and People with Intellectual and Developmental Disability*, https://doi.org/10.1007/978-3-030-98697-1_15

by an individual or by someone responsible for their care. Resuscitation typically involves cardiopulmonary resuscitation (CPR). CPR in non-hospital settings may take the forms of compression-only CPR and standard CPR—chest compressions and mouth-to-mouth rescue breaths. The introduction of compression-only CPR by community members in community settings has been found to increase the use of CPR overall (Riva et al., 2019). There is no data available on use of CPR by community members with people with intellectual disabilities. However, in many jurisdictions staff in settings for people with intellectual disabilities are required to be trained in CPR. Again, no data is available to confirm this, but it is likely that a person with an intellectual disability living in provider settings and/or receiving programming from providers will have more opportunity than other individuals to be offered CPR when in cardiopulmonary difficulties. Finally, no data appears to be gathered on use of compression-only CPR and/or standard CPR in persons with an intellectual disability living independently or with family. In hospital settings compression-only CPR is often mechanically supplemented to support breathing and to measure the effects of resuscitation efforts. Again, there is little data in different jurisdictions on its use in hospitals with people with an intellectual disability.

Although DNR orders may exist in writing, in the general population they are frequently only available in medical charts specific to a primary care practice and/or to the hospital where a person is receiving care. Some individuals carry a copy of their advance directives with them and increasingly there are electronic means of checking for advance directives and reading them. In some countries, these documents are a part of electronic health records that may be accessed by multiple providers.

For people with an intellectual disability in a supervised living situation or attending a day programme, their DNR desires may be a part of their record there and should therefore influence any use of CPR by staff. It is in the community where respect of the intent of an expressed or documented DNR request is most problematic. Most individuals generally do not carry a copy of an advance directive expressing their desires; not every country gives legal validity to advance directives. Written physician or other health professional (where permitted) orders tend to be limited to the place where they are given, and in emergency use of

CPR by community members they may not ask for, understand or feel bound by such documentation. A particular case is emergency response personnel whose primary goal is to get the person to more advanced care as quickly as possible and to take steps likely to increase the opportunity that the person will survive until they receive that care. In some jurisdictions but not all, such personnel are trained to ask for advance directives and DNR documentation and to respect a valid DNR request or order; and these personnel are clearly faced with competing imperatives. There is a growing practice of individuals wearing bracelets which express their do-not-resuscitate desire in circumstances where they are not able to speak for themselves, but the legal status of such devices also varies. To the extent that members of the public and emergency personnel who may seek to administer CPR are not trained in issues for people with intellectual disabilities and/or understand their ability to express their wishes or believe they are capable of making such decisions, then there is increased likelihood of use of CPR regardless of the individual with intellectual disability's expressed desires and documented orders.

For the purposes of this chapter, the discussion of use of DNR orders will largely but not exclusively concentrate on their use in in-patient situations. DNR orders may also be called a do-not-attempt-resuscitation (DNAR) order or allow-natural-death (AND) order (Wagemans et al., 2010) or a do-not-attempt-cardiopulmonary-resuscitation order (Care Quality Commission, 2021). From here on, the expression 'DNR/ DNAR order' will be used to maximise inclusivity of the range of definitions. The physician's or other health professionals (where permitted) order should be preceded by a discussion with the individual or with their proxy about their wishes when they might experience an event where CPR might be attempted and should reflect the individual's concerns and desires. Ideally, such wishes should be expressed in written advance directives and/or in the individual's assignment of authority to make decisions to another person, as a result of discussion with or directly created, by the individual (Wagemans et al., 2010). Therein lies the complexity of this issue for people with an intellectual disability. There are multiple steps then to creating and supporting a DNR/DNAR order, assumptions of discussion and understanding of the issues involved, and documenting the conveying of wishes to, and participation by, several parties.

The Decision Makers and the Basis for Decisions

There are potentially multiple levels of decision-maker. The usual participants are the individual with intellectual disabilities themselves, their assigned proxy and the physician/health professional (where permitted) overseeing care. Increasingly, there is also additional support by nurses, social workers and clergy, among others. Attorneys and legislated formal guardianship services may also become involved in situations where capacity for decision-making is being questioned. In emergency circumstances, this group may be expanded to family members who are present and interested, but who are not legal proxies (either by formal guardianship or by assignment through advance directive); others charged with implementing related policies within the jurisdiction or within the hospital or provider; staff members who see themselves as acting *in locus parenti* (although usually not a legal status) or as *knowing* the person's wishes and different physicians and other healthcare professionals faced with a situation where a response is required and feel guided by their professional knowledge and perceptions of their legal/licence obligations to act. In some rare cases, a close friend or spouse/partner of the individual with intellectual disability who may themselves have an intellectual disability may be consulted about what the person would have wanted. In a life and death situation and in the absence of a stated and documented wish of the person and a written physician order, for any one of these multiple decision-makers this process may be fraught and desired consultation/decision-making may not happen (see also Chapter 13). When the individual is not perceived by some or all as being competent to make this decision and/or when wishes/decisions are not documented (in both cases as often happens for people with an intellectual disability), situations arise where the appropriate actions in a cardiovascular or respiratory event appear less clear, and the presence of multiple opinions may encourage continuing valiant efforts to revive.

The decision to resuscitate itself is a complex one. In many studies, 30-day survival is the benchmark to be achieved, but there should also be concern for long-term impact. Riva et al. (2019), in a population-level study in Sweden of out-of-hospital cardiac arrest situations for the general population, found the chance of 30-day survival was doubled from any form of CPR compared with no CPR. These and similar statistics are widely cited and encourage community members to be trained in CPR. Looking at in-hospital use, research shows that people greatly overestimate the benefits of in-patient CPR. Adams and Snedden (2006) pointed out that in a survey of older adults in the U.S., approximately 81% believed that 50% of patients will survive in-patient CPR and later be discharged from hospital. The authors' review of the literature then countered that less than 5% of those who receive in-patient CPR actually survived to discharge. There have not been more recent published studies that contradict these findings. In-hospital outcomes therefore appear very different from community-based use of CPR.

Discussions about CPR and DNR/DNAR orders are likely being made on the basis of short-term outcomes, out-of-hospital event statistics and a poor understanding of both in-patient and longer term health findings post-CPR. Wagemans et al. (2010) largely concur with the Adams and Sneddon concerns, arguing in-patient CPR is unsuccessful in 70–98% of cases. Systematic reviews on the topic concluded that globally, survival rates of individuals post use of CPR are approximately 7% (Berdowski et al., 2010; Welbourn & Efstathiou, 2018) with positive effects argued to be unclear at best (Lippert et al., 2012). Adverse outcomes of attempted CPR included broken bones, ventilator dependence and damage to the brain and other organs (Welbourn & Efstathiou, 2018). The absence of agreed guidelines on the appropriate length of resuscitation efforts and the reality that CPR occurs in hospital and non-hospital settings and is administered by both health professionals and by trained community members, raises additional quality-of-implementation concerns in the general population (Welbourn & Efstathiou, 2018), and for people with intellectual disabilities (Wagemans et al., 2010). These varying circumstances, as well as potential for poor outcomes, raise another concept in the debate, futility, which Wagemans et al. (2010) describe as reflecting that many times decisions to pursue CPR are made to address a problem in a given moment, rather than with an eye to expected survival and positive outcomes (de Vos et al., 1999). This then means that decisions often do not sufficiently, or perhaps at all, consider likely long-term poor outcomes from the intervention. Such conclusions beg the questions of why then proceed with resuscitation, or at least are there circumstances where resuscitation should not be pursued?

An additional issue is the timing of DNR/DNAR orders. A DNR/DNAR order is frequently considered as occurring prior to a cardiovascular or respiratory life-threatening event, and to be based on a considered discussion, consent and a spoken or written statement, followed by a physician or other health professional (where permitted) order. In the absence of an expressed desire of the individual and/or their legal proxy and of described circumstances where they affirm CPR should not be attempted, the default would be that in emergency health circumstances the effort to revive should proceed.

In the case of the general population, there are circumstances, such as when a cardiovascular or respiratory incident occurs outside of a health-care setting, where even if a DNR/DNAR order has been agreed and issued, if the existence of the order is unknown then it is unlikely to be followed. This also occurs for people with an intellectual disability. Also, where people with intellectual disabilities are unable to express, or where they have not expressed their wishes or have an appointed proxy/guardian to inform those decisions, this may mean again that CPR will always be attempted.

Futility and Conflict in Use of DNR/DNAR Orders

The introduction of the idea that use of CPR may be futile raises a different ethical issue for physicians and other members of a health care team, including families. There are circumstances where a DNR/DNAR order may result primarily from an appraisal of current medical circumstances of the individual with an intellectual disability and while, where possible, this is discussed with the individual and family, in an emergency situation the order may be the result of the attending physician's appraisal of the emergent situation and will guide that physician's and other healthcare professional's actions. Wagemans et al. (2010) suggest that, at least in a European context, a healthcare professional, such as a physician cannot be required to offer an intervention they know will not work, or where they believe that risks and ill effects outweigh benefits, arguing such an intervention would be futile. They conclude this implies that there is basis for in-the-moment DNR/DNAR orders and actions initiated by the physician alone. They do acknowledge that decisions on futility are subjective, with uncertain outcomes, and encourage that such decisions should nevertheless be discussed with the individual with an intellectual disability, where possible, and if not, with the family.

DNR/DNAR orders issued or not issued may represent conflicts between decision-makers/stakeholders. In situations where there are multiple decision-makers and even when the person with an intellectual disability may speak for themselves or a family member on their behalf, the question arises of whose opinion should be primary. In a Dutch study surveying 94 intellectual disability care staff members, 273 intellectual disability physicians and 1000 general or primary care practitioners, Bekkema et al. (2014) reported that for complex medical decisions at the end of life, 43.6% of those surveyed thought that the wishes of the person with an intellectual disability about care should be primary, as compared to 72% believing that the person should be able to decide where they wished to die. As well as valuing the opinion of all staff, respondents also gave greater weight to the opinions of family as compared to the individual with an intellectual disability. The authors posit these twin findings may suggest that staff at all levels tend not to see people with intellectual disability as able to understand and make decisions about complex medical issues, but are more likely to agree they are capable of decisions around more practical issues influencing their end of life, including where they would like to die. The researchers are commended for raising these issues in their jurisdiction. It is likely that similar views are held in other countries but have yet to be documented.

It is understandable that questions around the appropriateness of DNR/DNAR orders represent similar complexity for staff and families, but it increases the likelihood that the involvement of the person with intellectual disability in decision-making is not always valued. In their systematic review, Voss et al. (2017) reported that three of five studies that met review criteria and were examined in detail indicated that do-not-resuscitate decisions were considered a critical component of advance care planning. However, they also found that people with intellectual disabilities were rarely involved in decision-making about this, or other end-of-life medical issues. None of the articles indicated that the views and decisions of people with intellectual disabilities themselves were even part of the data collected by study authors when considering these issues. In a field that continues to advance person-centred care, the first step is asking individuals with intellectual disabilities about their views and preferences. This first step appears to be largely absent in the discussion and decision-making around DNR/DNAR orders. Yet, that step is the basis for then considering whether the wishes of the person with an intellectual disability are respected and if subsequent care satisfies their needs.

Influencing DNR/DNAR Decisions

John's Story*

John, aged 61 years, a person with severe intellectual disability and limited verbal skills was admitted to hospital for the fourth time in as many months with heart failure. His decline over those four months included becoming increasingly bed-ridden, confusion, loss of appetite and incoherent verbalisations, interpreted by staff as expressions of pain and upset. There was no family available as his parents had died a number of years earlier, and his one older sibling was recently admitted to a nursing home herself due to advancing dementia. No one had been established as John's guardian, there were no advance directives on file and John was not able (and may have never been able) to, express his wishes for care.

John had been cared for by the same provider for many years and a group of staff who saw themselves as his family were insistent that John would get better over time and everything should be done to give him that chance. There had been two occasions in the home when John had cardiopulmonary difficulties where compression-only CPR had been applied by staff while awaiting arrival of an ambulance. Staff felt assured that their actions had helped preserve John's life and his subsequent discharge from hospital back to his group home was seen as confirming this.

On this fourth admission, the hospital's attending physician contacted the administrator of the intellectual disability provider and John's primary care physician in the community asking for instructions for John's care. She raised her concern that CPR was used both in his home and in previous hospitalisations, but John was continuing to decline and she did not feel continued use of CPR was recommended. She also pointed out that John had suffered a broken rib the last time CPR was initiated, and it was not yet completely healed. The attending physician asked them both to state that they wished her to issue a written DNR order, as in the absence of a guardian they were the persons responsible for John's care and well-being. The administrator and the community physician both stated that they did not have the authority to make that decision on John's behalf, that the staff who loved and cared for him wanted him to have the chance to get better, and that as the attending physician, she should decide if she wished to issue the DNR order.

*Assumed names are used in this story.

Understanding what DNR/DNAR orders and the related decision-making mean for the other actors and stakeholders is important. Sledge (2013), looking specifically at issues related to persons with intellectual disabilities on dialysis, argued for consideration of the Dubois (2008) framing of differences among stakeholders in three areas: volitional, cognitive and social differences.

Volitional differences arise when stakeholders have different and competing interests (DuBois, 2008). In John's story, some stakeholders wanted the preservation of life; others were insistent that the decision was not theirs and should be made by the attending physician; and the attending physician was concerned about the outcomes of the intervention and of being held responsible for those outcomes. *Cognitive differences* arise when there are different understandings of benefits and harms from procedures, and of the basic facts of the person's illness and the need for intervention. The attending physician saw that over time there were adverse outcomes for John from use of CPR. The staff saw hope for future recovery. Finally, *social differences* arise because of differences in norms and values. Here, the conflict was between beliefs that treatment is always preferable and may result in later improvement, and a view that treatment should not be attempted when the resulting quality of life will/may be poor or poorer.

Quality of Life and DNR/DNAR Decision-Making

Quality of life is often presented as a key issue in the decision on whether or not to resuscitate. Bekkema et al. (2014) in their survey of professionals, found that quality of life and well-being considerations were highly ranked by the professionals involved in determining and implementing end-of-life interventions including DNR/DNAR orders. The respondents held varying beliefs on whether people with intellectual disabilities should always be informed/consulted about medical interventions and whether their refusal of an intervention should be respected. Respondents were concerned about the potential for the person with an intellectual disability to adequately understand, and difficulties in presenting the pros and cons of an intervention so that the person themselves could make a fully determined decision. From their findings, it appeared that the wishes and quality of life desires of people with intellectual disabilities (especially of those with severe/profound intellectual disabilities) were often not considered. A slight majority (60.8%) of respondents believed that refusals of interventions should be respected, but 60.1% also believed that in such refusal, the decision-making capacity of the person with an intellectual disability should be tested. If it was determined that the person did not have sufficient capacity, a small majority thought that the family should be the decision-maker, but intellectual disability physicians in particular thought the final call should rest with the attending physician (Bekkema et al., 2014). If such strategies were to be adopted, decision-making would perhaps be more medically rather than quality of life driven. Yet, the language used by respondents was more focused upon quality of life concepts.

The *Caring at the End of Life* position statement of the American Association on Intellectual and Developmental Disabilities (2020) discusses quality of life at the end of life in terms of dignity:

Dignity: All persons are equally valuable, with or without disability, and deserve respect consistent with human dignity. The value of a person's life is not related to the type, degree or severity of disability. Inherent value must be distinguished from quality of life. Inherent value persists from birth to death, even though the quality of life may change as one approaches the end of life.

1. The quality of one's life must be assessed from a subjective viewpoint, that is, from the point of view of the person with a disability. Having a disability is not by itself a form of suffering. People without disability who fear becoming disabled must not assume that their feelings are shared by those who are living with a disability.
2. The mere presence of intellectual and developmental disabilities, or likelihood of having a disability in the future, does not make a person's life less valuable.
3. Decisions about care at the end of life must be made respectfully, consistent with the principle of dignity. Withdrawing or withholding care may be appropriate in some situations, although it should not itself imply a lack of respect for the importance of that person's life. Treatment should not be withdrawn or withheld only because a person has a disability (American Association on Intellectual and Developmental Disabilities, 2020).

Many other jurisdictions and professional groups take similar positions. The decision to either revive or to not resuscitate often references quality of life arguments. There is recognition of greater risk that the lives of people with disabilities are seen as less valuable, and there are also questions about whether people with disabilities experiencing chronic and life-threatening illness can be restored to an *acceptable* quality of life. There is considerable complexity and subjectivity involved in making a judgement about someone's quality of life, particularly if the person is dying and/or is not able to express their wishes.

In the debate on use of DNR/DNAR orders, approaches often do not begin with a view that lives with or without disability are equal. Judgements about the consequences of resuscitation for quality of life for people with intellectual disabilities by those in healthcare and when there are concerns about cost and distribution of resources, represent a denial of shared dignity or at least uncertainty about what quality of life means for people with intellectual disabilities (Wagemans et al., 2010). Too automatic a decision to institute a DNR/DNAR order has also been raised as an ethical concern.

The development of triage and other protocols to respond to shortages in care which occurred at times during COVID-19 has heightened these concerns (Oakley et al., 2020; Thomas, 2020). There have been reports of individuals with disabilities and with COVID-19 and their families not being consulted about DNR/DNAR decisions, and of being directly and indirectly excluded from consideration for lifesaving treatments (American Academy of Social Work and Social Welfare [AASWSW], 2020; Care Quality Commission, 2021). This is not a new concern. There has been historic exclusion of people with disabilities, in particular those with intellectual disabilities, from equitable access to healthcare. The COVID-19 pandemic has highlighted that judgements about the value of the lives of people with intellectual disabilities are sometimes renamed as so-called concerns for their experience of future quality of life. These judgements are held when making decisions about instituting DNRs/DNAR orders in the absence of the expressed wishes of those individuals (Care Quality Commission, 2021). Arguing that a decision is based upon quality of life considerations is not defensible when judgements are made by others and where the life of a person with an intellectual disability is not considered of equal value to that of others.

Self-Determination and DNR/DNAR Requests and Orders

Self-determination is seen as occurring when the person with an intellectual disability is the primary individual making decisions that influence and cause things desired and undesired to happen in their lives (Wehmeyer, 2011) meaning they are autonomous, self-regulating, empowered in initiating/responding to an event and self-realising (Wehmeyer & Abery, 2013). The exercise of self-determination is also influenced by individual motivation, knowledge, level of control, organisation/ community/family context, free will and legal and ethical rights within regulations (McCallion & Ferretti, 2017). Older age, illness and disability often become barriers to the exercise of self-determination, but Ekelund et al. (2014) argued that continuing exercise by the individual of self-determination is possible when there is commitment from professionals and caregivers, concerns are addressed (both the person's concerns and the beliefs of caregivers), purposeful efforts provide opportunities for the person's involvement in decision-making and evidence/experience that the views of the person matter (McCallion & Ferretti, 2017).

Bekkema et al. (2014) raised that genuine participation in decision-making, such as in a DNR/DNAR discussion, assumes a level of autonomy by the person with an intellectual disability where they are able to declare their personal care decisions without interference/involvement from others. Given that many people with an intellectual disability will have difficulties expressing themselves and may need the help of others to decide about medical interventions, Bekkema and colleagues instead argue for what they call a relational notion of patient autonomy, where autonomy is reached with the assistance of trusted others (see also Chapter 5). It could be argued that many other individuals in society also approach such decisions in a relational manner consulting with close family members, for example, about DNR/DNAR decisions before finalising them. Nevertheless, the finalisation of such a decision rests with the individual themselves, but this is infrequently the case for people with intellectual disabilities. Suzanne's story shows the complexity of relationships at play.

Suzanne's* Story

As part of an agency wide effort to better involve people with intellectual disabilities in decision-making about aged and end-of-life care, Suzanne, aged 68 and with a moderate level of intellectual disability had participated some years previously in a series of classes that addressed potential health issues in older years and decision-making about types of care desired, including end-of-life care.

The class was supported with a formal curriculum, some videos and easy read materials developed by a Center for Excellence in Developmental Disabilities. Participation in the class led to the supported development of Suzanne's own advance directive. Susanne was supported by an advocate she had chosen.

Suzanne particularly wanted to discuss her desire for a DNR/DNAR order because her mother had died previously after a long illness. Suzanne had been on a home visit when her mother experienced a heart attack. An ambulance was called, a next-door neighbour completed standard CPR until the ambulance arrived and Suzanne also watched emergency personnel take over carrying out CPR and placing her mother in the ambulance. Although her mother recovered and came home after a short time in hospital, Suzanne was very aware of the extensive bruising her mother experienced from CPR. Although the education materials available did not give a lot of information on DNRs/DNAR orders, as part of her own expressed desires for future healthcare, Suzanne stated in her advance directive that she never wanted to experience CPR.

Suzanne later had a major fall where she lost consciousness. Emergency personnel on arrival began CPR. The key worker informed the emergency personnel that Suzanne had an advance directive in which she refused CPR. The key worker did not have a copy of the advance directive, as health related documentation was managed by the nurse assigned to the home and her records were in a locked cabinet. The emergency personnel questioned whether as a person with an intellectual disability, Suzanne could fully understand what CPR was for and said that in the absence of a written physician's DNR/DNAR order they had an obligation to initiate the procedure and did. Suzanne was transported to hospital where she was treated and returned home after several days.

When she was feeling better, Suzanne complained to her case manager that no one had listened to her. The case manager convened a meeting involving Suzanne and her advocate, the key worker and her supervisor, and the nurse and her supervisor. She also invited Suzanne's older sister Marie who although not her guardian did attend and participate in her person-centred planning meetings. Suzanne's physician also joined the meeting virtually.

As the events were reviewed, the key worker raised that the nurse keeping Suzanne's advance directive locked up meant that it could not be shared with the emergency personnel. The nurse defended this decision as in keeping with privacy protections and the physician mentioned that an advance directive does not prevent a healthcare professional including emergency personnel making the best judgement for the circumstances to advance the best health outcomes for Suzanne. Marie offered that Suzanne really should listen to the doctor and nurses as they knew more about these things than she did and had her best interests at heart. The key worker said that Suzanne and she, as the person who knew Suzanne best, were better able to decide on and express Suzanne's wishes but they were not even asked. Others in the meeting did not see the key worker as having a role in representing Suzanne on health issue decision-making. The discussion became somewhat heated when the healthcare-related personnel all raised concerns that they were not going to do things that put their licences at risk. The key worker and the advocate argued that this was not about licences, that the most important thing was respecting Susanne's wishes, self-determination and rights. Suzanne interjected that she just did not want to be, as she said 'beaten up' like that again. Suzanne's words helped refocus the discussion.

The case manager asked for solutions that would respect Suzanne's wishes but not break laws or regulations and with Suzanne's permission a copy of the advance directive where she stated that CPR should not be attempted was placed in a locked area that could be accessed by her and her key worker. The physician agreed to note in Suzanne's electronic health record her desire not to be resuscitated and discussed with Suzanne how this would be reviewed with her at her annual physical examination in case she wished to change her instructions. The physician also insisted that Suzanne meet with the nurse to discuss the pros and cons of use of CPR. Suzanne was initially reluctant but her sister, Marie, asked her to do it for her. Suzanne agreed. After meeting with the nurse Suzanne did not change her mind about not wanting CPR and requested and did receive a do-not-resuscitate bracelet which she now always wears. With the nurse's support, Suzanne registered her advance directive with one of the national databases which hospitals can access. In addition, the provider administrator was approached by the two supervisors and the case manager to arrange an in-service by the provider for all first-responder emergency personnel in the municipality. The in-service purpose was about the abilities of people with intellectual disabilities to make decisions for themselves and also on what questions to ask and of whom, when there were emergency situations involving people with intellectual disabilities living in the provider's homes or attending their day programmes. The in-service was successfully scheduled and was well-attended.

*Assumed names are used in this story.

Reflection: Balancing Competing Needs, Regulations and Beliefs

Think about someone with an intellectual disability with chronic conditions which have necessitated several hospitalisations:

1. What were or are circumstances where you have or would question the value of continuing to resuscitate that person?
2. Do you think the person with an intellectual disability would agree with your viewpoint? How would you know?
3. Do you think other staff would agree with your viewpoint and/or the viewpoint of the person with an intellectual disability? How would you know? What would help create more agreement among everyone?

Suzanne's story illustrates that people with intellectual disabilities do develop personal advance directives and other end-of-life plans and documents and that there are tools to assist with this process (see Resources at the end of this chapter). However, implementation is not always well-supported. Pertinent to this particular chapter, as part of such efforts there must also be a strong emphasis on understanding and planning for use of DNR/DNAR orders. In Suzanne's situation, there was no plan for how other individuals would become aware of the advance directive and Suzanne's specific wishes regarding use of CPR. There is no self-determination if others choose to ignore one's desire to not be resuscitated. It becomes clear that even if a copy of the directive had been available there was poor understanding among the emergency personnel about whether Suzanne's wishes were valid given her disability. As noted in the earlier discussion, self-determination struggles to occur in situations where the person does not have a sufficient level of control and must contend with organisational, community and family contexts and differing interpretation of regulations (McCallion & Ferretti, 2017). Further, there was disagreement on Suzanne's decision, its communication and the extent to which it should be adhered to among those attending the so-called team meeting despite everyone seeing themselves as there to support Suzanne. Suzanne ended up directing the resolution but much of the discussion was driven by the desires, concerns and professional identities of the other team members. There must be greater attention by team members to Dubois' (2008) concerns about how issues that are volitional (different and competing interests in this case among professions), cognitive (understanding of benefits and harms among decision participants) and social (differences in norms and values including views of primacy of healthcare provider perspectives) create divisions and make it difficult to support self-determination.

In this particular case, Suzanne's own voice—not wanting to be 'beaten up' again—was heard but interestingly, the physical consequences of CPR which were driving Suzanne's decision were not acknowledged and discussed. This is not just a concern for people with intellectual disabilities alone. There is an absence of discussion of the potential negative consequences for everyone who may experience CPR. Suzanne, even in her success, was also still pressured to listen again to the healthcare professionals. There is a fine line between supporting the person with an intellectual disability with additional information in ways that are supportive of their self-determination process and seeking to further influence their decision away from what they desire.

Advance Care Planning and DNR/DNAR Orders

Advance care planning is discussed in more detail in Chapter 4. For this chapter's purposes, it is important to highlight that people with an intellectual disability are often able to participate in advance planning and will benefit from education on the different types of resuscitation, where and under what circumstances resuscitation might occur, the issues involved in deciding upon their desired position on resuscitation, the benefits and potential costs of any decision they make and their ability to change their minds and to revise their directives in this area. Discussion should address their desires in both hospital and in non-hospital circumstances. The decision-making process will be better informed by discussions with family members, peers and friends, who are also making decisions in this area, key workers and advocates as well as with healthcare staff. Discussions should include information on circumstances where attending physicians, other healthcare workers and staff and members of the general public trained in CPR may choose to overrule their decision, and steps that may be taken to reduce this likelihood. Most of all there should be assurance that final decisions about their directive rests with the person with an intellectual disability.

Resuscitation and Children with Intellectual Disabilities

Survival among children with intellectual disabilities has increased (Global Research on Developmental Disabilities Collaborators, 2018; Masquelier et al., 2018) including with health conditions that previously resulted in earlier mortality. Nevertheless, studies in several countries continue to confirm that mortality risk for children with intellectual disabilities to age 25 years are 5–15 times those of children without intellectual disabilities of the same ages, with children with severe intellectual disability at even greater risk (see e.g. Bourke et al., 2017). Survival of children with intellectual disabilities and significant health issues together with increased rates of mortality, mean some healthcare teams and families are facing and planning for the prospect of death while working to advance a productive and quality life for the child with an intellectual disability.

Adams and Levy (2017) argue for a shared decision-making approach towards daily care of individuals with intellectual disabilities with information exchanged in both directions between health professionals, families and their children (if able to be involved in the discussions) to address gaps in values, priorities and understanding of choices. In keeping with Bekkema et al. (2014), this approach fosters relational autonomy in decision-making, speaks of achieving balance and recognises that clinicians also have interests to be met in decision-making. The model comes from advanced disease care for older adults and has been piloted among families with children with autism (Strauss et al., 2015). Findings on its use suggest a need for significant training for the professionals who seek to practise this approach (Légaré et al., 2010). Looking specifically at DNR/DNAR orders, Adams and Levy (2017) argued that such an approach in families including a child with an intellectual disability means that everyone is better able to consider and address the related legal and ethical decisions. Two items to be noted are, (1) a belief in this model that the interests and desires of the clinician (usually the physician) are to be accommodated in decision-making and (2) while the child may be included, there is no guidance on when children should be included and on what weight should be given to their views.

In the case of children with intellectual disabilities, the decision-making process initially appears clearer given the legal status of families, parents in particular, as decision-makers are established in almost all jurisdictions. That said there has been an argument made that not only should children with intellectual disabilities be informed but they should give assent to procedures as this contributes to the development of autonomy and reflects an awareness of both their ability to understand and to contribute to decision-making on their own behalf (Miller & Nelson, 2006). In some jurisdictions, there is also legislative support for older teenagers in particular to provide assent. The potential to involve the individuals with intellectual disabilities themselves notwithstanding, there are often well-established patterns, sometimes since birth, of parents deciding on care. Work by Friedman and Gilmore (2007) established that the decision to support a DNR/DNAR order for the child with an intellectual disability is often influenced by the family/parent relationship with the attending physician, as well as by religious influences, family pressures and discussion with the physician. Others have acknowledged the role of religious belief and consultation (Nelson & Nelson, 1992) and of family member and physician discussions (Kearney & Griffin, 2001; Smith et al., 2001). Friedman and Gilmore (2007) in a sample of 60 families, documented that willingness to support DNR/DNAR orders more than doubled among families when information on the benefits and costs of resuscitation was provided. Conversely those who decided on full resuscitation were more likely to be influenced by family and religious leader discussions. They concluded that ambivalence about DNR/DNAR orders is more likely to lead to discussions and to being influenced by others, whereas an existing belief in the value of withholding resuscitation means the decision needs no further discussion.

It is of note that the literature on DNR/DNAR orders for children with intellectual disabilities has explored these issues and who and what is likely to influence decision-making by the family, but similar issues are largely unexplored for adults with intellectual disabilities.

Protocols for Approving and Monitoring DNR/DNAR Orders When There Are Concerns About Consent

Much of what has been discussed assumes that the person with an intellectual disability is able to participate in the discussions and decisions regarding resuscitation. There are many people with intellectual disabilities who will not be able to and who may not have a designated person to act on their behalf. For adults with intellectual disabilities in particular, there is a clash of values about decision-making ability. In the past, any decision-making often automatically rested with other family members or with the attending physician. Greater attention to consent issues, that the interests of the person may be different from those of family members and/or of the health or provider establishment, concerns about the consequences for the person with an intellectual disability and a greater recognition of the value of lives of people with intellectual disabilities, have all influenced different countries to create processes to both protect and support people with intellectual disabilities who are not able to speak for themselves. A critical issue in many countries is the extent to which, in the absence of legal adjudication of a need for formal guardianship and the actual appointment of such a guardian, people with intellectual disabilities must be viewed as adults and as competent to make their own decisions.

To address the complexity in decision-making and satisfy the requirements for consent approaches, many countries are taking an approach with similarities to the following New York State example.

1. Strong encouragement of the assumption that most people with intellectual disabilities are able to convey their wishes and the provision of support to develop advance directives including wishes about resuscitation and/or to designate a decision-maker consistent with guidance/legislation available for proxy decision-makers.
2. Appraisal in writing, usually by a physician or a psychiatrist (when there are mental health concerns) after review of records and consultation with family and members of the interdisciplinary team (if assigned) of why capacity to participate in decision-making should not be assumed.
3. Independent appraisal, also in writing, by another physician or psychiatrist (when mental health issues are believed to influence capacity to make decisions) to confirm the approval of the original evaluator.
4. Placement of all related appraisals in the person with an intellectual disability's record.
5. Development of DNR/DNAR orders for non-hospital and hospital-based CPR with both person-centred and chronic illness specific justification.
6. Periodic review of the DNR/DNAR orders—every 90 days is one recommendation.

7. Periodic evaluation by relevant bodies of individual physician DNR/DNAR practices and recommendations for people with intellectual disabilities.
8. A willingness and mechanism to recognise that there may be changes in the person with an intellectual disability's capacity to make their own decisions or to incorporate information or prior statements and decisions should they subsequently come to light. This may result in person-informed changes to the DNR/DNAR decision.
9. The legal appointment of a guardian may but will not always supersede the decision-making achieved through this process (New York Codes, Rules & Regulations, 2020).

The approach outlined here is to be recommended, but underlying each country's response will be their legislative approach to incapacity and supported decision-making, the legal status of advance directives, living wills and other instruments such as DNR/DNAR orders, and any additional legal or legislative requirements specifically for people with intellectual disabilities. It may be that these additional issues need to be addressed first to set up circumstances where people with intellectual disabilities do not receive CPR when it is unlikely to result in benefit.

More challenging are the circumstances where people with intellectual disabilities may be denied (overtly or covertly) access to resuscitation. As noted previously, the delivery of CPR is widespread and may actually be more available to people with intellectual disabilities given requirements in many countries that staff of services for people with intellectual disabilities be certified in CPR. Unexplored among provider staff, emergency response personnel and the CPR trained general public is their level of willingness to provide CPR to people with intellectual disabilities as compared to others in distress. COVID-19 has highlighted that there are circumstances where professional judgements are also being made about the *usefulness* of CPR for this population.

Conclusion

As is true for others, there are circumstances, particularly in non-hospital situations where CPR may make the difference for people with intellectual disabilities in being able to survive to receive the care that they need. However, as is also true for others, there are concerns resulting from repeated use of CPR in terms of ultimate outcomes for the person with an intellectual disability and the potential for injury and other negative consequences. People with intellectual disabilities should have the opportunity to make decisions about their willingness to receive CPR and/or should be able to appoint a proxy to make those decisions for them. Where people with intellectual disabilities are not able to make those decisions, there must also be a mechanism to reduce the likelihood that there will be unnecessary and unhelpful use of CPR in the absence of a DNR/DNAR order.

Conversely, there is recent COVID-19 related evidence that persons with intellectual disabilities may also receive DNR/DNAR orders without consultation or permission that instead reflect a devaluing of their lives as compared to others, and/or assumptions about the quality and potential quality of their lives. Mechanisms are also needed to ensure that DNR/DNAR orders are not imposed in these circumstances.

As noted, legislative approaches to incapacity and supported decision-making, advance directives, living wills and other instruments such as DNR/DNAR orders and legal or legislative requirements specifically for people with intellectual disabilities may be required to achieve both of these goals. However, person-centred approaches and consideration of the wishes of the person with an intellectual disability must always be the starting point. Education of people with intellectual disabilities, and of staff in non-hospital and hospital settings, availability of consultation from both self-advocates and service experts, and genuine discussion of the issues present are also critical.

Resources

1. *Do not attempt resuscitation, end-of-life decision-making and associated directives and their implications for people with intellectual disabilities.* An article addressing how legal provisions for people choosing where and how they die, such as advance directives and lasting powers of attorney for health and welfare, affect those with intellectual disabilities. http://www.intellectualdisability.info/changing-values/articles/do-not-attempt-resuscitation,-end-of-life-decision-making-and-associated-directives,-and-their-implications-for-people-with-intellectual-disabilities
2. *Advanced care planning tools and resources.* A range of planning ahead tools for future medical needs. https://coalitionccc.org/CCCC/Resources/Advance-Care-Planning-Resources/CCCC/Resources/Advance-Care-Planning-Tools-and-Resources.aspx?hkey=a0a83455-ff0f-4a89-b700-2572ef44f077
3. *Video library: People with developmental disabilities.* A series of videos featuring people with developmental disabilities thinking ahead about the end of life (English and Spanish) https://coalitionccc.org/CCCC/Resources/Video-Library/CCCC/Resources/Video-Library.aspx?hkey=48e1b795-499e-43c5-b9b9-9703c39686dd
4. *Do Not Resuscitate? DNACPR & people with learning disabilities.* Webinar presented by Professor Irene Tuffrey-Wijne about medical decision-making plans for the end of life. https://www.youtube.com/watch?v=ePJKT9MZRXA.

References

Adams, D. H., & Snedden, D. P. (2006). How misconceptions among elderly patients regarding survival outcomes of inpatient cardiopulmonary resuscitation affect do-not-resuscitate orders. *Journal of the American Osteopathic Association., 106*(7), 402–404. PMID: 16912338.

Adams, R. C., & Levy, S. E. (2017). Shared decision-making and children with disabilities: Pathways to consensus. *Pediatrics, 139*(6), e20170956. https://doi.org/10.1542/peds.2017-0956

American Academy of Social Work and Social Welfare (AASWSW). (2020). *Policy of the American Academy of Social Work and Social Welfare (AASWSW) on the right of all Americans to potential life-saving vaccinations and care.* https://aaswsw.org/aaswsw-policy-on-equal-right-to-vaccination-and-care/

American Association on Intellectual and Developmental Disabilities (AAIDD). (2020). *Position statement: Caring at the end of life*. https://www.aaidd.org/news-policy/policy/position-statements/caring-at-the-end-of-life

Bekkema, N., de Veer, A. J. E., Wagemans, A., Hertogh, C. M. P. M., & Francke, A. L. (2014). Decision making about medical interventions in the end-of-life care of people with intellectual disabilities: A national survey of the considerations and beliefs of GPs, ID physicians and care staff. *Patient Education and Counseling, 96*(2), 204–209. https://doi.org/10.1016/j.pec.2014.05.014

Berdowski, J., Berg, R. A., Tijssen, J. G. P., & Koster, R. W. (2010). Global incidences of out-of-hospital cardiac arrest and survival rates: Systematic review of 67 prospective studies. *Resuscitation, 81*(11), 1479–1487. https://doi.org/10.1016/j.resuscitation.2010.08.006

Bourke, J., Nembhard, W. N., Wong, K., & Leonard, H. (2017). Twenty-five year survival of children with intellectual disability in Western Australia. *The Journal of Pediatrics, 188*, 232–239. https://doi.org/10.1016/j.jpeds.2017.06.008

Care Quality Commission. (2021). *Protect, respect, connect—Decisions about living and dying well during COVID-19: CQC's review of 'do not attempt cardiopulmonary resuscitation' decisions during the COVID-19 pandemic: Final report.* https://www.cqc.org.uk/sites/default/files/20210318_dnacpr_printer-version.pdf

de Vos, R., de Haes, H. C., Koster, R. W., & de Haan, R. J. (1999). Quality of survival after cardiopulmonary resuscitation. *Archives of Internal Medicine, 159*(3), 249–254. https://doi.org/10.1001/archinte.159.3.249

DuBois, J. M. (2008). Solving ethical problems: Analyzing ethics cases and justifying decisions. In J. M. DuBois (Ed.), *Ethics in mental health research: Principles, guidance, and cases* (pp. 46–57). Oxford University Press.

Ekelund, C., Dahlin-Ivanoff, S., & Eklund, K. (2014). Self-determination among older people: A concept analysis. *Scandinavian Journal of Occupational Therapy, 21*(2), 116–124. https://doi.org/10.3109/11038128.2013.853832

Friedman, S., & Gilmore, D. (2007). Factors that impact resuscitation preferences for young people with severe developmental disabilities. *Intellectual and Developmental Disabilities, 45*(2), 90–97. https://doi.org/10.1352/1934-9556(2007)45[90:FTIRPF]2.0.CO;2

Global Research on Developmental Disabilities Collaborators. (2018). Developmental disabilities among children younger than 5 years in 195 countries and territories, 1990–2016: A systematic analysis for the Global Burden of Disease Study 2016. *Lancet Global Health, 6*(10), e1100–e1121. https://doi.org/10.1016/S2214-109X(18)30309-7

Kearney, P. M., & Griffin, R. (2001). Between joy and sorrow: Being a parent of a child with developmental disability. *Journal of Advanced Nursing, 34*(5), 582–592. https://doi.org/10.1046/j.1365-2648.2001.01787.x

Légaré F., Ratté S., Stacey, D., Kryworuchko, J., Gravel, K., Graham, I. D., & Turcotte, S. (2010). Interventions for improving the adoption of shared decision making by healthcare professionals. *Cochrane Database Systematic Reviews* (5), CD006732. https://doi.org/10.1002/14651858.CD006732.pub2

Lippert, F. K., Raffay, V., Georgiou, M., Steen, P. A., & Bossaert, L. (2012). European Resuscitation Council Guidelines for Resuscitation 2010 Section 10. The ethics of resuscitation and end-of-life decisions. *Resuscitation, 81*(10), 1445–1451. https://doi.org/10.1016/j.resuscitation.2010.08.013

Masquelier, B., Hug, L., Sharrow, D., You, D., Hogan, D., Hill, K., Liu, J., Pederson, J., & Alkema, L. (2018). Global, regional, and national mortality trends in older children and young adolescents (5–14 years) from 1990 to 2016: An analysis of empirical data. *The Lancet Global Health, 6*(10), E1087–E1099. https://doi.org/10.1016/S2214-109X(18)30353-X

McCallion, P., & Ferretti, L. A. (2017). Understanding, supporting and safeguarding self-determination as we age. In M. L. Wehmeyer, K. A. Shogen, T. D. Little, & S. J. Lopez (Eds.), *Development of self-determination through the life-course* (pp. 145–158). Springer. ISBN: 9789402410402.

Miller, V. A., & Nelson, R. M. (2006). A developmental approach to child assent for nontherapeutic research. *The Journal of Pediatrics., 149*(1), S25–S30. https://doi.org/10.1016/j.jpeds.2006.04.047

Nelson, L. J., & Nelson, R. M. (1992). Ethics and the provision of futile, harmful, or burdensome treatment to children. *Critical Care Medicine, 20*(3), 427–433. https://doi.org/10.1097/00003246-199203000-00022

New York Codes, Rules and Regulations. (2020). *Part 633—Protection of individuals receiving services in facilities and services operated and/ or certified by Office for People with Developmental Disabilities; Section 633.18—Procedures governing do not resuscitate (DNR) orders.* https://casetext.com/regulation/new-york-codes-rules-and-regulations/title-14-department-of-mental-hygiene/chapter-xiv-office-for-people-with-developmental-disabilities/part-633-protection-of-individuals-receiving-services-in-facilities-and-services-operated-andor-certified-by-opwdd

Oakley, B., Tillmann, J., Ruigrok, A., Baranger, A., Takow, C., Charman, T., Jones, E., Cusack, J., Doherty, M., Violland, P., Wroczyńska, A., Simonoff, E., Buitelaar, J. K., Gallagher, L., & Murphy, D. G. M. (2020). COVID-19 health and social care access for autistic people: European policy review. *BMJ Open, 11*(6). https://doi.org/10.1136/bmjopen-2020-045341

Riva, G., Ringh, M., Jonsson, M., Svensson, L., Herlitz, J., Claesson, A., Djärv, T., Nordberg, P., Forsberg, S., Rubertsson, S., Nord, A., Rosenqvist, M., & Hollenberg, J. (2019). Survival in out-of-hospital cardiac arrest after standard cardiopulmonary resuscitation or chest compressions only before arrival of emergency medical services: Nationwide study during three guideline periods. *Circulation, 139*(23), 2600–2609. https://doi.org/10.1161/CIRCULATIONAHA.118.038179

Sledge, R. (2013). Do not resuscitate orders for adults with developmental disabilities: Ethical considerations in the dialysis center. *National Kidney Foundation Journal of Nephrology Social Work, 37*, 8–14.

Smith, T. B., Oliver, M. N. I., & Innocenti, M. S. (2001). Parenting stress in families of children with disabilities. *American Journal of Orthopsychiatry, 71*(2), 257–261. https://doi.org/10.1037/0002-9432.71.2.257

Strauss, K., Benvenuto, A., Battan, B., Siracusano, M., Terribili, M., Curatolo, P., & Fava, L. (2015). Promoting shared decision making to strengthen outcome of young children with autism spectrum disorders: The role of staff competence. *Research in Developmental Disabilities, 38*, 48–63. https://doi.org/10.1016/j.ridd.2014.11.016

Thomas, R. (2020, April 25). *Unprecedented' number of DNR orders for learning disabilities patients.* https://www.hsj.co.uk/coronavirus/unprecedented-number-of-dnr-orders-for-learning-disabilities-patients/7027480.article

Voss, H., Vogel, A., Wagemans, A. M. A., Francke, A. L., Metsemakers, J. F. M., Courtens, A. E., & de Veer, A. J. E. (2017). Advance care planning in palliative care for people with intellectual disabilities: A systematic review. *Journal of Pain and Symptom Management, 54*(6), 938–960. https://doi.org/10.1016/j.jpainsymman.2017.04.016

Wagemans, A. M. A., van Schrojenstein Lantman-de Valk, H. M. J., Tuffrey-winjne, I., Widdershoven, G., & Curfs, L. (2010). End-of-life decisions: An impotant theme in the care for the people with intellectual disabilities. *Journal of Intellectual Disability Research, 54*(6), 516–524. https://doi.org/10.1111/j.1365-2788.2010.01276.x

Wehmeyer, M. L. (2011). Empowerment. In A. C. Michalos (Ed.), *Encyclopedia of quality of life and wellbeing research.* Springer. https://doi.org/10.1007/978-94-007-0753-5

Wehmeyer, M. L., & Abery, B. H. (2013). Self-determination and choice. *Intellectual and Developmental Disabilities, 51*(5), 399–411. https://doi.org/10.1352/1934-9556-51.5.399

Welbourn, C., & Efstathiou, N. (2018). How does the length of cardiopulmonary resuscitation affect brain damage in patients surviving cardiac arrest? A systematic review. *Scandinavian Journal of Trauma, Resuscitation & Emergency, 26*(77). https://doi.org/10.1186/s13049-018-0476-3

16

Positioning the Issues: An Agenda for Future End-of-Life Research, Policy and Practice

Michele Y. Wiese, Roger J. Stancliffe, Philip McCallion, and Mary McCarron

Thou know'st 'tis common; all that lives must die.
—William Shakespeare, *Hamlet*, Act I, Scene 2, Line 72

M. Y. Wiese (✉)
School of Psychology, Western Sydney University, Penrith, NSW, Australia
e-mail: m.wiese@westernsydney.edu.au

R. J. Stancliffe
Centre for Disability Research and Policy, University of Sydney, Sydney, NSW, Australia
e-mail: roger.stancliffe@sydney.edu.au

P. McCallion
School of Social Work, Temple University, Philadelphia, PA, USA
e-mail: philip.mccallion@temple.edu

M. McCarron
Trinity Centre for Ageing and Intellectual Disability, Trinity College Dublin, Dublin, Ireland
e-mail: mccarrm@tcd.ie

R. J. Stancliffe et al. (eds.), *End of Life and People with Intellectual and Developmental Disability*, https://doi.org/10.1007/978-3-030-98697-1_16

Regardless of disability, race or creed, the inevitability of death is an undeniable part of our common humanity. Recognising this shared reality, one theme of this book is awareness of end-of-life issues in the general community and how they apply to people with intellectual and developmental disability. Chapter 1 focuses on end of life for everyone, and many subsequent chapters mention the general-community context. In this present chapter, we may mention issues current in the broader end-of-life literature that have not yet received attention in the intellectual and developmental disability field and are therefore not represented elsewhere in this book. We also refer to topics presented in the disability-specific chapters. However, our intention is not to summarise points made elsewhere, as the respective chapter authors have presented their concerns clearly. Instead, we consider several cross-cutting or overlooked issues to put forward important parts of a practical research, policy and practice agenda intended to assist people with intellectual and developmental disability to have opportunities they want for active participation, social inclusion and self-determination in all end-of-life dimensions.

Research, Research Gaps and Future Research

In Chapter 2, we identified several overarching issues and gaps in existing end-of-life research:

- Very limited involvement by people with intellectual and developmental disability in research as participants, and as researchers, apparently reflecting widespread avoidance and protection by others (caregivers, professionals and seemingly researchers) about involving people with disability in end-of-life issues. However, research consistently indicates that such avoidance and protection is often unnecessary and counterproductive (Stancliffe et al., 2021; see also Chapter 2). We therefore expect that these findings will help reduce paternalistic protection and lead to future research involvement by people with intellectual and developmental disability.

- The primary focus on problems and inequities, with little research attention given to policies, practices and interventions aimed to prevent or ameliorate these problems.
- With a few notable exceptions (e.g. suicide and people with autism, see Chapter 3), the paucity of research on the end-of-life experiences of people with developmental disability, except those with intellectual disability. Diagnostic groups without co-occurring intellectual disability but with specific support needs, such as people with autism or those with cerebral palsy, are largely absent from end-of-life research and therefore too often missing in this book.
- Limited development of disability-specific measurement instruments (see below).

Many chapters in this book have noted important topics that need attention in the future, and other publications have offered important ideas for the future research agenda (e.g. Stancliffe, Wiese, & Read, 2017; Todd et al., 2013). Below we discuss in greater detail four key future research topics: Relationships, social inclusion, people with disability providing care and support, and end-of-life assessment instruments, followed by briefly listing other research gaps.

Relationships, Social Support and End of Life

High quality, individually appropriate health and palliative care are undoubtedly important components of end of life but underpinning the quality of life of the dying person and their caregivers is human contact, especially involving relationships with trusted people who can offer support and emotional care. Chapters in this book have emphasised that relationships with trusted, familiar people are central to quality end-of-life experiences (see Chapters 5, 7 and 13). However, these relationships should be understood within a broader context of social networks throughout life. Research consistently finds that adults with intellectual and developmental disability have smaller social networks than people without disability and that their most common relationships are with peers with disability (usually fellow service users), staff and family (Amado et al., 2013).

The relative presence of these groups in the network of a person is affected by individual life circumstances. Those living with family usually have more family-focused social networks (Amado et al., 2013). By contrast, McCausland et al. (2021) found that among older (40+ years) Irish adults with intellectual disability, living mostly in disability service settings (40% group homes, 44% institutions), co-resident peers with intellectual disability were the most common friends, closely followed by staff, but with 8% having no friends. Friendship quality was a common issue, with only 52% of people having a best friend (typically a non-co-resident person with intellectual disability). Given this situation, social and emotional support from network members is particularly valuable, especially from those with deep relationships such as best friends and close family. Yet end-of-life research has focused disproportionately on only one group—disability staff—in attempting to understand end-of-life care and social support. Likewise, most research attention is given to dying people who live in disability settings. These practices have resulted in obvious gaps, with little or no research on:

- the perspective and experiences of peers with disability as carers and sources of social support, especially best friends
- the practical and emotional support needed to enable friends with disability to provide support to a dying person
- dying at home for people living in their own home, some with a partner as the primary carer
- people dying in their family home, including the experiences and support needs of family caregivers
- the benefits of, barriers to, and extent to which members of the person's social network (staff, peers, family, friends) are able to provide social and practical support to a person with intellectual and developmental disability dying in a hospital or hospice.

Such research should provide examples of good practice, as well as identifying problems.

With the knowledge that key social supports tend to be family and staff, there is some evidence of tensions between these groups, pronounced when the dying person is living in supported accommodation. While tensions in end-of-life care may be expected, the extent to which these influence the experience for all involved is only partially understood. To date the focus tends to have been disability staff (Ryan et al., 2010; Todd, 2013), perhaps a residual of the relative ease of locating staff for research participation. Staff have reported limited consultation afforded them in the last days about day-to-day care, medical decisions and funeral arrangements, all particularly intensified for those staff in long-term caring relationships with the dying individual. Meanwhile, the family voice in this context is not well-understood and demands equal research attention. In Chapter 7, *Joan's story* provides a poignant good-practice example of staff-family collaboration, where the family were invited to stay in Joan's home while she was dying. The extent, however, to which this scenario reflects typical practice is unclear.

Scoping reviews in the general population end-of-life literature have identified the complexity of family issues at the end of life (Coombs, 2015). Social and emotional issues, for example, acknowledging long-held grievances or reconciling fractured relationships, are perceived as often more important than day-to-day care needs (Toseland et al., 2002). For families with a member with intellectual and developmental disability these issues may be intensified at the end of life. For example, the history in some jurisdictions of child institutionalisation and severing family ties can have profound impact on now ageing individuals (Lightfoot & McCallion, 2016). This personal history may well colour who is important to the individual when they are dying in older age. It may also be an issue for siblings who may not have been party to the decision to separate.

Family relationships are important. With the institutional era overwhelmingly in the past and less separations of these kinds, family relationships will likely be more consistent throughout life and therefore very important when dying. Future research needs to engage more fully with these family issues.

Social Inclusion, Place of Death and Compassionate Communities

People with intellectual and developmental disability continue to experience higher levels of social isolation and loneliness, and as noted in the previous section, have fewer friends and smaller social networks, typically family, disability staff and other disability service users (Amado et al., 2013). With restricted social inclusion throughout life, people likewise probably also have very limited sources of social and emotional support at the end of their life. General-community research shows that end of life is often accompanied by social isolation and *social death* where the dying person's social existence deteriorates and ultimately ceases before actual bodily death (Králová & Walter, 2018), acutely witnessed in the COVID pandemic with people dying alone in hospital. This circumstance contrasts with the focus on relationship-based care set out in Chapter 7. In addition, people with intellectual disability identify the importance of practical, emotional, social and spiritual support at end of life provided by trusted, familiar caregivers (Bekkema et al., 2016; Cithambarm et al., 2021). Although these issues are demonstrably important, research has yet to analyse and describe the nature and extent of social support available to dying individuals with intellectual and developmental disability, let alone compare this to their preferred supports or to the experience of those dying in the general community. Further, the support needs of those providing care and those surrounding the individual are also poorly researched.

As noted in Chapter 6, of the few with intellectual and developmental disability who have indicated a preference about where they wish to die, individuals in the UK (Hunt et al., 2020) and Ireland (McCarron, Burke et al., 2017) all preferred to die at home. Importantly, these preliminary findings are consistent with the place-of-death wishes of the general community (Ali et al., 2019). People with intellectual disability strongly endorse the importance of honouring end-of-life wishes (Bekkema et al., 2016). More research is needed to document individual wishes about preferred place of care and death, to determine the circumstances when their wish is met and to probe their understanding of the implications of home death as compared to dying in hospital/hospice (Segerlantz et al., 2020).

In the broader community, there is an increasing focus on involving friends and community members in end-of-life care, a movement often referred to as *compassionate communities* (Kellehear, 2013; Sallnow et al., 2016), representing a return to the norms of family and community care that characterised dying in previous eras. These ideas are part of a public health approach aiming to support the quality of life of the dying person and those around them by involving and developing the broader community's capacity in caregiving and supporting caregivers (Grindrod & Rumbold, 2018; Kellehear, 2013). Support from community members can come from existing social networks, volunteer community members or in other ways. Such support is thought to benefit the well-being of the dying person and family caregivers through reducing social isolation, loneliness, stigma and caregiver fatigue. There is substantial interest in and uptake of this approach in the UK and elsewhere (Grindrod & Rumbold, 2018; Kellehear, 2013). To date, research on outcomes has been limited, although there is evidence that support from friends as caregivers in addition to family members is associated with being more likely to have a home death and greater use of palliative care (Sallnow et al., 2016).

These approaches have not yet been directly considered by the disability community but are very compatible with the field's emphasis on social inclusion and social support throughout life (Amado et al., 2013) and with the emphasis on relationship-based care at the end of life (see above and Chapters 7 and 11). The compassionate communities approach means that informal networks have a continued role in supporting the dying person. For example, the dying person who had worked in supported employment could (re)connect with work friends, who might provide companionship through visits or phone calls. The Irish example in Chapter 10 (see Funeral Rituals in Ireland) is pertinent, where in preparation for the wake a housemate helped a staff member lay out the body of a community-living-service resident who had died. Such approaches reflect relationship-based care for those in the immediate situation but are not yet inclusive of all potential helpers which, if achieved, will enhance the social capital so important in building compassionate communities (Rosenberg et al., 2015).

Aspects of the available research show that valued end-of-life outcomes involve the participation of and support from compassionate members of the dying person's social network. For example, McKenzie et al. (2017) researched the situations of adults with intellectual disability in New Zealand who knew they were dying and participated in their own advance-care planning. The strongest theme was focussing on the life they continued to live, so it was as full, active and socially connected as possible with events with others to look forward to. At times, this required adaptations of favourite activities to accommodate declining health. Likewise, Bekkema et al.'s (2016) participants with intellectual disability highlighted dying people being supported to live life and have positive experiences. McCarron, Burke et al. (2017) emphasised as far as possible maintaining normal activities, routines and social connections even in the person's last weeks.

Taken together, these ideas could inform future research. As suggested previously, studies could describe the extent of social contact and support from families, staff, friends and familiar community members for people dying in various care settings. If such contact was minimal, potential intervention could involve reconnecting with family members, finding old friends and former staff and supporting them to visit, with research evaluating the feasibility and benefits.

People with Disability Providing Care and Support

Involvement by people with intellectual and developmental disability in compassionate communities can encompass being a provider and a recipient of support—receiving support from one's community towards the end of your own life or bereavement, or giving support to others as a member of a local community or friendship network.

Little research attention has focussed on what people with intellectual and developmental disability can do to support others who are dying and what effect this has on the support provider and recipient. One of Bekkema et al.'s (2016) participants with intellectual disability stated:

> Take somebody outside if the weather is nice, and go for a walk. Try to make them think of other things, distract them. When my sister was ill, we took her on a trip in a wheelchair. On a boat with music. She really enjoyed that. That's something you could do for somebody. (p. 630)

Such actions probably have benefits and challenges for both parties, but we know of no research examining the issue from the perspective of people with intellectual and developmental disability. Truesdale et al. (2021) recently reported on the experiences, support needs of and impact on people with intellectual disability caring for a co-resident older family member with health problems. This methodology could also be applied to end-of-life care.

End-of-Life Assessment Instruments

Measurement is fundamental to research. Few assessments are designed for people with intellectual and developmental disability, especially those involving *self-reports* (see Resources at the end of Chapters 2 and 3). This gap encompasses issues as varied as suicidality, funeral satisfaction and understanding do-not-resuscitate orders. This situation constrains research and practice about end of life. Valid and reliable instruments enable studies of prevalence, disparity with other populations, risk factors and intervention efficacy. Instruments facilitate good practice through, for example, identifying individuals with clinically important difficulties who may benefit from support/intervention. The lack of self-report instruments further disenfranchises people in areas where their views are often either not sought or ignored (e.g., Voss et al., 2017, 2019). More good quality self-report instruments would help increase people with intellectual and developmental disability's involvement in research as participants in their own right (Stancliffe, Wiese, Read et al., 2017).

Assessments should meet the comprehension and communication needs of people with intellectual and developmental disability to enable them to self-report (see section below about accessible information). Well-designed disability-specific assessments allow substantially more people with intellectual and developmental disability to respond validly and express their views compared to using instruments designed for the general community (Stancliffe et al., 2014). Unmodified general-population instruments often use wording that is too difficult for people with intellectual and developmental disability to understand, and/or response methods that are too complex. End of life is a sensitive topic with the potential to remind people of unpleasant experiences or even upset them (O'Keeffe et al., 2019), so instrument developers should report data on acceptability, including participant discomfort, withdrawal, refusals etc. as indices of potential distress when answering the questions. This data can reassure instrument users, and gatekeepers like research ethics committees, that the tool is not distressing for participants to use (Stancliffe, Wiese, Read et al., 2017, 2021).

Although we have emphasised the development of self-report instruments, there are some people with severe disability and very limited conventional communication who are currently unable to self-report, so proxy reports may be necessary. In this event, the responsibility of the proxy reporter to best represent the person's values, rather than their own, cannot be over-emphasised.

The Example of Complicated Grief. Measurement of complicated grief (also known as prolonged grief, or persistent complex bereavement disorder) experienced by people with intellectual disability provides an example of applied research that is particularly important because it is (a) distressing to the individual, (b) associated with mental health morbidity and (c) may be more common among people with intellectual disability (Dodd et al., 2008). Understanding risk factors, prevention and treatment requires measurement of the presence, severity and persistence of complicated grief. Irish researchers developed proxy and self-report versions of a complicated grief scale by extensively modifying an existing general-community scale (Dodd et al., 2008, 2021; Guerin et al., 2009; O'Keeffe et al., 2019). O'Keeffe et al. (2019) described the self-report version's development process in detail, providing a method for developing assessments of other sensitive topics.

Practice guidelines were developed by Ballan and Sormanti (2006), but apart from Dowling et al. (2006), we know of no research on treatments of grief experienced by people with intellectual and developmental disability. Dowling et al. used proxy-report instruments assessing challenging behaviour and mental health to evaluate the effects of their intervention. The availability of a self-report instrument offers a potentially more sensitive and valid way to evaluate grief interventions in future research.

Other Research Gaps

These studies of complicated grief assessment show how disability researchers can build on mainstream research. Many important issues have been examined in general-population studies but remain little considered in intellectual and developmental disability. Among the areas for further research are:

- The experiences of the dying and their caregivers, family, staff and peers
- Assisted dying legislation
- Real-life experience of implementing do-not-resuscitate orders
- Parents with disability whose children die
- Ethical issues in end-of-life care.

For example, on the first point listed above, mainstream studies of farewelling behaviour of the dying (Kellehear & Lewin, 1988–1989) have no equivalent in intellectual and developmental disability research. The nature, timing and other characteristics of these farewells could be examined and compared to the person's wishes.

Research areas more specific to people with intellectual and developmental disability include:

- The need for disability identifiers in population-based data on dying and death to enable comparisons with the general community and among disability subgroups

- The effects (intended and unintended), scope, processes and participants' experience of official scrutiny of deaths of people with disability in care, through critical incident and coronial processes
- The extent that the funeral industry is inclusive of and makes accommodations for people with intellectual and developmental disability.

Finally, there are newer topics to explore:

- Pandemic dying—experiences of the dying and their caregivers
- Pandemic funerals—the online experience and its ability to offer a positive collective mourning.

There is vast scope for future research involving people with intellectual and developmental disability. These enquiries can build on relevant general-population research, address important little-studied topics, involve people with disability as active participants and researchers and supplement the current research focus on people with intellectual disability by involving participants with other types of developmental disability. Greater involvement of families, staff and peers in the identification and investigation of additional research questions will also expand research agendas.

Policy and Practice Issues: A Future Agenda

In this chapter, we consider *policy* as social in nature and intended to impact the real and every-day lives of people (Bochel & Daly, 2020). *Practice* is essentially a procedure, or way of doing something. With a few minor exceptions to assist clarity, we will customarily use the term *policy*. We will use *practice* only when there are specific issues arising from a policy that have procedural implications for end of life.

Using Bochel and Daly's (2020) policy definition we observe that each chapter includes suggestions addressing policy gaps or identifies existing policy which does not include people with intellectual and developmental disability. These findings speak for themselves. Here, we deal with some fundamental principles that cut *across* these findings to propose policy design that is inclusive of people with intellectual and developmental disability, addresses their particular needs and enacts currently available best evidence. This involves policy on multiple levels, at the national (or international) level, guiding the overall direction of society's response to end-of-life issues and disability inclusion, as well as at the local level within individual services (both mainstream and disability specific).

These principles require clear and explicit representation in policy. We propose too that they are not exclusive to individuals with intellectual and developmental disability and fit equally well in end-of-life policy and practice for all members of society.

Community Inclusion

In earlier sections, we presented a research agenda to address how people with intellectual and developmental disability are included in community, with respect to both their own and others' end of life. This agenda is positioned within the compassionate communities' movement and reflects current palliative-care approaches outlined in Chapters 1, 8 and 14. Thus, we propose that end-of-life policy needs to reflect the person's inclusion within their community and provide an explicit commitment to maintaining identity and presence therein. That community should be understood not simply as bricks and mortar. Instead, a true understanding of inclusion observes community as an ecology involving the physical community together with the social relationships and interactions the individual experiences.

Honouring the principle of community inclusion means not only explicit commitment in policy, but also identification of practices about *how* this commitment might be enacted. Taking the person's own dying as an example, community inclusion would likely be reflected by a policy commitment to care in the person's chosen own community, typically their own home surrounded by familiar people. This policy should also include practices that address caregiver support, capacity and necessary training, support of the local community palliative-care team, equipment and all the other practices that comprise quality in-home care. Because community is seen as a social ecology, practice must ensure the maintenance of valued relationships, like the visits of friends from the dying person's former workplace previously described.

Advancing community inclusion within policy about the person's own dying should not be interpreted as care at home at all costs. Aiming solely for this outcome ignores other critical dimensions of quality end-of-life care, including health care. If the health care needs necessitated a transfer away, even temporarily, from the physically familiar community to hospital or hospice, the practice would be about transferring with the person the portable elements like a familiar pillow, quilt, music and loved objects. Valued social relationships with familiar people could also be transferred with visits, bringing with them familiar objects like photographs that could assist shared conversations and reminiscence. Equally, hospitals and hospices must continue to be challenged to deliver care in ways that honour the person's wishes, and de-medicalise and de-institutionalise a person's last days.

The Right to Active Participation in Decision-Making

Active participation in decision-making reflects a commitment to autonomy and self-determination. While decision-making may be more challenging for individuals with learning problems associated with their disability, the pattern of avoidance and protection about end-of-life topics (see Chapter 2) means that many are currently excluded from any meaningful participation. Decision-making is a right that must be reflected in all policy.

Acknowledging that legal jurisdictions differ, we propose that policy should, as both Chapters 5 and 15 conclude, have self-determination at the forefront—that is, individual ownership of decision-making, free from undue influence or coercion. Where support for decision-making is needed, this approach must be adopted with integrity and substitute decision-making used only as an absolute last resort. In Chapter 15, Suzanne's story demonstrates a policy position that respected her decision to develop her own advance directive. The story also listed a range of practices that enabled this right. Suzanne learnt about end-of-life decisions through classes and easy-read materials, meetings were held to address key personnel's concerns, agreements were made about where to keep the advance directive and the people with responsibility to honour the directive were trained.

Access to Full Range of Planning Options Requiring Decision-Making

The right of the person with intellectual and developmental disability to act as decision-maker requires that the full range of planning options be present in policy. Many chapters offer examples of these options including, funeral wishes (Chapter 10), involvement of a doula (Chapter 11), advance-care planning (Chapter 4), preferred place of care (Chapter 6), health and palliative-care planning (Chapter 8), do-not-resuscitate orders (Chapter 15) and the appointment of proxy decision-makers (Chapters 4 and 15). Other topics not covered elsewhere in the book should equally be included. Bequeathing (leaving possessions to others after death), organ and tissue donation, and plans for care when dying (including preferred carers) round out the range of planning options that might routinely be offered to the general community, and therefore also to people with intellectual and developmental disability (Wiese & Tuffrey-Wijne, 2020).

End-of-life planning and decision-making should not wait for a life-limiting illness. Many of us make some plans when we are well, for example, decisions about bequeathing, or organ and tissue donation (Wiese & Tuffrey-Wijne, 2020). And, although perhaps not with a clearly defined plan, most of us privately ponder or discuss these future issues with our loved ones. The same approach should be accorded to people with intellectual and developmental disability. The evidence showing continuing patterns of protection and avoidance by caregivers (Chapter 2), however, indicates that policy should include practices to help ensure these opportunities *are* actually afforded.

One practice to consider is *when* to plan. For the general community, this usually occurs somewhere in the adult life stage. In funded accommodation services with person-centred planning, adulthood could signal when the range of planning options described above could begin to be introduced. The individual's choice to make (or choose not to make) decisions could be facilitated, and periodically revisited throughout life. There are many practical resources that caregivers could use with the individual. One is the *When I Die* booklet (see Resources at the end of this chapter) in which Tony makes choices about a range of end-of-life plans including preferred visitors when dying, funeral wishes and bequeathing. In some jurisdictions such conversations are already encouraged as part of person-centred care planning. Opportunities to discuss death and to participate in death rituals will ensure such planning is meaningful and possible.

Although these suggestions may appear to mechanise an experience that for most of us is organic, without explicit focus the opportunity for people with intellectual and developmental disability to self-determine about planning options and to participate in these decisions will be compromised (Kirkendall et al., 2017; Voss et al., 2017, 2019). Worse, many will be confronted with the issues only when death is imminent, stress and emotions are heightened, and the opportunity to learn over time will have been lost.

Accessible and Complete Information

In this section, we address two components of information. Firstly, that it is understandable (accessible), and secondly that all information is offered (complete). First to the principle of information accessibility. Active participation in decision-making and accessing a range of planning options both require sufficient information. Genuine commitment to providing accessible information requires specific skills, including support to the individual to find, think through, understand and use the content (Chinn & Homeyard, 2017). Committing to this principle maximises the chance that the person can make informed decisions. Further, having information means that the person can exercise their informed right to *not* make decisions about end-of-life issues, just as many in the general community do.

There is a growing range of readily available accessible information across a number of end-of-life topics, many listed as Resources at the close of the book's chapters. In Chapter 4, McGinley and Waldrop emphasise an individualised approach to information provision about the complex subject of advance-care planning and offer strategies to do so. The *Books Beyond Words* website (see Resources at the end of this chapter) offers a range of picture stories, both in hard copy and in a downloadable app, that can be used to provide information and start conversations about topics like *Getting on with cancer* and *When someone dies from coronavirus.* Further, Tuffrey-Wijne (2013) provided practical strategies about disclosing bad news, including about death, with methods like breaking the information down into small chunks, giving information sequentially, and adding information as the person's understanding grows. Maximising the accessibility of information need not always rely on formal resources. A creative individual solution is described in Paul's funeral story (Chapter 10), where the headstone featured a coloured photograph of him so people with reading difficulties could easily identify his grave.

The evidence in this book highlights significant gaps in the completeness of information offered about end of life. Healthcare information (Chapters 2 and 14) and advance-care planning (Chapter 4) are notable. Death disclosure (discussed in Chapters 2 and 6) is a pivotal pre-requisite to much subsequent information. A person cannot share in the collective ritual mourning offered by a funeral if the death of a loved one is not disclosed (Chapter 10). Further, unless fully informed about a life-limiting illness, participation in subsequent healthcare decisions is not based on complete information about prognosis (Chapter 14). Therefore, organisational-level policy which addresses information provision about the end of life must make clear the principle of offering all information, commencing with a commitment to honesty about disclosing death. This disclosure might be about the upcoming or actual death of a loved one or friend, or the individual's own life-limiting illness. In some jurisdictions, decisions to share such information rest with an appointed guardian who must be consulted.

The human right to know is fundamental, even about the most distressing information. The individual's choice to further engage or make decisions is then based on having complete information. Tuffrey-Wijne et al. (2020) summarise the many issues, including readiness (by both the disclosure-giver and recipient), the disclosure-giver's skill, the accessibility of the information provided, and the disclosure-giver's ability to pick-up the recipient's emotions and capacity to engage. Several practice recommendations arise which respond to these issues, principle among which is support and training to those charged with the responsibility of disclosure, be they health professionals, disability staff or family (Clayton et al., 2012; Tuffrey-Wijne et al., 2020). The use of accessible resources discussed earlier can play an important role.

Accessible information requires extra time and resources, but this is not sufficient reason to abrogate our responsibility to do so (Chinn & Homeyard, 2017). Any lesser commitment is tokenistic and does little to authentically engage people with intellectual and developmental disability in understanding, planning and decision-making about end of life.

Actively and Inclusively Supporting Caregivers

End of life highlights the critical and intimate position of caregivers. Here, the term *caregivers* encompass parents, siblings, partners, friends and peers, as well as paid caregivers like disability support professionals.

Any policy about caregivers should reflect the principle of active and inclusive support for that role. Support may include emotional support to grieving caregivers, or training to ensure that supported decision-making can be implemented with integrity. While not discounting the principal position of the person with intellectual and developmental disability as key decision-maker, the preferences and needs of caregivers should also be heard and respected about end-of-life issues. As is true for the general population, it becomes challenging when there are multiple people with caregiving, familial and affectional links to the person to determine whose opinions should be given more weight.

Caregiver choice, willingness and capacity to care for the person who wishes to receive care and die at home should be actively sought rather than presumed. A range of caregiver supports should be made available, including palliative-care expertise, available on-call advice and medical and equipment needs. Within disability accommodation services, caregiving may have implications for staff deployment, with some staff finding the prospect of end-of-life care daunting or stressful. Available evidence indicates that for many who express initial hesitation, support and training is an effective solution (Clayton et al., 2012), with the added bonus of building future workforce capacity. These issues can also apply to family caregivers and similar supports must be made available.

An Early-Intervention Approach—Supporting Conversations and Literacy About Death

One policy imperative involves taking an early-intervention approach to death literacy, defined as a "set of knowledge and skills that make it possible to…understand and act upon end-of-life and death care options" (Noonan et al., 2016, p. 31). For people with intellectual disability, a characteristic feature of which is difficulties with learning, building death literacy requires a long-term approach (Wiese & Tuffrey-Wijne, 2020). The lesson is to start early. Research also shows that engaging with end-of-life information is not emotionally or psychologically harmful (Stancliffe et al., 2021) (see Chapter 2), so there is no reason to fear that helping an individual build their death literacy is harmful, so long as usual care and sensitivity is adopted. It might also be speculated that death literacy has a protective emotional and psychological role. For example, in Chapter 1, Deliens describes the benefits of information about palliative care early in disease progression.

Early-intervention practice reflecting death literacy could include supporting active participation at funerals, sending sympathy cards, reminiscing about loved ones who have died, tending their graves, visiting the dying, and initiating conversations about various end-of-life topics. These activities should ideally begin in childhood and occur throughout life (see Resources at the end of this chapter). The aim of start-early practice is to assist the person to gradually build death literacy in a natural, manageable way. These practices have the additional benefit of mitigating the protection and avoidance discussed earlier, as well as not delaying engagement with these topics till too late.

The early-intervention principle also focusses on reducing the incidence of potentially avoidable deaths and ensuring sustained quality health care (O'Leary et al., 2018; see also Chapter 6). Early intervention in health care can be implemented by ensuring quality individual health care and through public health approaches. Better attention to both primary care and coordinated care would facilitate early diagnosis, lower hospitalisations, reduce deaths from preventable conditions and decrease caregiver stress (Moro & McGinley, 2021; Segerlantz et al., 2020). Consistent policy mandating annual health checks is an urgent need. For some, a more individualised approach is needed. For example, individuals with Down syndrome have higher risk for dementia, with an average diagnosis age of 55 years (McCarron, Burke et al., 2017) (see Chapter 7), suggesting that annual health checks inclusive of falls-risk assessments from mid adulthood would be helpful. Memory assessments would also be useful, but may require adaptation. Those with intellectual disability may not be able to be assessed with standard memory tests, however, adaptive behaviour instruments which measure changes in participation in daily living activities offer an alternative indicator of memory changes (McCarron, McCallion et al., 2017). Practice guidelines are increasingly available to assist quality health care for particular diagnostic groups, including dementia care (Dodd et al., 2018).

Regarding public health policy, the need for caregiver awareness and training, and similar accessible training for people with intellectual and developmental disability themselves, is critical. Public health approaches aim to address complex health needs, limit the incidence of chronic disease and reduce mortality. Though not an exhaustive list, practices to assist this aim include awareness and training in falls prevention, epilepsy management, safe swallowing practices, nutrition and exercise. These education programmes would make a substantial contribution to mitigating some potentially avoidable illnesses and death experienced by people with intellectual and developmental disability, including heart disease, epilepsy and cerebrovascular disease (Glover et al., 2017; Hosking et al., 2016).

Commitment to Collaboration

Perhaps the greatest need for policy commitment to collaboration is between the disability and palliative-care sectors. Research shows that the limited collaboration between both sectors is currently a barrier to quality care at the end of life. Adam et al. (2020), in a systematic review of the palliative-care needs of adults with intellectual disability, identified the following collaboration barriers: poor referral to specialist palliative care, lack of established relationships between local disability and palliative-care services, poor understanding of each other's role, mistrust, conflicts about ownership of the patient and a clash between social and medical models; this clash sometimes translating to a focus on the social aspects of care to the neglect of physical care needs. These barriers have been previously identified (McCallion et al., 2012) and remain largely unaddressed.

There are, however, some international efforts to improve collaboration. The European Association of Palliative Care includes an endorsed reference group on intellectual disability (see Resources at the end of this chapter). That group's White Paper of thirteen palliative-care consensus norms explicitly identifies collaboration and the early identification of *all* experts, including health professionals, disability staff, family carers and spiritual leaders (Tuffrey-Wijne et al., 2016). The principles of community inclusion and active participation in decision-making identified in this chapter suggest that the dying person with intellectual and develop mental disability should also be included wherever possible, as their own expert.

Collaboration must also include partners like the funeral industry and religious leaders. In Chapter 10, the story of Paul's funeral is a demonstration of collaboration. The funeral civil celebrant customised the service to include active roles for Paul's friends, with and without disability, including the opportunity for each of his housemates to place a flower on his coffin. Future policy and practice could be directed to a collaboration between the disability and funeral sectors to jointly assist community inclusion and accessibility. The collaboration could build practice protocols about how people with disability could be actively included in a funeral and what an accessible funeral looks like. Similarly, the funeral sector could offer expertise to help develop educational opportunities to assist people with intellectual and developmental disability understand things like the difference between burial and cremation, and what to expect at a funeral.

We conclude this section on policy and practice with a comment about the need for disability to be included in mainstream end-of-life policy. While policy, particularly at international and national levels, may include the term disability (e.g. in Australia the *National Palliative Care Standards*; Palliative Care Australia, 2018), the nuts-and-bolts of what authentic inclusion looks like remains largely unspecified. This lack of top-down drive means policy and practice solutions are developed instead by creative individuals locally, which constrains sustainability and reach. In the U.S., civil rights language with which all policy documents need to comply ensures inclusion of people with disability in end-of-life care practices and standards. The specifics of such inclusion are often addressed in cross-discipline training packages and in additional guidance on how policies may be implemented in inclusive ways (see Resources at the end of this chapter).

Authentic inclusion of people with intellectual and developmental disability in mainstream policy is not easy, due in part to the undeniable complexity of the intellectual and developmental disability experience and society's response to it. Further, the breadth and depth of end-of-life topics, discussed throughout this book, add further complexity. The principles we have outlined offer a small and hopefully practical contribution to address the many challenges.

Conclusion

We began this chapter with a Shakespearean quote about the universality of death. This reality is undeniable, but the individual experience of end of life is too often compromised for people with intellectual and developmental disability through poor practice. Well-meaning but unhelpful avoidance and protection in turn denies individuals the level of death literacy needed to make informed personal decisions. This approach has resulted in exclusion from end-of-life decision-making, lack of information and consequent limited understanding of the issues, with increased risk for consequences such as complicated grief. Excellent examples of good practice exist, and many accessible end-of-life resources are available to help address these issues. We have highlighted research gaps, including the need for greater understanding of the social networks around the person dying and their related needs, but growing research has considerably expanded our knowledge and should inform future practice. Fundamental to continued progress are participation and empowerment of people with intellectual and developmental disability at all levels, supported by integration of policy and practice across disability, health and other sectors involved in end of life. The motivation to continue this work is clear. For each individual, we don't have a second chance to enable a good death, so we need to do the very best we can the first, and only, time.

Resources

1. *Books Beyond Words*. Wordless picture stories for people who have trouble reading, to provide information and open up conversations about death. https://booksbeyondwords.co.uk/
2. *End of life care: A guide for supporting older people with intellectual and developmental disabilities*. Helpful guidance for families and professionals. Purchasable by order from NYSARC Trust Services. https://www.nysarctrustservices.org/contact/
3. *Intellectual Disability reference group*. A group of world experts who contribute information about the needs of people with intellectual disability to the European Association for Palliative Care. https://www.eapcnet.eu/eapc-groups/reference/intellectual-disability/
4. *Talking End of Life...with People with Intellectual Disability [TEL]*. Australian website with 12 modules on end of life and people

with intellectual and developmental disability. Designed for disability support workers but also helpful for families, health professionals, and educators. TEL includes a suite of topics on (1) Teaching how to understand the end of life, and (2) Teaching the planning options. https://www.caresearch.com.au/TEL/

5. Tuffrey-Wijne, I. (2013). *How to break bad news to people with intellectual disabilities: A guide for carers and professionals*. London: Jessica Kingsley. Guidelines for practitioners, families and carers to ease the process of explaining bad news. http://www.breakingbadnews.org.
6. *When I Die: The choices that Tony has made for the end of his life.* An easy-read template for decision-making about a range of end-of-life planning options. From Palliative Care for Persons with Learning Disabilities Network, UK. http://www.pcpld.org/wp-content/uploads/when_i_die_2_0.pdf.

References

Adam, E., Sleeman, K. E., Brearley, S., Hunt, K., & Tuffrey-Wijne, I. (2020). The palliative care needs of adults with intellectual disabilities and their access to palliative care services: A systematic review. *Palliative Medicine, 34*(8), 1006–1018. https://doi.org/10.1177/0269216320932774

Ali, M., Capel, M., Jones, G., & Gazi, T. (2019). The importance of identifying preferred place of death. *BMJ Supportive & Palliative Care, 9*, 84–91. https://doi.org/10.1136/bmjspcare-2015-000878

Amado, A. N., Stancliffe, R. J., McCarron, M., & McCallion, P. (2013). Social inclusion and community participation of individuals with intellectual/developmental disabilities. *Intellectual and Developmental Disabilities, 51*(5), 360–375. https://doi.org/10.1352/1934-9556-51.5.360

Ballan, M., & Sormanti, M. (2006). Trauma, grief and the social model: Practice guidelines for working with adults with intellectual disabilities in the wake of disasters. *Review of Disability Studies: an International Journal, 2*(3), 78–97.

Bekkema, N., de Veer, A. J., Hertogh, C. M., & Francke, A. L. (2016). Perspectives of people with mild intellectual disabilities on care relationships at the end of life: A group interview study. *Palliative Medicine, 30*(7), 625–633. https://doi.org/10.1177/0269216316640421

Bochel, H., & Daly, G. (2020). Conclusions. In H. Bochel & G. Daly (Eds.), *Social policy* (4th ed., pp. 568–572). Routledge.

Chinn, D., & Homeyard, C. (2017). Easy read and accessible information for people with intellectual disabilities: Is it worth it? A Meta-Narrative Literature Review. *Health Expectations, 20*(6), 1189–1200. https://doi.org/10.1111/hex.12520

Cithambarm, K., Duffy, M., & Courtney, E. (2021). What constitutes good quality end-of-life care? Perspectives of people with intellectual disabilities and their families. *Journal of Policy and Practice in Intellectual Disabilities, 18*(3), 207–216. https://doi.org/10.1111/jppi.12376

Clayton, J., Adler, J., O'Callaghan, A., Martin, P., Hynson, J., Butow, P., Laidsaar-Powell, R., Arnold, R., Tulsky, J., & Back, A. (2012). Intensive communication skills teaching for specialist training in Palliative Medicine: Development and evaluation of an experential workshop. *Journal of Palliative Medicine, 15*(5), 585–591. http://online.liebertpub.com/doi/pdfplus/10.1089/jpm.2011.0292

Coombs, M. (2015). A scoping review of family experience and need during end of life care in intensive care. *Nursing Open, 2*(1), 24–35. https://doi.org/10.1002/nop2.14

Dodd, K., Watchman, K., Janicki, M., Coppus, A., Gaertner, C., Fortea, J., Santos, F. H. D., Keller, S. M., & Strydom, A. (2018). Consensus statement of the international summit on intellectual disability and Dementia related to post-diagnostic support. *Aging & Mental Health, 22*(11), 1406–1415. https://doi.org/10.1080/13607863.2017.1373065

Dodd, P., Guerin, S., McEvoy, J., Buckley, S., Tyrrell, J., & Hillery, J. (2008). A study of complicated grief symptoms in people with intellectual disabilities. *Journal of Intellectual Disability Research, 52*(5), 415–425. https://doi.org/10.1111/j.1365-2788.2008.01043.x

Dodd, P., McEvoy, J., Lockhart, K., Burke, T., O'Keeffe, L., & Guerin, S. (2021). An exploratory study of self-reported complicated grief symptoms in parentally bereaved adults with intellectual disability. *Journal of Intellectual Disability Research, 65*, 297–305. https://doi.org/10.1111/jir.12812

Dowling, S., Hubert, J., White, S., & Hollins, S. (2006). Bereaved adults with intellectual disabilities: A combined randomized controlled trial and qualitative study of two community-based interventions. *Journal of Intellectual Disability Research, 50*(4), 277–287. https://doi.org/10.1111/j.1365-2788.2005.00759.x

Glover, G., Williams, R., Heslop, P., Oyinlola, J., & Grey, J. (2017). Mortality in people with intellectual disabilities in England. *Journal of Intellectual Disability Research, 61*(1), 62–74. https://doi.org/10.1111/jir.12314

Grindrod, A., & Rumbold, B. (2018). Healthy End of LIfe Project (HELP): A progress report on implementing community guidance on public health palliative care initiatives in Australia. *Annals of Palliative Medicine, 7*, S73–S83. https://doi.org/10.21037/apm.2018.04.01

Guerin, S., Dodd, P., Tyrell, J., McEvoy, J., Buckley, S., & Hillery, J. (2009). An initial assessment of the psychometric properties of the Complicated Grief Questionnaire for People with Intellectual Disabilities (CGQ-intellectual disability). *Research in Developmental Disabilities, 30*(6), 1258–1267. https://doi.org/10.1016/j.ridd.2009.05.002

Hosking, F. J., Carey, I. M., Shah, S. M., Harris, M., DeWilde, S., Beighton, C., & Cook, D. (2016). Mortality among adults with intellectual disability in England: Comparisons with the general population. *AJPH Research, 106*(8), 1483–1490. https://doi.org/10.2105/AJPH.2016.303240

Hunt, K., Bernal, J., Worth, R., Shearn, J., Jarvis, P., Jones, E., Lowe, K., Madden, P., Barr, O., Forrester-Jones, R., Kroll, T., McCarron, M., & Todd, S. (2020). End-of-life care in intellectual disability: A restrospective cross-sectional study. *BMJ Supportive & Palliative Care, 10*(4), 469–477. https://doi.org/10.1136/bmjspcare-2019-001985

Kellehear, A. (2013). Compassionate communities: End-of-life care as everyone's responsibility. *QJM: An International Journal of Medicine, 106*(12), 1071–1075. https://doi.org/10.1093/qjmed/hct200

Kellehear, A., & Lewin, T. (1988–1989). Farewells by the dying: A sociological study. *Omega: Journal of Death and Dying, 19*(4), 275–292. https://doi.org/10.2190/HWJC-D0PD-5ADA-32U9

Kirkendall, A., Linton, K., & Farris, S. (2017). Intellectual disabilities and decision making at end of life: A literature review. *Journal of Applied Research in Intellectual Disabilities, 30*(6), 982–994. https://doi.org/10.1111/jar.12270

Králová, J., & Walter, T. (2018). *Social death: Questioning the life-death boundary*: Routledge.

Lightfoot, E., & McCallion, P. (2016). Older adults with intellectual and developmental disabilities. In D. Kaplan & B. Berkman (Eds.), *Handbook of social work in health and aging* (2nd ed., pp. 489–499). Oxford University Press.

McCallion, P., McCarron, M., Fahey-McCarthy, E., & Connaire, K. (2012). Meeting the end of life needs of older adults with intellectual disabilities. In E. Chang & A. Johnson (Eds.), *Contemporary and innovative practice in palliative care* (pp. 255–267). Intech. https://doi.org/10.5772/29438

McCarron, M., Burke, E. M., White, P., O'Dwyer, C., Ryan, K., O Farrell, J., & McCallion, P. (2017). *"He'd mind you, you mind him": Experiences of end of life care for people with an intellectual disability as perceived by staff carers.* Trinity College Dublin. https://www.tcd.ie/tcaid/research/endoflifepublications/ReportExperiencesofendoflife.pdf

McCarron, M., McCallion, P., Reilly, E., Dunne, P., Carroll, R., & Mulryan, N. (2017). A prospective 20-year longitudinal follow-up of dementia in persons with Down syndrome. *Journal of Intellectual Disability Research, 61*(9), 843–852. https://doi.org/10.1111/jir.12390

McCausland, D., McCallion, P., Carroll, R., & McCarron, M. (2021). The nature and quality of friendship for older adults with an intellectual disability in Ireland. *Journal of Applied Research in Intellectual Disabilities, 34*, 763–776. https://doi.org/10.1111/jar.12851

McKenzie, N., Mirfin-Veitch, B., Conder, J., & Brandford, S. (2017). "I'm still here": Exploring what matters to people with intellectual disability during advance care planning. *Journal of Applied Research in Intellectual Disabilities, 30*(6), 1089–1098. https://doi.org/10.1111/jar.12355

Moro, T., & McGinley, J. (2021). End-of-life care for adults with intellectual disabilities. In M. Putnam & C. Bigby (Eds.), *Handbook on ageing with disability* (pp. 310–319). Routledge Taylor & Francis.

Noonan, K., Horsfall, D., Leonard, R., & Rosenberg, J. (2016). Developing death literacy. *Progress in Palliative Care, 24*(1), 31–35. https://doi.org/10.1080/09699260.2015.1103498

O'Keeffe, L., Guerin, S., McEvoy, J., Lockhart, K., & Dodd, P. (2019). The process of developing self-report measures in intellectual disability: A case study of a complicated grief scale. *British Journal of Learning Disabilities, 47*(2), 134–144. https://doi.org/10.1111/bld.12261

O'Leary, L., Cooper, S.-A., & Hughes-McCormack, L. (2018). Early death and causes of death of people with intellectual disabilities: A systematic review. *Journal of Applied Research in Intellectual Disabilities, 31*(3), 325–342. https://doi-org.ezproxy.uws.edu.au/10.1111/jar.12417

Palliative Care Australia. (2018). *National palliative care standards.* https://palliativecare.org.au/standards

Rosenberg, J. P., Horsfall, D., Leonard, R., & Noonan, K. (2015). Informal caring networks for people at end of life: Building social capital in Australian communities. *Health Sociology Review, 24*(1), 29–37. https://doi.org/10.1080/14461242.2014.999400

Ryan, K., McEvoy, J., Guerin, S., & Dodd, P. (2010). An exploration of the experience, confidence and attitudes of staff to the provision of palliative care to people with intellectual disabilities. *Palliative Medicine, 24*(6), 566–572. https://doi.org/10.1177/0269216310371413

Sallnow, L., Richardson, H., Murray, S. A., & Kellehear, A. (2016). The impact of a new public health approach to end-of-life care: A systematic review. *Palliative Medicine, 30*(3), 200–211. https://doi.org/10.1177/0269216315599869

Segerlantz, M., Axmon, A., & Ahlström, G. (2020). End-of-life care among older cancer patients with intellectual disability in comparison with the general population: A national register study. *Journal of Intellectual Disability Research, 64*(5), 317–330. https://doi.org/10.1111/jir.12721

Shakespeare, W. (1992). *Hamlet, prince of Denmark*. In C. Watts & K. Carabine (Eds.), *Wordsworth editions* (Original work published 1599).

Stancliffe, R. J., Wiese, M. Y., & Read, S. (2017). Editorial: End of life and people with intellectual disability. *Journal of Applied Research in Intellectual Disabilities, 30*(6), 977–981. https://doi.org/10.1111/jar.12383

Stancliffe, R. J., Wiese, M. Y., Read, S., Jeltes, G., & Clayton, J. M. (2017). Assessing knowledge and attitudes about end of life: Evaluation of three instruments designed for adults with intellectual disability. *Journal of Applied Research in Intellectual Disabilities, 30*(6), 1076–1088. https://doi.org/10.1111/jar.12358

Stancliffe, R. J., Wiese, M. Y., Read, S., Jeltes, G., Barton, R., & Clayton, J. M. (2021). Does talking about end of life with adults with intellectual disability cause emotional discomfort or psychological harm? *Journal of Applied Research in Intellectual Disabilities, 34*(2), 659–669. https://doi.org/10.1111/jar.12835

Stancliffe, R. J., Wilson, N. J., Bigby, C., Balandin, S., & Craig, D. (2014). Responsiveness to self-report questions about loneliness: A comparison of mainstream and intellectual disability-specific instruments. *Journal of Intellectual Disability Research, 58*(5), 399–405. https://doi.org/10.1111/jir.12024

Todd, S. (2013). 'Being there': The experiences of staff in dealing with matters of dying and death in services for people with intellectual disabilities. *Journal of Applied Research in Intellectual Disabilities, 26*, 215–230. https://doi.org/10.1111/jar.12024

Todd, S., Bernal, J., & Forrester-Jones, R. (2013). Death, dying and intellectual disability research. *Journal of Applied Research in Intellectual Disabilities, 25*(6), 571–583. https://doi.org/10.1111/jar.12027

Toseland, R., McCallion, P., Gerber, T., & Banks, S. (2002). Predictors of health and human services use by persons with dementia and their family caregivers. *Social Science & Medicine, 55*, 1255–1266. https://doi.org/10.1016/s0277-9536(01)00240-4

Truesdale, M., Taggart, L., Ryan, A., & McConkey, R. (2021). Experiences of reciprocal caring among adults with an intellectual disability caring for an older family member. *Journal of Policy and Practice in Intellectual Disabilities, 18*(3), 240–248. https://doi.org/10.1111/jppi.12380

Tuffrey-Wijne, I. (2013). *How to break bad news to people with intellectual disabilities: A guide for carers and professionals*. Jessica Kingsley.

Tuffrey-Wijne, I., Finlayson, J., Bernal, J., Taggart, L., Kar Kei Lam, C., & Todd, S. (2020). Communicating about death and dying with adults with intellectual disabilities who are terminally ill or bereaved: A UK-wide survey of intellectual disability support staff. *Journal of Applied Research in Intellectual Disabilities, 33*(5), 927–938. https://doi.org/10.1111/jar.12714

Tuffrey-Wijne, I., McLaughlin, D. M., Curfs, L., Dusart, A., Hoenger, C., McEnhill, L., Read, S., Ryan, K., Satgé, D., Strasser, B., Westergård, B.-E., & Oliver, D. (2016). Defining consensus norms for palliative care of people with intellectual disabilities in Europe, using Delphi methods: A White Paper from the European Association of Palliative Care. *Palliative Medicine, 30*(5). https://doi.org/10.1177/0269216315600993

Voss, H., Vogel, A., Wagemans, A. M., Francke, A. L., Metsemakers, J. F., Courtens, A. M., & de Veer, A. J. E. (2017). Advance care planning in palliative care for people with intellectual disabilities: A systematic review. *Journal of Pain and Symptom Management, 54*, 938–960. https://doi.org/10.1016/j.jpainsymman.2017.04.016

Voss, H., Vogel, A., Wagemans, A. M. A., Francke, A. L., Metsemakers, J. F. M., Courtens, A. M., & de Veer, A. J. E. (2019). Advance care planning in the palliative phase of people with intellectual disabilities: Analysis of medical files and interviews. *Journal of Intellectual Disability Research, 63*(10), 1262–1272. https://doi.org/10.1111/jir.12664

Wiese, M. Y., & Tuffrey-Wijne, I. (2020). End-of-life choices. In R. J. Stancliffe, M. L. Wehmeyer, K. A. Shofren, & B. H. Abery (Eds.), *Choice, preference, and disability: Promoting self-determination across the lifespan* (pp. 317–335). Springer.

17

End-of Life Resources

Michele Y. Wiese

At the close of most chapters in this book a list of resources is given to assist practice across a range of end-of-life issues for people with intellectual and developmental disability. In this chapter the full set is brought together as a handy compendium. The following table lists the resources organised by topic, with topics arranged alphabetically. Within topics each resource is listed in alphabetical order by title and cross-referenced to the chapter/s where it is listed. Some resources are listed more than once because they are relevant to more than one topic. Almost all the resources listed have been designed specifically for people with intellectual and developmental disabilities. Resources intended for the general community are identified as such.

M. Y. Wiese (✉)
School of Psychology,
Western Sydney University, Penrith, NSW, Australia
e-mail: m.wiese@westernsydney.edu.au

R. J. Stancliffe et al. (eds.), *End of Life and People with Intellectual and Developmental Disability*, https://doi.org/10.1007/978-3-030-98697-1_17

Topic	Resource	Chapter/s
Ageing	*Trinity Centre for Ageing and Intellectual Disability (TCAID)*—Information and reports about the ageing process for people with an intellectual disability https://www.tcd.ie/tcaid/	9
Assessment	*Complicated Grief Questionnaire for People with Intellectual Disabilities (CGQ-intellectual disability)* (Dodd et al., 2021, https://doi.org/10.1111/jir.12812; O'Keeffe et al., 2019, https://doi.org/10.1111/bld.12261). This 23-item self-report scale is administered by interview. It complements the existing proxy version of the CGQ-intellectual disability (Guerin et al., 2009, https://doi.org/10.1016/j.ridd.2009.05.002). The authors propose that the scale is suitable for individuals with mild-moderate intellectual disabilities. Questions and response options are both supported with visual aids to enhance comprehension and to enable non-verbal responding (pointing). The CGQ-intellectual disability includes a practice pre-test intended to aid comprehension and minimise response bias. Enquiries should be directed to the authors	2
Assessment	*Concept of Death Questionnaire (CODQ)*—developed by McEvoy et al. (2012, https://doi.org/10.1111/j.1365-2788.2011.01456.x). This 13-item scale evaluates understanding of the five components of death (Causality, Finality, Non-functionality, Universality, and Inevitability). Items relate to three simply worded vignettes concerning a person's death. The interviewer reads each vignette aloud then asks the questions about the components of death. CODQ internal consistency and inter-rater agreement are satisfactory (McEvoy et al., 2012). The CODQ's utility and psychometrics have also been evaluated by an independent team and found to be sound (see Stancliffe et al., 2016, https://doi.org/10.1016/j.ridd.2015.11.016; 2017, https://doi.org/10.1111/jar.12358). Contact the CODQ authors for more information about availability	2

(continued)

(continued)

Topic	Resource	Chapter/s
Assessment	*Distress and Discomfort Assessment Tool (DisDat).* A resource for understanding distress in people with communication difficulties. https://www.stoswaldsuk.org/how-we-help/we-educate/education/resources/disability-distress-assessment-tool-disdat/	14
Assessment	*End-of-Life Planning Scale (EOLPS)*—developed by Stancliffe et al. (2017, https://doi.org/10.1111/jar.12358). This scale is designed to measure a person with intellectual disability's understanding and self-determination about end-of-life planning. It is a not a tool to do such planning. The EOLPS has seven subscales: (a) Bequeathing, (b) Preferred place of care, (c) Funeral wishes, (d) Preferred carers, (e) Advance care planning, (f) Things to take with me, and (g) Organ donation. Each subscale has a simple vignette that the interviewer reads aloud before asking questions related to who decides, and what the person can do now (i.e., while alive) so that their wish is known. Each vignette is supported by one or two colour photos. Comprehensibility, reliability and validity data were mostly satisfactory or better (Stancliffe et al., 2017, https://doi.org/10.1111/jar.12358). A copy of the EOLPS is available from the authors	2
Assessment	*Fear of Death Scale 3.0-Intellectual Disability (FODS3.0-Intellectual Disability).* This scale was adapted by Stancliffe et al. (2017, https://doi.org/10.1111/jar.12358) from an existing fear-of-death scale for the general population (Lester & Abdel-Khalek, 2003, http://dx.doi.org/10.1080/07481180302873). Adaptation involved simplifying the wording and response options. The FODS 3.0-intellectual disability has 28 items in four 7-item factors: (a) others' dying, (b) others' death, (c) your own dying and (d) your own death. The 3-point response scale ranges from 1 (not frightened), to 3 (very frightened). It is supported by a 3-point pictorial scale to enable pointing. Comprehensibility, reliability and validity were sound (Stancliffe et al., 2017, https://doi.org/10.1111/jar.12358). A copy of the FODS3.0-intellectual disability is available from the authors	2

(continued)

(continued)

Topic	Resource	Chapter/s
Assessment	*Paediatric Palliative Screening Scale (PaPaS).* A tool for screening paediatric patients for palliative care needs. Useful for primary care clinicians unfamiliar with palliative care. Though designed for the general community, the information could be helpful for people with intellectual and developmental disabilities. https://bmcpalliatcare.biomedcentral.com/articles/10.1186/s12904-021-00765-8	12
Assessment	*Patient Health Questionnaire-9* (PHQ-9) includes a single question concerning potential for self-harm/suicidal ideation. In the absence of alternative instruments, the PHQ-9 provides a useful, brief screening tool for suicide risk in autistic people that has been used in research settings (Arnold et al., 2019, https://doi.org/10.1007/s10803-019-03947-9). The PHQ-9 is widely available online	3
Assessment	*Suicide Behaviours Questionnaire-Revised* (SBQ-R) is a four question self-report instrument designed to assess risk factors for suicide. The instrument provides cut-off scores to indicate risk of suicidal behaviour in the general population and in people with clinical diagnoses. The original SBQ-R has been used in research settings with autistic people and a revised version for autistic people has been developed (SBQ-ASC). The revised version is only recommended for use in research settings and is not recommended for assessing risk of future suicide attempts or self-harm. The SBQ-ASC has not been tested in people with co-occurring intellectual disability. Readers interested in the SBQ-ASC are referred to Cassidy et al. (2020, https://doi.org/10.1016/S2215-0366(14)70248-2)	3
Children	*Paediatric Palliative Screening Scale (PaPaS).* A tool for screening paediatric patients for palliative care needs. Useful for primary care clinicians unfamiliar with palliative care. Though designed for the general community, the information could be helpful for people with intellectual and developmental disabilities. https://bmcpalliatcare.biomedcentral.com/articles/10.1186/s12904-021-00765-8	12

(continued)

(continued)

Topic	Resource	Chapter/s
Children	*Together for short lives.* A website of resources to support children who are dying and their families. While not purposefully designed for children with intellectual disability, the information is helpful. https://www.togetherforshortlives.org.uk/	12
Caring for the dying	Butler, K. (2020). *The art of dying well: A practical guide to a good end of life*. Though designed for the general community, the information could be helpful for end-of-life care of people with intellectual and developmental disabilities. Scribner. ISBN: 9781501135477; ISBN-10: 1501135473	11
Caring for the dying	Cochran, D. (2019). *Accompanying the dying: Practical, heart-centred wisdom for end-of-life doulas and health care advocates.* Though designed for the general community, the information could be helpful for end-of-life care of people with intellectual and developmental disabilities. Sacred Life Publishers. ISBN: 9780989659352, ISBN-10: 0989659356	11
Caring for the dying	*Decision-making about the best place of palliative care for people with intellectual disabilities.* A guide for care staff and healthcare professionals providing palliative care for people with intellectual disabilities (Bekkema, Tuffrey-Wijne, et al., 2015, ISBN 978-94-6122-333-3, https://www.nivel.nl/sites/default/files/bestanden/Handreiking-decisionmaking-palliative-care-disabilities.pdf)	13
Caring for the dying	*End of life and palliative care for people living with a disability*. Fact sheets about planning ahead and palliative care for people with disability. https://www.betterhealth.vic.gov.au/health/ServicesAndSupport/End-of-life-and-palliative-care-for-people-living-with-a-disability	11
Caring for the dying	*End of life care: A guide for supporting older people with intellectual and developmental disabilities*. Helpful guidance for families and professionals. Purchasable by order from NYSARC Trust Services. https://www.nysarctrustservices.org/contact/	8, 16

(continued)

(continued)

Topic	Resource	Chapter/s
Caring for the dying	*Family carers' experiences of caring for a person with intellectual disability.* Report by the National Disability Authority, Ireland. Looks at coping distress and caring in family carers of people with an intellectual disability. http://nda.ie/nda-files/family-carers%E2%80%99-experiences-of-caring-for-a-person-with-intellectual-disability.pdf	9
Caring for the dying	Glenn, A. W. (2017). *Holding space: On loving, dying, and letting go.* Though designed for the general community, the information could be helpful for end-of-life care of people with intellectual and developmental disabilities. Parallax Press. ISBN: 9781941529799	11
Caring for the dying	*Moral distress in the health professions.* Book edited by Connie Ulrich and Christine Grady. This book addresses moral distress and dealing with moral distress in the workplace. ISBN 978-3-319-64626-8	9
Caring for the dying	*Moral resilience: Transforming moral suffering in healthcare.* Book edited by Cynda Hylton Rushton. This book looks at how moral resilience can be cultivated in individuals and argues for a new architecture to support ethical practice. https://doi.org/10.1093/med/9780190619268.001.0001	9
COVID-19	*Health Services Executive, Ireland.* A range of accessible resources for care staff, including information on staff protections, preparation for working during the pandemic, psychological support and how to keep healthy. https://healthservice.hse.ie/staff/coronavirus/?source=banner-hse-staff	9
COVID-19	*Trinity Centre for Ageing and Intellectual Disability (TCAID).* Information about how to support people with an intellectual disability with a diagnosis of COVID-19. https://www.tcd.ie/tcaid/about/webinars.php	9, 14

(continued)

(continued)

Topic	Resource	Chapter/s
Death fear	*Fear of Death Scale 3.0-Intellectual Disability (FODS3.0-Intellectual Disability).* This scale was adapted by Stancliffe et al. (2017, https://doi.org/10.1111/jar.12358) from an existing fear-of-death scale for the general population (Lester & Abdel-Khalek, 2003, http://dx.doi.org/10.1080/07481180302873). Adaptation involved simplifying the wording and response options. The FODS 3.0-intellectual disability has 28 items in four 7-item factors: (a) others' dying, (b) others' death, (c) your own dying and (d) your own death. The 3-point response scale ranges from 1 (not frightened), to 3 (very frightened). It is supported by a 3-point pictorial scale to enable pointing. Comprehensibility, reliability and validity were sound (Stancliffe et al., 2017, https://doi.org/10.1111/jar.12358). A copy of the FODS3.0-intellectual disability is available from the authors	2
Death understanding	*Books beyond words.* Wordless picture stories for people who have trouble reading, to provide information and open up conversations about death. https://booksbeyondwords.co.uk/	6, 10, 16
Death understanding	*Concept of Death Questionnaire (CODQ)*—developed by McEvoy et al. (2012, https://doi.org/10.1111/j.1365-2788.2011.01456.x). This 13-item scale evaluates understanding of the five components of death (Causality, Finality, Non-functionality, Universality, and Inevitability). Items relate to three simply worded vignettes concerning a person's death. The interviewer reads each vignette aloud then asks the questions about the components of death. CODQ internal consistency and inter-rater agreement are satisfactory (McEvoy et al., 2012). The CODQ's utility and psychometrics have also been evaluated by an independent team and found to be sound (see Stancliffe et al., 2016, https://doi.org/10.1016/j.ridd.2015.11.016; 2017, https://doi.org/10.1111/jar.12358). Contact the CODQ authors for more information about availability	2

(continued)

(continued)

Topic	Resource	Chapter/s
Death understanding	*Death over dinner: Let's have dinner and talk about death:* A guide to planning a conversation about death over a meal. Though designed for the general community, the information could be helpful for people with intellectual and developmental disabilities. https://deathoverdinner.org/	11
Death understanding	*Glancing back, planning forward: Facilitating end of life conversations with persons with an intellectual disability: A guide for carers*. Findings from the end of life interviews of the Intellectual Disability Supplement to the Irish Longitudinal Study on Ageing (IDS-TILDA). http://www.professionalpalliativehub.com/sites/default/files/Carers%20Guide.pdf	9
Death understanding	Gordon, S., & Kacandes, I. (2015). *Let's talk about death: Asking questions that profoundly change the way we live and die*. Though designed for the general community, the information could be helpful for people with intellectual and developmental disabilities. Prometheus Books. ISBN 781633881129	11
Death understanding	Hebb, M. (2018). *Let's talk about death (over dinner): An invitation and guide to life's most important conversation*. Though designed for the general community, the information could be helpful for people with intellectual and developmental disabilities. Orion Publishing. ISBN: 9781841883014	11
Death understanding	*Let's talk about death*. A free booklet from Down's Syndrome Scotland in easy read format about death and funerals for adults with intellectual and developmental disability. https://www.dsscotland.org.uk/wp-content/uploads/2019/05/DSS_LETS-TALK-ABOUT-DEATH_WEB.pdf	8, 10
Death understanding	*Talking End of Life … with People with Intellectual Disability (TEL)*. Australian website with 12 modules on end of life and people with intellectual and developmental disability. Designed for disability support workers but is also helpful for others including families, health professionals, and educators. https://www.caresearch.com.au/TEL/	10, 11, 16

(continued)

(continued)

Topic	Resource	Chapter/s
Death understanding	Tuffrey-Wijne, I. (2013). *How to break bad news to people with intellectual disabilities: A guide for carers and professionals*. London: Jessica Kingsley. Guidelines for practitioners, families and carers to ease the process of explaining bad news. http://www.breakingbadnews.org	4, 6, 10, 14, 16
Decision-making	*Listening to those rarely heard—A guide for supporters*. A training package to guide supporters of people with severe to profound disability through the supported decision-making process (Watson & Joseph, 2015, https://shop.scopeaust.org.au/shop/listening-rarely-heard-guide-supporters/)	13
Decision-making	*Support for decision-making: A practice framework: La Trobe University Living with Disability Research Centre*. A guide to ensuring people with cognitive disability have support to make decisions. https://www.firah.org/upload/notices3/2015/support-for-decision-making.pdf	5
Dementia	Fahey-McCarthy, E., McCallion, P., Connaire, K., McCarron, M. (2008). *Supporting persons with intellectual disability and advanced dementia: Fusing the horizons of care. An introductory education and training programme*. Trainer's Manual. Dublin: Trinity College Dublin. The curriculum emerged from an effort to understand staff experiences in supporting persons with intellectual disability and advanced dementia. Intellectual disability service providers and a specialist palliative care provider in Ireland were involved. Their experiences were interpreted to gain an understanding of their education and training needs. This information was then the basis for a pilot educational intervention which was designed, delivered and evaluated with these services. A partnership approach which involved the Trinity College School of Nursing and Midwifery research team, intellectual disability service providers and a specialist palliative care service was crucial to success. Contact: https://nursing-midwifery.tcd.ie/	8

(continued)

(continued)

Topic	Resource	Chapter/s
Dementia	*Glancing back, planning forward: Facilitating end of life conversations with persons with intellectual disability: A guide for carers.* Findings from the end-of-life interviews of the Intellectual Disability Supplement to the Irish Longitudinal Study on Ageing (IDS-TILDA). https://www.tcd.ie/tcaid/accessibleinformation/carersguide.pdf	7
Dementia	*"He'd mind you, you mind him": Experiences of end-of-life care for people with intellectual disability as perceived by staff carers.* Findings from the end-of-life interviews of the Intellectual Disability Supplement to the Irish Longitudinal Study on Ageing (IDS-TILDA). https://www.tcd.ie/tcaid/research/endoflifepublications/ReportExperiencesofendoflife.pdf	7
Dementia	*People living with dementia supports: Trinity Centre for Ageing and Intellectual Disability.* A range of resources and links that may be of assistance to health and social care professionals during the COVID-19 outbreak: https://www.tcd.ie/tcaid/about/dementia.php	7
Dementia	*The National Intellectual Disability Memory Service.* Trinity Centre for Ageing and Disability site describing a dementia assessment service and easy-read support materials about dementia. https://www.tcd.ie/tcaid/research/NIDMS.php	7
Dementia	*Understand together.* A campaign designed for the general population by the Health Service Executive, Ireland aimed at combatting stigma around dementia. https://www.understandtogether.ie/	7
Do-Not-Resuscitate (DNR)	*Advanced care planning tools and resources.* A range of planning ahead tools for future medical needs. Though designed for the general community, the information could be helpful for end-of-life care of people with intellectual and developmental disabilities. https://coalitionccc.org/CCCC/Resources/Advance-Care-Planning-Resources/CCCC/Resources/Advance-Care-Planning-Tools-and-Resources.aspx?hkey=a0a83455-ff0f-4a89-b700-2572ef44f077	15

(continued)

(continued)

Topic	Resource	Chapter/s
Do-Not-Resuscitate (DNR)	*Do not attempt resuscitation, end-of-life decision-making and associated directives, and their implications for people with intellectual disabilities.* An article addressing how legal provisions for people choosing where and how they die, such as advance directives and lasting powers of attorney for health and welfare, affect those with intellectual disabilities. http://www.intellectualdisability.info/changing-values/articles/do-not-attempt-resuscitation,-end-of-life-decision-making-and-associated-directives,-and-their-implications-for-people-with-intellectual-disabilities	15
Do-Not-Resuscitate (DNR)	*Do not resuscitate? DNACPR & people with learning disabilities*. Webinar presented by Professor Irene Tuffrey-Wijne about medical decision-making plans for the end of life. https://www.youtube.com/watch?v=ePJKT9MZRXA	15
Do-Not-Resuscitate (DNR)	*Video library: People with developmental disabilities.* A series of videos featuring people with developmental disabilities thinking ahead about the end of life (English and Spanish).https://coalitionccc.org/CCCC/Resources/Video-Library/CCCC/Resources/Video-Library.aspx?hkey=48e1b795-499e-43c5-b9b9-9703c39686dd	15
Doula	Cochran, D. (2019). *Accompanying the dying: Practical, heart-centred wisdom for end-of-life doulas and health care advocates.* Though designed for the general community, the information could be helpful for end-of-life care of people with intellectual and developmental disabilities. Sacred Life Publishers. ISBN: 9780989659352, ISBN-10: 0989659356	11
Doula	*End of life planning*. The work of a death doula, training opportunities and resources. Though designed for the general community, the information could be helpful for end-of-life care of people with intellectual and developmental disabilities. https://endoflifeplanning.com.au/portfolio/beth-obrien/	11
Doula	*International End-of-life Doula Association.* A professional association with training and events about the doula role. https://www.inelda.org/	11

(continued)

(continued)

Topic	Resource	Chapter/s
Families	*End of life care: A guide for supporting older people with intellectual and developmental disabilities*. Helpful guidance for families and professionals. Purchasable by order from NYSARC Trust Services. https://www.nysarctrustservices.org/contact/	16
Families	*Talking End of Life…with People with Intellectual Disability [TEL]*. Australian website with 12 modules on end of life and people with intellectual and developmental disability. Designed for disability support workers but also helpful for families, health professionals, and educators. TEL includes a suite of topics on (1) Teaching how to understand the end of life, and (2) Teaching the planning options. https://www.caresearch.com.au/TEL/	10, 11, 16
Families	*Together for short lives*. A website of resources to support children who are dying and their families. While not purposefully designed for children with intellectual disability, the information is helpful. https://www.togetherforshortlives.org.uk/	12
Funerals	*Let's talk about death*. A free booklet from Down's Syndrome Scotland in easy read format about death and funerals for adults with intellectual and developmental disability. https://www.dsscotland.org.uk/wp-content/uploads/2019/05/DSS_LETS-TALK-ABOUT-DEATH_WEB.pdf	10
Funerals	*Talking End of Life…with People with Intellectual Disability [TEL]*. Australian website with 12 modules on end of life and people with intellectual and developmental disability. Designed for disability support workers but also helpful for families, health professionals, and educators. TEL includes a module on *Funeral Wishes*. https://www.caresearch.com.au/TEL/	10, 11, 16
Funerals	*When I die*. An example of a completed end-of-life plan by a person with intellectual and developmental disability, with a section on funeral wishes. http://www.pcpld.org/wp-content/uploads/when_i_die_2_0.pdf	4, 10, 16

(continued)

(continued)

Topic	Resource	Chapter/s
Grief	*Bereavement and loss: Learning resource pack and training*. From PAMIS: Promoting a More Inclusive Society, guidance for those supporting bereaved people with profound or multiple learning disabilities and their parents and carers. http://pamis.org.uk/resources/bereavement-and-loss/	10, 13
Grief	*Books beyond words: When Mum died* and *When Dad died*. UK wordless books designed for people with intellectual and developmental disability that include only pictures so the story can be tailored to the individual. https://booksbeyondwords.co.uk	6, 10, 16
Grief	*Breaking Bad News* This UK website contains information and guidelines for caregivers of people with intellectual and developmental disability about how to break bad news. http://www.breakingbadnews.org/	4, 6, 10, 14, 16
Grief	*Complicated Grief Questionnaire for People with Intellectual Disabilities (CGQ-intellectual disability)* (Dodd et al., 2021, https://doi.org/10.1111/jir.12812; O'Keeffe et al., 2019, https://doi.org/10.1111/bld.12261). This 23-item self-report scale is administered by interview. It complements the existing proxy version of the CGQ-intellectual disability (Guerin et al., 2009, https://doi.org/10.1016/j.ridd.2009.05.002). The authors propose that the scale is suitable for individuals with mild-moderate intellectual disabilities. Questions and response options are both supported with visual aids to enhance comprehension and to enable non-verbal responding (pointing). The CGQ-intellectual disability includes a practice pre-test intended to aid comprehension and minimise response bias. Enquiries should be directed to the authors	2
Palliative care	*Decision-making about the best place of palliative care for people with intellectual disabilities* A guide for care staff and healthcare professionals providing palliative care for people with intellectual disabilities (Bekkema, Tuffrey-Wijne, et al., 2015, ISBN 978-94-6122-333-3, https://www.nivel.nl/sites/default/files/bestanden/Handreiking-decisionmaking-palliative-care-disabilities.pdf)	13

(continued)

(continued)

Topic	Resource	Chapter/s
Palliative care	*End of life and palliative care for people living with a disability*. Fact sheets about planning ahead and palliative care for people with disability. https://www.betterhealth.vic.gov.au/health/ServicesAndSupport/End-of-life-and-palliative-care-for-people-living-with-a-disability	11
Palliative care	*Intellectual Disability reference group*. A group of world experts who contribute information about the needs of people with intellectual disability to the European Association for Palliative Care. https://www.eapcnet.eu/eapc-groups/reference/intellectual-disability/	16
Palliative care	*Palliative Care and People with Learning Disabilities (PCPLD) Network*. A network of people, including those with intellectual disability, professionals and families with the aim of ensuring quality palliative care. This website includes webinars, podcasts and resources on a range of palliative care topics. https://www.pcpld.org	4, 6
Planning	*Advanced care planning tools and resources*. A range of planning ahead tools for future medical needs. Though designed for the general community, the information could be helpful for end-of-life care of people with intellectual and developmental disabilities. https://coalitionccc.org/CCCC/Resources/Advance-Care-Planning-Resources/CCCC/Resources/Advance-Care-Planning-Tools-and-Resources.aspx?hkey=a0a83455-ff0f-4a89-b700-2572ef44f077	15
Planning	*Advance care planning*. A range of guides to help people talk about their wishes for care. Though designed for the general community, these could be adapted for people with intellectual and developmental disabilities. https://www.nhdd.org/public-resources#where-can-i-get-an-advance-directive	4

(continued)

(continued)

Topic	Resource	Chapter/s
Planning	*Bonn Lighthouse (Future planning at the end of life: What I want).* Living will in plain language Original of website in German: https://bonn-lighthouse.de/pv/ Translation to English: https://bonn--lighthouse-de.translate.goog/pv/?_x_tr_sl=de&_x_tr_tl=en&_x_tr_hl=en&_x_tr_pto=nui,sc	5
Planning	Butler, K. (2020). *The art of dying well: A practical guide to a good end of life*. Though designed for the general community, the information could be helpful for end-of-life care of people with intellectual and developmental disabilities. Scribner. ISBN: 9781501135477; ISBN-10: 1501135473	11
Planning	*End of life and palliative care for people living with a disability*. Fact sheets about planning ahead and palliative care for people with disability. https://www.betterhealth.vic.gov.au/health/ServicesAndSupport/End-of-life-and-palliative-care-for-people-living-with-a-disability	11
Planning	*End-of-Life Planning Scale (EOLPS)*—developed by Stancliffe et al. (2017, https://doi.org/10.1111/jar.12358). This scale is designed to measure a person with intellectual disability's understanding and self-determination about end-of-life planning. It is a not a tool to do such planning. The EOLPS has seven subscales: (a) Bequeathing, (b) Preferred place of care, (c) Funeral wishes, (d) Preferred carers, (e) Advance care planning, (f) Things to take with me, and (g) Organ donation. Each subscale has a simple vignette that the interviewer reads aloud before asking questions related to who decides, and what the person can do now (i.e., while alive) so that their wish is known. Each vignette is supported by one or two colour photos. Comprehensibility, reliability and validity data were mostly satisfactory or better (Stancliffe et al., 2017, https://doi.org/10.1111/jar.12358). A copy of the EOLPS is available from the authors	2
Planning	*Five wishes.* An easy-to-use legal document to help express wishes ahead of a serious illness. https://www.agingwithdignity.org/five-wishes	4

(continued)

(continued)

Topic	Resource	Chapter/s
Planning	*Future planning*. Learning about and building a plan for the future. Includes resources and short videos on why and how individuals have made plans. https://futureplanning.thearc.org/	4
Planning	*Glancing back, planning forward, A guide for planning end-of-life care with people with Intellectual disability*. Trinity Centre for Ageing and Intellectual Disability has developed an accessible planning tool for people with an intellectual disability to help them, their families and their carers plan ahead so that their wishes can be respected at the end of their life. https://www.tcd.ie/tcaid/accessibleinformation/index.php	10
Planning	*How to break bad news to people with intellectual disabilities: A guide for carers and professionals.* London: Jessica Kingsley. Guidelines for practitioners, families and carers to ease the process of explaining bad news. http://www.breakingbadnews.org	4, 6, 10, 14, 16
Planning	*Making my own healthcare decisions for people with disabilities.* A guide to writing a letter to the doctor about preferred healthcare decisions. https://thearc.org/wp-content/uploads/forchapters/Making-My-Own-Health-Care-Decisions--A-Letter-for-to-My-Doctors.pdf	4
Planning	*Palliative Care and People with Learning Disabilities (PCPLD) Network*. A network of people, including those with intellectual disability, professionals and families with the aim of ensuring quality palliative care. On the Resources tab there is a section on planning ahead/ advance care planning. https://www.pcpld.org	4, 6
Planning	*People planning ahead*. Guidance to ensure loved ones receive care that conforms to their personal, cultural and religious beliefs. Contact American Association on Intellectual and Developmental Disabilities (AAIDD) for more information. https://www.aaidd.org	4

(continued)

(continued)

Topic	Resource	Chapter/s
Planning	*Persönliche Zukunftsplanung (Network Personal Future Planning).* A network of people and organisations from Germany, Luxembourg and Austria to promote future planning about end of life. Original of website in German: https://www.persoenliche-zukunftsplanung.eu/neuigkeiten.html Translation to English: https://www.persoenliche-zukunftsplanung.eu/neuigkeiten.html	5
Planning	*PiCarDi (Palliative Care for People with Intellectual and Multiple Disabilities).* A research project conducted in Germany about providing the best possible palliative care and hospice support for people with intellectual and multiple disabilities. Includes easy-read information and resources about advance care planning. Original of website in German: https://www.picardi-projekt.de Translation to English: https://www-picardi--projekt-de.translate.goog/?_x_tr_sl=de&_x_tr_tl=en&_x_tr_hl=en&_x_tr_pto=nui,sc	5
Planning	*Respecting choices: Person-centred care.* Guidance for organisations and communities about evidence-based practices to ensure individual healthcare choices are known about and honoured. https://respectingchoices.org/	4
Planning	*Talking End of Life…with People with Intellectual Disability [TEL].* Australian website with 12 modules on end of life and people with intellectual and developmental disability. Designed for disability support workers but also helpful for families, health professionals, and educators. https://www.caresearch.com.au/TEL/	10, 11, 16

(continued)

(continued)

Topic	Resource	Chapter/s
Planning	*The conversation project starter kits**. A tool to help have a conversation with a family member, friend or loved one about wishes regarding end-of-life care. Though designed for the general community, these could be adapted for people with intellectual and developmental disabilities. Available in several languages. https://theconversationproject.org/starter-kits/	4
Planning	*Thinking ahead: My way, my choices, my life at the end*. A workbook to make decisions about life support treatment and other end-of-life choices. https://mn.gov/mnddc/honoring-choices/Thinking-Ahead-English-web.pdf	4
Planning	*When I die: The choices that Tony has made for the end of his life.* An easy-read template for decision-making about a range of end-of-life planning options. From Palliative Care for Persons with Learning Disabilities Network, UK. http://www.pcpld.org/wp-content/uploads/when_i_die_2_0.pdf	4, 10, 16
Planning	*Video library: People with developmental disabilities.* A series of videos featuring people with developmental disabilities thinking ahead about the end of life (English and Spanish).https://coalitionccc.org/CCCC/Resources/Video-Library/CCCC/Resources/Video-Library.aspx?hkey=48e1b795-499e-43c5-b9b9-9703c39686dd	15
Profound intellectual and multiple disability	*Bereavement and loss learning resource pack and training*. For those supporting bereaved people with PIMD. Pamis Scotland. http://pamis.org.uk/resources/bereavement-and-loss/	13
Profound intellectual and multiple disability	*Decision-making about the best place of palliative care for people with intellectual disabilities* A guide for care staff and healthcare professionals providing palliative care for people with intellectual disabilities (Bekkema, Tuffrey-Wijne, et al., 2015, ISBN 978-94-6122-333-3, https://www.nivel.nl/sites/default/files/bestanden/Handreiking-decisionmaking-palliative-care-disabilities.pdf)	13

(continued)

(continued)

Topic	Resource	Chapter/s
Profound intellectual and multiple disability	*Listening to those rarely heard—A guide for supporters*. A training package to guide supporters of people with severe to profound disability through the supported decision-making process (Watson & Joseph, 2015, https://shop.scopeaust.org.au/shop/listening-rarely-heard-guide-supporters/)	13
Profound intellectual and multiple disability	*Listening to those rarely heard*. A video that is part of a training package for people who know someone with a severe to profound intellectual disability. Scope Victoria Australia. https://vimeo.com/21176882	13
Profound intellectual and multiple disability	*Representing the rights of persons with disabilities*. TED Talk by Joanne Watson (2016). https://www.youtube.com/watch?v=qMerG7CULJE	13
Spirituality	*Spiritual support professional certification*. A certification process that seeks to ensure that individuals whose work focuses on providing and/ or facilitating ministry with people with intellectual and developmental disabilities is of a competent standard. Run by the American Association on Intellectual and Developmental Disabilities (AAIDD) Religion and Spirituality Interest Network (https://aaiddreligion.org/certification/)	8
Suicide	*Autism speaks—Eight critical measures to counter suicide*. A list of tips for talking about and preventing suicide in autistic people. www.autismspeaks.org/blog/8-critical-measures-counter-suicide	3
Suicide	*International Association for Suicide Prevention (IASP)* provides a worldwide directory of resources and hotlines for suicide support, as well as a forum for academics, mental health professionals, crisis workers, and suicide survivors. www.iasp.info	3
Suicide	*National Autistic Society—Suicidality in autism: Risk and prevention*. Practical suggestions for assessment and prevention of suicide for autistic people. network.autism.org.uk/good-practice/evidence-base/suicidality-autism-risk-and-prevention	3

(continued)

(continued)

Topic	Resource	Chapter/s
Suicide	*Patient Health Questionnaire-9* (PHQ-9) includes a single question concerning potential for self-harm/suicidal ideation. In the absence of alternative instruments, the PHQ-9 provides a useful, brief screening tool for suicide risk in autistic people that has been used in research settings (Arnold et al., 2019, https://doi.org/10.1007/s10803-019-03947-9). The PHQ-9 is widely available online.	3
Suicide	*Suicide Behaviours Questionnaire-Revised* (SBQ-R) is a four question self-report instrument designed to assess risk factors for suicide. The instrument provides cut-off scores to indicate risk of suicidal behaviour in the general population and in people with clinical diagnoses. The original SBQ-R has been used in research settings with autistic people and a revised version for autistic people has been developed (SBQ-ASC). The revised version is only recommended for use in research settings and is not recommended for assessing risk of future suicide attempts or self-harm. The SBQ-ASC has not been tested in people with co-occurring intellectual disability. Readers interested in the SBQ-ASC are referred to Cassidy et al. (2020) (reference available at: https://doi.org/10.1016/S2215-0366(14)70248-2)	3

(continued)

(continued)

Topic	Resource	Chapter/s
Suicide	*Suicidal Ideation Attributes Scale* (SIDAS) is a five question web-based measure of the severity of suicidal ideation and the *Suicide Assessment Kit* (SAK) is a comprehensive assessment and policy package designed to assess and manage suicide risk, which incorporates an 11-question suicide risk screen. Both instruments are included here as they have been modified for use by autistic people by the chapter authors in collaboration with the authors of the original instruments. The original versions of these instruments are available through the Australian National University (SIDAS; https://rsph.anu.edu.au/research/tools-resources/suicidal-ideation-attributes-scale-sidas) and the University of New South Wales (SAK; https://ndarc.med.unsw.edu.au/suicide-assessment-kit). Because SIDAS and SAK are currently being evaluated for use with autistic people, caution is recommended if they are to be used in this population. At the time of writing, the modified instruments are not yet available for general use (enquiries can be directed to the lead author of Chapter 3).	3
Suicide	*Suicide Prevention Australia* is the peak body for suicide prevention in Australia. Provides information on suicide prevention policy and advocacy, resources, webinars, research and publications. www.suicidepreventionaust.org	3
Suicide	*Suicide Response Project* is an evidence-based free suicide prevention educational toolkit and resource that provides information about how to detect and respond to people at risk of suicide, including neurodivergent and sex/gender diverse people. www.suicideresponseproject.com	3

Afterword

Allan Kellehear

I have been a public health sociologist working in palliative care for the last 25 years. In that time, I have critiqued the lack of recognition that the longer part of dying, caregiving, and grieving occurs away from the face-to-face encounters with professionals. The longer part of the end-of-life experience is living, dying, caregiving, and grieving inside communities. The compassionate community movement has been a response to the growing recognition that professional services have a limited role to play and that they must partner—not only with service users—but also with everyone in the wider community to forge quality and sustainable forms of caring.

Every book edited and written by academics and practitioners tends to emphasise—and rightly so—the need for further research and policy change. The final chapter (see Chapter 16) in this collection has summarised and highlighted gaps and ambiguities within our current knowledge and for our policy needs. There are daunting but also exciting challenges ahead for those of us dedicated to a more inclusive world

A. Kellehear
University of Vermont, Burlington, VT, USA

R. J. Stancliffe et al. (eds.), *End of Life and People with Intellectual and Developmental Disability*, https://doi.org/10.1007/978-3-030-98697-1

for everyone. The findings and observations in this book are abundantly clear. Armed with those insights we can move forward in a clear direction and act now. For why do we write and research if not to shed greater light on our next practical steps?

Clearly, our advocacy organisations and our professional associations in intellectual and developmental disability have a major role to play for the communities they represent. A warm and practical invitation can and should be extended by these organisations, to affiliate with hospice and palliative care associations. In these new scenarios, we will be partner organisations for mutual recruitment to conferences, research, policymaking, and for the mutual exchange of volunteers or board members.

Our advocacy and support organisations can identify and participate in local compassionate communities as these are now led by palliative, primary care, ageing, or hospital organisations around the world. We can employ the *Compassionate Community Charter* to revise, amend, and market to our own and the wider communities what this charter will involve (Kellehear, 2016).

Our advocacy and support organisations should integrate the best public health/health promotion information on end-of-life care, ensuring that this helps balance a more traditional clinical idea about dying as terminal care, or caregiving and grief and loss. Dying is not the last 24 hours or even the last week. People live with life-threatening and life-limiting illness for months and even years. And grief and loss in life, often compounded by a bereavement, is forever. Our abilities to develop prevention, harm reduction, and early actions towards anxiety, depression, social isolation or rejection, job loss or school refusal, early death, or suicide depend significantly on greater public education efforts. Public education about health and safety is both normal and routine for bodily, mental, spiritual, and social well-being. This educational principle is identical for matters to do with our experiences of dying, caregiving, and grief and loss. Ignorance about these matters lights the way for paternalism, division, disenfranchisement, and poor outcomes.

End-of-life care education is therefore crucial—not merely for the terminal phase of life but for *all* phases of life—because dying can occur at any age, as do caregiving responsibilities, and living with grief and loss. Bereavement must not be viewed by anyone as a sequestered experience unconnected from life's broader and inevitable tapestry of other separation and loss experiences.

The end-of-life policy and practice world is increasingly beginning to recognise that this form of care, like health care itself, is everyone's responsibility. End-of-life care is not merely something that health professionals do *on* us, as patients, clients, or customers. Health and social care at the end of life must work *with* us. Everyone has a role to play. Just as workplaces and schools have health and safety policies, so too should they have end-of-life care policies. Neighbours have a duty to care also—one of the main inspirations for the neighbourhood watch scheme or the child safe house schemes that are two excellent examples of neighbours looking out for neighbours. Local compassionate communities mean you and I have a role. The public health principles of prevention and harm reduction mean that we do not wait for crisis and illness to act. No-one is exempt, and this means the intellectual and developmental community must play their part in concert with others.

Steps to Better End-of-Life Care: An Intellectual and Developmental Disability Action Plan for End-of-Life Care (DID Act)

- Our associations will affiliate with palliative and hospice care associations.
- Our associations will contribute to existing compassionate communities.
- We will advocate for, or support moves to establish new compassionate communities.
- Death, dying, and end-of-life care education should be integrated into all life course/human development education.
- Intellectual and developmental disability education should be integrated into all death, dying, and end-of-life care education.
- Act and remind others that end-of-life care is everyone's responsibility.
- Always remember that information *is* support. Participation is power.

Finally, information is support, not simply the existence of helpful people. An informed person is the difference between a helpful person and a hindrance. There is much information and research already in existence about end-of-life needs, challenges, and characteristics for people living with disabilities. These insights require a greater public profile and a greater integration into professional and community practice wisdom in the broader end-of-life care world. Those living, or who have lived with dying, caregiving, and or grief and bereavement in their lives as disabled people or their caregivers, have something of critical value and importance to give to others. Participation in end-of-life care will help transform those insights about the human condition into a more positive legacy for us all.

Allan Kellehear, Ph.D. is a Medical and Public Health Sociologist working in end-of-life care. He has been a major advocate of the public health approach to palliative care and has held former professorial posts in Australia, Japan and the United Kingdom.

Reference

Kellehear, A. (2016). The compassionate City Charter: Inviting the cultural and social sectors into end-of-life care. In K. Wegleitner, K. Heimerl & A. Kellehear (Eds.), *Compassionate communities: Case studies from Britain and Europe* (pp. 76–87). Routledge. https://doi.org/10.4324/9781315735801

Correction to: Suicide and Autism: A Lifespan Perspective

Darren Hedley, Susan M. Hayward, Alison Clarke,
Mirko Uljarević, and Mark A. Stokes

Correction to:
Chapter 3 in: R. J. Stancliffe et al. (eds.), *End of Life and People with Intellectual and Developmental Disability*, https://doi.org/10.1007/978-3-030-98697-1_3

The original version of Chapter 3 was inadvertently published with wrong middle initial for the co-author, which has now been changed from "Susan H. Hayward" to "Susan M. Hayward". The correction to the chapter has been updated with the changes.

The updated version of this chapter can be found at
https://doi.org/10.1007/978-3-030-98697-1_3

R. J. Stancliffe et al. (eds.), *End of Life and People with Intellectual and Developmental Disability*, https://doi.org/10.1007/978-3-030-98697-1_18

Index

R. J. Stancliffe et al. (eds.), *End of Life and People with Intellectual and Developmental Disability*, https://doi.org/10.1007/978-3-030-98697-1

E

O

P

GPSR Compliance
The European Union's (EU) General Product Safety Regulation (GPSR) is a set of rules that requires consumer products to be safe and our obligations to ensure this.

If you have any concerns about our products, you can contact us on

ProductSafety@springernature.com

In case Publisher is established outside the EU, the EU authorized representative is:

Springer Nature Customer Service Center GmbH
Europaplatz 3
69115 Heidelberg, Germany

www.ingramcontent.com/pod-product-compliance
Ingram Content Group UK Ltd.
Pitfield, Milton Keynes, MK11 3LW, UK
UKHW020258051026
15470UKWH00021B/1323
9783030986964